Streaming Media Demystified

Michael Topic

McGraw-Hill
New York   Chicago   San Francisco   Lisbon
London   Madrid   Mexico City   Milan   New Delhi
San Juan   Seoul   Singapore   Sydney   Toronto

Library of Congress Cataloging-in-Publication Data

Topic, Michael.
    Streaming media demystified / Michael Topic.
       p.   cm.
    Includes bibliographical references.
    ISBN 0-07-138877-X (alk. paper)
    1. Streaming technology (Telecommunications)   I. Title.

  TK5105.386 .T67   2002
  006.7'876—dc21

                         2002067793

McGraw-Hill

A Division of The McGraw-Hill Companies

1 2 3 4 5 6 7 8 9 0   DOC/DOC   0 9 8 7 6 5 4 3 2

ISBN 0-07-138877-X

The sponsoring editor for this book was Stephen S. Chapman, the editing supervisor was David E. Fogarty, and the production supervisor was Sherri Souffrance. It was set in New Century Schoolbook by Patricia Wallenburg.

Printed and bound by R. R. Donnelley & Sons Company.

McGraw-Hill books are available at special quantity discounts to use as premiums and sales promotions, or for use in corporate training programs. For more information, please write to the Director of Special Sales, Professional Publishing, McGraw-Hill, Two Penn Plaza, New York, NY 10121-2298. Or contact your local bookstore.

In memory of Daniel Lewin—innovator and visionary.

McGRAW-HILL
TELECOMMUNICATIONS

CONTENTS

Contents

PREFACE

This book, my first, was written under fairly difficult circumstances. The project coincided with the birth of my second child and first daughter, high-pressure project deadlines relating to the Aqua streaming media encoder, the building up of a business with only bootstrap finance, a near-death experience and dealing with the negative consequences of the September 11, 2001 atrocity, the U.S. high-tech recession and the Nimda virus. I am particularly pleased to have completed this book during all of that.

The book's content may strike the reader accustomed to technical books as somewhat unusual, since there are chapters dealing with the social, political and business issues relating to the technology. The reason for including such peripheral information is that I strongly believe technologists ought to understand their technologies within a human context. If a technology does not serve humanity and improve people's lives, what justification is there for its existence?

Scientists and engineers must take responsibility for what they thrust upon humanity. I have never agreed with technologists who hide behind the beauty of their creations in order to avoid having to confront the problems their technology creates. I also cringe at business decisions made solely on the basis of technical argument, without some basic understanding of the people the technology is for, what it will do for them and why they might buy it. Only by understanding the context within which a technology will exist can technologists make sound judgments about how to shape their products, fashion the features and create solutions that are relevant to people's real lives. Too many bad applications and products get made because the designers don't take into account the context of their work.

Technical books that failed to give the "big picture" have always tended to bewilder rather than clarify. The old adage of not being able to see the wood for the trees always applied. I have endeavored to set my own explanations of streaming media technology against a background of the issues surrounding the technology. I hope that my peculiar and particular viewpoint serves to illumine the process of demystification.

The book could have been very much longer, since there is a lot of ground to cover in explaining everything about streaming media. Consequently I have, in places, reluctantly resorted to sketches rather than detailed examinations of various aspects of the technology and medium.

This is a fast-moving technology, so I expect much of the book to date very rapidly. By concentrating on the underlying principles rather than the specifics of various current solutions, however, I hope that the work will serve the reader for many years to come.

MICHAEL TOPIC
Ripley, Surrey
michael.topic@imag-eng.com

ACKNOWLEDGMENTS

I used to read the acknowledgments in other books with a good deal of skepticism, but having now written a book of my own, I have come to appreciate just how essential the efforts of other people are in the process and what a great debt they are owed. I, therefore, offer my humble thanks and appreciation to the following people:

Peter Symes for suggesting that I write this book and for making the necessary introductions. I hope my book is half as good as yours. Steve Shepard—thanks for the helping hand and encouragement when I was in over my depth on telecommunications protocols and optical networking. Laura Clemons—thank you for your insight and for taking the time to write down your thoughts on where the industry was going and where it had been. Your input was not only authoritative, but also inspirational. Nancy Arculin Jandorf—I am eternally indebted to you for your critiques of the first drafts, suggestions for added clarity, unfailing encouragement, and for continually reminding me that I could do this project. John Portnoy—thank you for being there over the years and for your generous information on the film industry and its internal workings. Ben Roeder—thanks for introducing me to the Kendra project and for giving me some understanding of what it is like to be a broadband service provider. Ross Summers—thanks especially for the insights into the streaming media industry and for tramping through all the trade shows with me, while protecting "the project" like a lion. Julian Medinger—thank you for clear-headed thought and for explaining digital rights management to me in words of one syllable. Ray Baldock, Mike Cronk, Beth Bonness and Tim Thorsteinson—thank you for granting me access to the streaming media industry from close quarters and for your thoughts and ideas on the future of the industry. Without your help and support, this book could never have been written. Mark Leonard and Rob Charlton—thanks for helping to debate and clarify various issues concerning data synchronization on mobile computing devices and for having the intelligence to play a part in shaping the industry according to what you can see. There are many people in the industry, who I met at trade shows and conferences, or else interviewed as part of my consultancy work, that generously spent time debating various hot issues with me, adding insight and opinions to my partly formed views. Listing everybody by name would be an impossible task and I am sure to overlook somebody vital. Please accept my sincere and heartfelt thanks for shaping and forming my ideas.

Thanks are due to Steve Chapman, Jessica Hornick, and all the wonderful "behind-the-scenes" people at McGraw-Hill, for acting as calm, collected and patient midwives to this project, even when I was struggling to get the book written. Extra special thanks are due to Patty Wallenburg of TypeWriting for making my work look so darn good and for providing me with a much needed "buffer;" to Marion Brady for patiently reconstructing my awkward sentences and mending all my split infinitives, and to Joann Woy for proofreading with the eyes of a hawk and for indexing this diverse, free-ranging subject matter intelligently.

I offer special gratitude and thanks to my staff at Imaginative Engineering, especially Anne Elliott and Ewan Smith, for keeping things running while I was deeply immersed in the writing of this book and for all the trade show support and sheer excellence. I feel truly privileged to have worked with such outstanding people in my lifetime.

To my parents, I will never be able to adequately thank you or repay you for working like slaves so that I could get a decent education. Dad, I especially thank you for teaching me that the world is really very simple to understand, once you figure out how it works. Nothing is too complicated to attempt to grasp.

Finally, to Clare, Alexander and Elise, thank you for understanding when I couldn't be at places I should have been with you, for accepting fewer hugs while I was busy writing, locked away in my office, or in some far-off city working out how streaming media works, for creating precious peace and quiet when I needed it and for tolerating this obsession that overtakes you when you begin to write a book. I've been working on this book for your entire life, Elise. Now that it's done, daddy has returned to normal. I'm afraid this as normal as it gets, sweetheart.

CHAPTER **1**

Introduction

I undertook the task of trying to demystify streaming media because I personally believe that streaming media will be the biggest thing since television. It might even have a greater impact on the world than books! I am certain that the earliest manifestations of streaming media on the Internet and the World Wide Web will appear as primitive to our children as the earliest days of wireless broadcasting seem to us today. In terms of using streaming media to capture and disseminate our cultural artifacts, we are at the same stage of evolution as the painstakingly hand-copied ecclesiastical manuscripts were, compared to modern mass-produced airport paperbacks and e-books.

This book takes the long view, therefore. We survey the technology and medium independently of the dot-com crash that overtook stock markets during 2001. This cataclysmic economic event, significant as it is to the present business environment for streaming media, will have very little lasting impact on the technology or the medium.

Streaming media, at the time of writing, is in the chasm between being a product that appeals to early adopters, the "techno-enthusiasts," and one that is ready for the early majority of consumers (Figure 1.1). In the terminology of high-tech marketing guru Geoffrey Moore, the "whole-product offering" is not complete. Niche applications must be exploited before streaming media is truly ready for prime time. The next phase in the technology's development will see companies creating offerings that even grandma can use.

Figure 1.1
The product chasm.

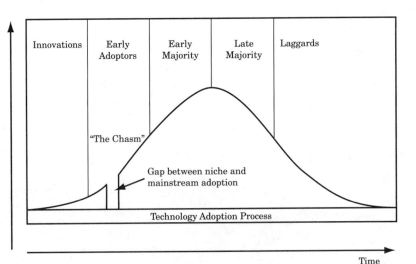

In the progress of the technology from its emergence to its present state, dozens of companies have come and gone. There have been many more failures than successes. Some business models patently didn't work. Consumer uptake has been slow, hampered by the lack of "killer applications," a paucity of imagination, and disappointing infrastructure buildout. The problem is that the industry is still very much built on vision, but vision is hard to realize and harder yet to sell. When the technology becomes a practical proposition for even the most casual of users, the medium will undoubtedly achieve widespread acceptance. These setbacks are temporary. Part of the process of demystifying streaming media is, then, an exploration into why the medium is still in its infancy and why it will inevitably change in the future.

This book will guide the reader through the maze of acronyms, proprietary and open technologies, business models, and related communications and digital media technologies used to create and deliver streaming media. We unravel the medium itself, including its component technologies. We examine the audience for streaming media, to understand what is driving demand for the medium. The business of streaming media is also investigated, to provide a fiscal context for its supply and demand. Finally, we look at the upsides and downsides of streaming media. Most importantly, this is a book about my hopes for the future of the medium. It is a medium with limitless possibilities. Even though there are obstacles, this new media technology is compelling, for reasons that will become apparent throughout the book. We will examine not only how things are, but, crucially, how they could be.

A survey of the literature available on streaming media will reveal that there has been very little written on the subject offline. This is partly because of the speed of evolution of the technology, but also because this medium is in something of a "blind spot" for traditional media commentators and analysts. With only a view of what exists, it is difficult for them to extrapolate the technology and infrastructure, in order to explore what the ultimate impact of the medium will one day be. Media analysts and commentators do not always have the technical insight to be able to project the technology that far ahead. Most of the information available on streaming media has been written by technology vendors with their own particular worldview, either in white papers and application guides, or else as press releases. Little serious independent analysis or exploration of the enormous potential of the medium is available. Vendors overwhelmingly present their solutions as complete and the ultimate state of the art, yet consumers remain largely unimpressed, asking themselves "is this it?" We will offer a vendor-agnostic analysis of the medium as a new way of communicating to mass and individual audiences.

Another of the main theses of this book is the evolution of the delivery of home entertainment. The death of television, as we know it, will eventually take place and streaming media is what will be there instead: a new kind of television, if you will. However, it would be a mistake to cast an understanding of streaming media merely in television replacement terms. It is television for a new century, but much more than that. The same news, information, and entertainment programs will be made, but enhanced using streaming media technology and delivered in entirely new ways. Television favorites will still be available, but on demand, via streaming channels, not just broadcast when the network controllers decide to air them. Already, shows like "Big Brother" are becoming media events, exploiting streaming technology to great effect and drawing unprecedented audiences as a result. In addition, every country's televisual output will be available to the entire world, creating an unprecedented cultural and creative impact. We will highlight the significant characteristics and fundamental properties of streaming media that will make it more than television ever can be. Then, we'll look into the business obstacles that have prevented streaming media from superseding television, to date, in the expectation that enterprises will come forth to overcome them.

This book is also written for the general public: people who have heard of streaming media, but don't understand what it is, how it works, what it is good for, and how significantly it might affect our lives in the future. To that end, I have described technical concepts by way of simple analogy and resisted using special language and jargon, where possible. Where specialist terms give exact meanings, I have endeavored to explain clearly both the meaning and why use of the special term was most appropriate.

I am also unashamedly aiming this book at industry professionals who can make a difference to the development of the medium, given a good overview of the landscape. Other audiences I am targeting include media regulators, since they have the power to make the medium truly great. I am addressing content creators who already have excellent material and could exploit this medium's unique creative potential most effectively. Compelling content will drive adoption and create unexpected new media stars. Finally, I am speaking to telecommunications companies who, so far, seem to have been less than entirely successful at universal deployment of broadband.

Streaming media has the potential to create a world very different from the one we know. With streaming media technology, rich information content achieves a wider reach. People can learn, at their own pace, from the most brilliant experts, in any field of inquiry they care to choose. News reporting can be, if not completely unbiased, at least open to verification and scrutiny at the source, with an editorial agenda that is more democratically set, not distorted to agree with some powerful media magnate's particular point of view, as has regrettably been the case with earlier media. When news programs are no longer subject to scheduling constraints, there is always enough air time left to allow the interviewee to answer the crucial questions, preventing the time-honored trick of talking about nothing until the available time is used up. Indeed, the questions may be posed directly by the audience. Streaming media can be used to ensure that public figures are always called to account.

Streaming media makes physical distance irrelevant. It can entertain you, wherever you happen to be, any time of the day or night, in your own language, and with subject matter guaranteed to appeal to you. With streaming technology, you can not only talk to your distant relatives and friends, but also see and interact with them in real time, without the relative user-hostility of current videoconferencing technologies. Streaming media technology will also enable cinema-quality presentations to be routinely available to the most remote, least urbanized populations on the planet. Streaming game play will one day be photo-realistic, three-dimensional, richly interactive, and totally immersive, comparable in quality to the computer-generated special effects used in Hollywood feature films. The expense and discomfort of business travel can be replaced by easy-to-use, better-than-television-quality video conferencing. This application, alone, will have a significant impact on economic growth.

The societal impact of streaming media should not be underestimated. In a world of wide-reaching, rich information, it is much more difficult to remain ignorant or prejudiced. Democratic choices can be informed choices and tyranny and oppression harder to sustain. You don't have to take anybody's word for it. You can check sources and conflicts of interests. If you have something to say, you can say it to anybody and everybody who cares to listen, uncensored, immediately. If you didn't understand what was said, you can play it again or ask directly for clarification. The message can even be translated into your native tongue instantaneously. If someone is foolish enough to steal your car or mistreat your children, images of them, caught in the act, can be relayed to you instantly (and perhaps to the authorities as well). Streaming media is a technology that can significantly contribute to the security of every citizen.

Streaming media technology holds the promise of making some of this very different world come true soon. However, streaming media is in its infancy. If you had asked the average person, at the turn of the twentieth century and into its early years, about the importance of the motorcar, few, if any, could have envisaged its full social and economic impact; its pervasiveness into all aspects of life, just a hundred years hence. How many could have imagined two hundred-mile per hour sports cars being available to the public, just for fun, and a motor vehicle breaking the speed of sound? Even if they had understood where the technology might lead, who would build the roads and how would they be financed? Would there ever be a road between where you lived and where you wanted to go and could you ever afford to use it anyway? What would a car be for? Yet, just as with the automobile last century, all the key technologies and systems that enable streaming media have already been designed. What's missing is a Henry Ford.

When I started writing this book, I began asking ordinary people I knew, from all walks of life and backgrounds, what they knew about streaming media. Most, if not all, had heard of it. Some had even tried it. Hardly anyone knew what the big deal with it was. It hadn't impressed them. The general public does not yet see the potential, just the jerky, postage-stamp-sized pictures with warbling, poorly synchronized sound. Very few people, other than streaming media industry professionals, know how it works. This remains a major challenge for the streaming media industry. Until the general public "gets it," the medium will appeal only to specialists and the business will not grow to the size it has the potential of reaching.

Those skeptical of the importance of streaming media's vast potential could be forgiven for asking why I chose to write a book on the subject, rather than present my views via streaming media. Books, as a format, present information in a highly available way. Books don't crash. As an information delivery device, a book also uses very little energy, dissipates almost no heat, produces insignificant noise and interference, is available in high resolution, does not have limited battery life, is relatively benign to the environment when disposed of, is widely available, relatively cheap, provides the user with rapid random-access browsing capabilities, is relatively lightweight, highly portable, does not require that the user read a manual as a prerequisite to accessing the information it contains and has very good viewing angles. Such are the requirements and the quality bar already set for designers of streaming media receivers! However, interactivity, animation, hyperlinking, and information currency have all been sacrificed. One day, that trade-off may be very much harder to justify. The intended audience of this book also includes people who might not already be fully conversant with streaming media, so the traditional medium of the book is necessary as a means of awakening the imagination of the reader to the creative possibilities and social impact of streaming technology. After all, even rocket scientists learn from books!

The Medium

What Is Streaming Media?

Ask a dozen people what streaming media is and you are likely to get a dozen different answers. In its most basic definition, the only difference between streaming media and media that you have in its entirety before accessing, is that with streaming, you can begin to access the media before you have received it all. In other words, while you are watching, the rest is arriving (Figure 2.1).

Figure 2.1
How streaming works.

 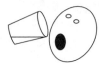 

Downloading is like pouring the
glass of milk, then drinking it.

Streaming is like drinking
straight form the carton.

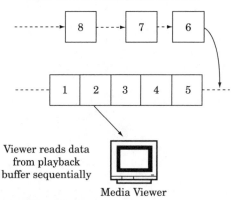

In the world of terrestrial television and radio, media always did stream. Even where analog transmission is used, your receiver is continuously receiving "data," while simultaneously displaying what it already has (Figure 2.2). With an analog television, what is displayed is, for all intents and purposes, pretty much what is currently being received. This has changed a little with digital television, because the data consists of binary numbers, which can, theoretically, be stored; but the average digital television receiver doesn't store very much. Again, what you are viewing is pretty much what is currently being received. Streaming is the default.

Figure 2.2
Analog streaming.

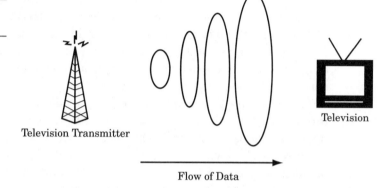

Television Transmitter

Television

Flow of Data

Why does streaming matter? Who cares if you get to see what you are downloading before you have it all? The answer lies in the rich array of digital media types that have been invented, the vastly expanded range of available "channels" for distribution of that digital media provided by the Internet, and the myriad ways that this digital media can be received and rendered for consumption. With a streaming model, the consumer of streaming media has the freedom to shift attention to another stream almost instantaneously. You can effectively channel surf through all the kinds of streaming media that exist, just as you can flip channels on a television.

Once, media consisted of radio, television, books, newspapers, magazines, records, and videotapes. All of these were essentially *analog*, in that when you made copies, there was always some loss of quality involved. Also, the product was manufactured and distributed through a vast, expensive and slow network, owned and controlled by a handful of powerful media companies. With analog media, the producer of the content had great control and the most profitable products were those that appealed to a mass audience, since economies of scale applied to the manufacturing and distribution of media products.

The advent of the desktop computer created an opportunity for the invention of new media types—*digital media*. These include digital versions of all the old media types we've already talked about, plus new ones like three-dimensional interactive multi-player games, virtual reality, live video chat, synchronized multimedia, animated vector graphics, and computer-generated imagery. Digital media, when copied, can perfectly reproduce the original and distribution is via communications networks, not as physical freight. Suddenly, the economics of production and distribution of digital media and access to the means of mass distribution have swung in favor of the small producer. Now it is possible to address niche audiences profitably and logistically, not just the mass market. Not only that, but new types of digital media are being invented all the time, creating new user experiences and applications; spawning new business opportunities and growth potential.

The only way to get old media was to tune into a broadcast (either through the air or on a cable network) or by purchasing a physical object (tape, newspaper, or record). This meant that consumer choice was limited to what the owner of the broadcast medium wanted to show. The broadcaster dictated what would be shown and when. Consumer choice was also limited to whatever the manufacturers of physical media wanted to make (often, these were the same companies). Distribution of media was a non-trivial financial undertaking and the media was geared toward serving mass audiences, as we have already noted. People like what they know, so vast amounts of money were spent making sure that the public knew about products that the producers wanted to ship in volume. The more they knew about them, the more they bought. Broadcasting was harnessed in the service of promoting the sale of physical media products.

Digital media can be served on the Internet, via phone lines and modem, a cable modem, a digital television broadcast (DTV) carrier, cell phone networks ADSL (Asynchronous Digital Subscriber Line), optical fiber connection, or a corporate Ethernet, via satellite, using infrared networks or via wireless LAN (Local Area Network) connections. Remember, these distribution paths are only the ones that carry TCP/IP (Transport Control Protocol/Internet Protocol) traffic, in which the digital media are carried in small packets conforming to this widely used communications protocol. There are many other networks and protocols which can stream media.

The fact that streaming media payloads are broken into tiny packets of data for delivery presents some significant problems. The Internet was designed to be resilient. Data is broken up into small packets and each packet finds its own route to the end-user. The Internet was designed this way so that if part of the network was destroyed or busy, subsequent packets could follow other routes. The lost packets could also be resent, the packets all assembled back into the correct order by the receiver, and the payload recovered in its entirety with no loss. But if packets get lost and must be resent, you cannot stream. The loss of even one of these packets causes the pictures to freeze (Figure 2.3). Buffering can help, so that there is enough time to resend and recover data lost during transmission before anybody notices, but the longer the buffer, the less responsive the streaming player feels. It starts to feel less like changing a television channel and more like waiting for a download.

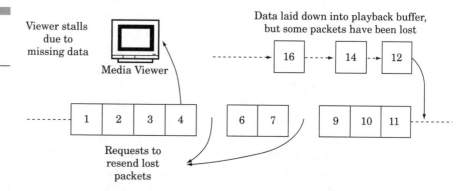

Figure 2.3
Packet loss halts streaming.

Part of the answer to this fundamental problem has been the invention of protocols to ensure quality of service. These protocols, like RTSP and MPLS, will be discussed later in the book. Suffice it to say that they attempt to maintain the integrity of the stream and hence allow streaming to take place uninterrupted.

In addition to these new network and communications technologies, there are new digital techniques for transmitting television and radio, as well as a plethora of new and emerging optical techniques to service the telecommunications industry. Add to these digital tape formats and CD (Compact Disk) and DVD (Digital Video Disk or Digital Versatile Disk, depending on who you ask and when) optical storage technologies and you soon discover that there are more ways to get a stream of digits to the end user, at varying costs, than ever before. The provision of ever-increasing amounts of bandwidth to each and every consumer does not yet seem to have reached any discernible limit. Companies are laying cables, installing satellite receivers, and making offers to consumers to give them higher-bandwidth connections. There also doesn't seem to be any decrease in the number of new digital distribution schemes being invented. Digital media can be delivered via the Internet, using terrestrial digital television channels, on a cell-phone network, through cable television connection, or via satellite. These distribution schemes use different techniques for carrying the data along the physical medium (*modulation* is the technical term) and different protocols to ensure the data has been delivered successfully. In other words, the future will be one of more data paths with higher capacity, capable of delivering digital media in more and varied ways, almost anywhere you are, anytime.

Analog electronic media were experienced by turning on a receiver and selecting the program, or else by placing a physical tape or record into a player. You had to buy a special machine to render each of those electronic media products into sound and pictures. With the vast array of digital media types, there are a greater number of potential user experiences, delivered via a multitude of different delivery paths. Does the consumer buy a special receiver for every conceivable combination of digital media type and delivery path? Right now, yes, unless you accept the limitations of using a general-purpose personal computer to receive and render all the digital media experiences available. If you have a DVD, you put it into a DVD player. If you watch digital video, you do that with a set-top box. For digital audio, you need a portable CD or MP3 player. Because of the cost and inconvenience of having to own a different device to receive different kinds of streaming digital media, people have had to make choices and ignore some of the digital media available. However, as the digital media content production industry evolves and consumer electronics manufacturers begin better to understand digital media, new types of receivers, capable of receiving digital media in different ways and rendering it wherever the consumer happens to be, will be designed. The home PC is probably the most versatile digital media receiver yet made, since it can handle almost all the digital media types available, given the right hardware interfaces. However, as a consumer appliance that must live at the other end of the living room opposite a sofa, it falls short of the ideal. The idea of a home streaming media gateway has much more appeal, since it can interface to all of the data pipes entering the home, locally store content of interest and serve the household's lightweight screens and handheld wireless devices directly (Figure 2.4). The potential for new and exciting consumer appliances is vast.

The same holds true for authoring and delivering digital media. Today, the media producer must author separately for each individual medium and delivery channel. In the future, it will be an economic necessity for the digital multimedia producer to repurpose all media assets automatically, regardless of the delivery mechanism or the rendering capabilities of the player receiving them (Figure 2.5). Similarly, media-receiving appliances will need to cope with more delivery paths and a wider variety of media types. The design of digital media receivers

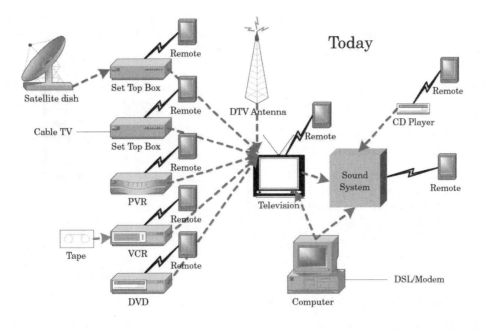

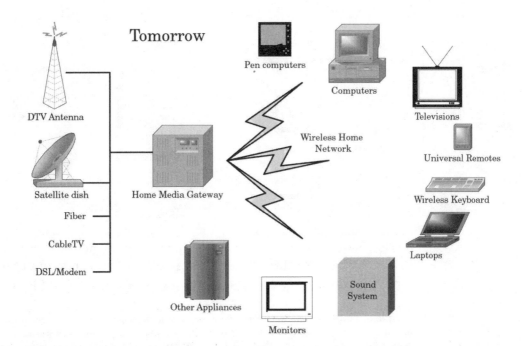

Figure 2.4   Home media gateway solves consumer appliance nightmare.

is the most backward aspect of streaming media today, and possibly the main reason that digital media have not yet reached critical mass in user acceptance.

Figure 2.5
Streaming media production workflow.

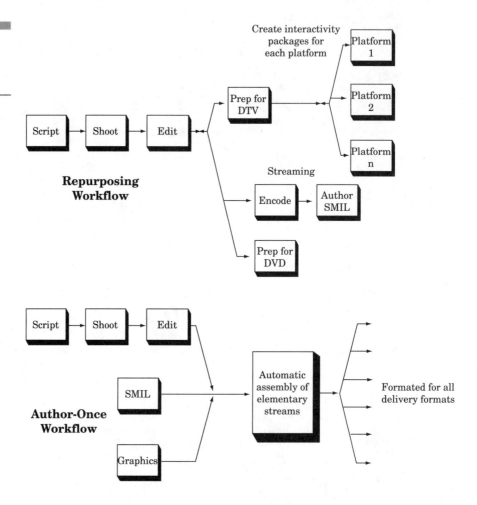

This book, will examine only streaming *digital* media. A revolution driven by Moore's law* has taken place in the capacity of cheap receivers to render complex multimedia presentations in real time. The

---

*In 1965, Gordon Moore, one of the founders of microprocessor manufacturer intel, predicted that the number of transistors that could be placed on a silicon chip would double every eighteen months, thereby giving the consumer geometrically increasing computing power, for the same money. Moore's law has held for over three decades so far. It may hold for another two decades. See http://www.intel.com/research/silicon/mooreslaw.htm.

combination of cheap, powerful computing machinery, the invention of new digital media types, and the existence of broadband networks has made possible a new kind of medium superior to any that went before. Because it can both stream and be stored economically, it is a flexible medium suitable for many new applications. Many of these unique applications will be enumerated in this chapter, by way of illustration. In addition, because the distribution networks that can deliver streaming digital media include the broadband Internet, the power structures of existing media empires are under threat. These are sociologically and economically significant characteristics of the streaming medium.

So, streaming media refers to the near-instantaneous delivery of various kinds of digital media, carried to the consumer via a multitude of distribution paths and received on a variety of rendering devices (by rendering, we mean machines that are capable of converting digital media data into something you can see, hear, experience, etc.).

As new digital media types, new distribution infrastructure, and new receivers are designed and deployed, streaming media will remain a moving target, changing identity as technology advances. For this reason, the treatment of streaming media in this book will concentrate not only on existing embodiments, but also on those that might happen in the near future. We'll stick to the fundamental characteristics of the medium, rather than debate which current system will prevail.

The following sections in this chapter will define streaming media by denoting some characteristics and applications. By describing what the media will be like, we'll answer the question "what is streaming media?"

A New Distribution Channel

The best way to think about streaming is as a new way of delivering digital media to an audience. Even though digital television, in fact, streams digital media to a receiver, it is merely aping the characteristics of the analog channel it replaces. A streaming channel can be much more flexible. It can be both a broadcast infrastructure, in competition with traditional broadcast channels, as well as an extension to the Internet, with added media types, interactivity, and speed.

There are a bewildering number of digital media types. The total of all the standards bodies, proprietary and open, that want to define digital media types for home delivery, including those that want to add interactivity, numbers well over a dozen. Even though many of these standards for media types offer approximately the same end-user experience, they require their own authoring processes and often tie themselves to particular distribution standards. There are overlaps in what many of the standards can do and a great deal of disunity, at present.

The different delivery methods for streaming media also number well over a dozen and some media types can be delivered via multiple distribution channels, whereas others cannot. For example, the video payload of digital television, compressed according to the rules of the Motion Picture Experts Group (MPEG) can be sent through the airwaves directly to a set-top box on a television as a Digital Video Broadcast (DVB) bit stream. However, that same video payload can be wrapped in TCP/IP packets and delivered to the same set-top box, via a cable modem, using Hyper Text Transport Protocol (HTTP). In fact, you could even wrap the video payload, wrapped in TCP/IP packets as the IP payload of a satellite DVB stream, so that the set-top box would receive a DVB stream that contained Internet packets, laden with MPEG compressed video! At the bottom of the heap, it's just data representing moving images. How it gets to you is a mish-mash of complexity and competing standards.

Finally, streaming media is presented to the end-user by a variety of appliances. We've already mentioned the set-top box, but we must also mention PCs (Personal Computers) of various flavors (Windows, Linux and Macintosh, for example) and DVD players. Add to that list game consoles (Sony Playstation, Nintendo, and Microsoft Xbox, for example), wireless handsets and Web tablets. Many of these appliances receive their media streams and render them in particular, often-proprietary ways, using only one delivery method, whereas others can receive streaming media from multiple carriers. It follows, then, that not all receivers can render all possible digital media types. Once again, the array of possible combinations is bewildering.

With streaming digital media, it is the end-user who experiences the largest amount of complexity and confusion, with the possible permutations and combinations of program, media type, delivery channel, delivery protocol, and receiver left to his or her choice. How does granny make sense of all this?

The flip side of complexity is flexibility. The ways in which streaming media can add flexibility are illustrated by comparing and contrasting the characteristics of the most flexible streaming channel compared to, say, television. A good way to begin to imagine the most flexible streaming delivery method possible is to think of the World Wide Web, but with high-quality video and sound, as shown in Figure 2.6.

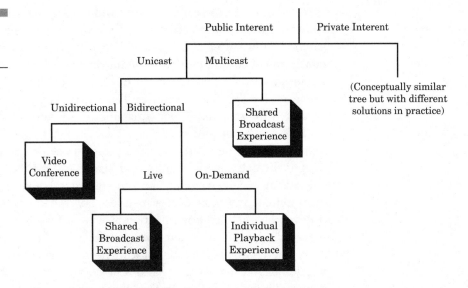

Figure 2.6
Streaming media
delivery options.

As with broadcast television, streaming media can "push" programming to the viewer according to a schedule, deliver an immersive and high-quality experience, and address a large viewing audience. Programs can be broadcast live and program makers can use traditional program-making techniques to create a streaming media presentation. Access to the program can be controlled using conditional access technologies (smart cards, such as might be found on a cable television service, for example). Today, you can literally broadcast a live rock concert direct from a venue to the entire world, while it happens, using streaming media technology.

Unlike broadcast television, however, with flexible streaming media you can access the same video globally, without the need for a different player or any standards conversion. The viewer can request the program on-demand and "pull" it from a server. A person can interact with the program or program producer and with other people currently watching the program or even those who watched it in the past. You can follow links, some of which may be embedded in the video image itself, to other related media or even contact advertisers and find out more about what they are selling. Other multimedia can be synchronized to the video. You have the freedom to replay sections whenever you want, even if the stream is of a live event, and then continue watching from where you left off, or jump to the live stream. It is possible to serve niche audiences with specialist programming even though they might be distributed across the globe. Many more conditional access options are available, so that the producer can control who watches, when and where they watch, and how much they pay to watch. The producer of the program can even let you watch for the next month and then no more. The media can be delivered on either wired or wireless networks and the video quality can potentially be much higher than even HDTV (High Definition Television). Because of the existing open standards for streaming media, compatibility and interoperability issues are less prevalent and everybody has the opportunity to create for the medium. Most importantly, streaming media programs can be located using dedicated search engines and streaming media syndication agents, rather than placing the onus on the viewer to trawl through endless program guides and broadcast schedules. For scheduled events, the viewer can even be e-mailed a notification of when to watch. Finally, if you want to publish a program of your own, you don't need to convince the gatekeepers of the channels, the network owners, to schedule your program for broadcast; you merely post it to a streaming media server and register your media with a search engine. For a program maker, streaming media provides a way to get powerful, emotionally engaging content to a massive number of individuals, either as mass media or as mass customized media. Conversely, it also is economical and feasible to serve niche audiences without having to compete for mass media channel space. Streaming media changes the economics of production and distribution profoundly.

As a new distribution medium, streaming can be both the same as television and yet at the same time completely different. It is a truly flexible way of delivering digital media content to audiences.

No More Downloads

It is important to remember that digital media has been distributed for quite some time, both physically and electronically. In 1969, when CompuServe first started as an online community, digital media artifacts like text documents and later, digital images were already in circulation. Indeed, the act of sending files on a floppy disk to a friend through the mail constituted a primitive form of digital media distribution. When the CD emerged in the early 1980s, it became possible to distribute larger amounts of digital media (music) relatively cheaply, albeit on physical media. Distribution of digital media is not new.

The difference between those early digital media distribution methods and streaming is that in the past, it was an all-or-nothing event. You either had the digital media in its entirety, or you had nothing. If you used a computer connected to a network service, like CompuServe, you had to download the entire file before you could view it or listen to it. For long songs, downloading could take quite a while, given the modem speeds that were once the norm, so if you didn't like what you were downloading, there was no way to tell until you had downloaded all of it.

Let's contrast downloading to streaming, by examining how the process of streaming actually works. Streaming is, in some senses, just a trick of the light. It's really just downloading while playing. The trick is to hide that fact from the end-user.

In *multi-threaded software*, where the computer effectively appears to do two things at once, software engineers can make one part of a computer do the job of obtaining the digital media from a remote source, packet by packet, while another software task is simultaneously sending the data that has already been received to some form of digital-to-analog converter, so that a human can view or listen to the media. So long as the software task that is getting the data from the remote source never gets caught holding nothing by the task that renders the data into user experience, the person experiencing the media cannot tell whether the entire digital media asset is in the computer, or just the parts of it that have been received so far. To guarantee that the data-gathering task is never caught by the data-rendering task, the data gatherer usually starts some time before the data-rendering task begins to render the data already received. This time difference is called the *buffering latency*. It is the reason why, when you click on a streaming media link on the Web, for example, the streaming media player often tells you it is buffering.

Another prerequisite for streaming is that the data-gathering task can provide the data to the data-rendering task at least as quickly as it can consume it. For stereo digital audio at CD quality, that means that the data must flow at a rate of 176,400 bytes per second (just under 1.4 megabits per second). A modem is a device that connects a computer to a phone line, in order to transmit data using the telephone network. Most computers have a modem. At first glance, it would appear that you should not be able to get digital audio to stream at CD quality using a modem that can only deliver a maximum of 56 kilobits of information per second (such as popular modems can do today). There just isn't enough speed to deliver data at 1.4 megabits per second, which the digital to analog converters require in order to create CD-quality audio. The answer lies in compression technology. The audio data is coded in such a way that it can be transmitted in a smaller amount of data, which can then be reconstructed, at least approximately, at the receiver, using the right recipe to decode the compressed data into its native raw format.

A final prerequisite for streaming to continue uninterrupted is that the remote source of the data be capable of delivering the data to the data-gathering task at the same rate as it is consumed. Clearly, if the network connection between the data source and the data-gathering task is interrupted, the stream will be interrupted and will stop. In this case, the end-user of the media becomes painfully aware that the entire digital media asset is not present on his or her computer. If the user were listening to music, it would simply suddenly stop.

Streaming digital media, then, is akin to turning on any other utility, like water, gas, or electricity. When you want it, you turn it on. No waiting. When you don't want any more, you can turn it off just as easily as turning it on. It can be metered by usage, or else made freely available on an "all-you-can-use" payment plan. Unlike those utilities, however, today the provider pays, not the consumer. This is, of course, an over-simplification. The provider must pay for the capacity to serve streams to an intended audience, in both the cost of servers and the bandwidth to connect those servers to end-users. Advertisers sometimes subsidize these costs, provided they think they can reach an audience that will ultimately buy something from them. Finally, the consumers pay in terms of connection charges and on-line charges levied by their Internet Service Providers (ISPs), not to mention the cost of buying a PC to receive streaming media in the first place. If streaming digital media becomes the commodity which delivers our news, entertainment, and information, will that model be sustainable or will it change?

Audio/Visual Web Stuff

When most people think about streaming media, they think about audio and video delivered to the desktop of their personal computer. In fact, this kind of streaming media only came into existence after 1995, when companies like RealNetworks were started, to pioneer the creation, delivery and playback of rich media via the Internet.

Web browsers have pieces of software, called *streaming media players*, which can be installed as plug-ins (or are already built in), that make it possible to play audio or video. To date, most of the players have been available free, as a downloaded component. They use proprietary technology, and player vendors have expended a great deal of effort to make their particular players the de facto standard.

The data required by the streaming media player is delivered to the computer, often using the same transfer protocols that deliver ordinary

Web pages (i.e., HTTP). In fact, there are other protocols used for delivering streaming media, like RTP (Real Time Protocol) and RTSP (Real Time Streaming Protocol), which can allow the delivery of streaming media with more control.

The development of streaming media has been and will continue to be heavily influenced by the development of Internet protocols and companies that use the IP networks to deliver digital data. New traffic management, media synchronization, and quality-of-service protocols will greatly enhance the end-user's experience of streaming media.

Today, a typical home computer with a 56k modem can receive quarter-frame video at a frame rate of around five frames per second (this temporal resolution results in motion that looks jerky), with a picture size of 320 × 240 pixels (very poor spatial resolution, which makes the pictures look impressionistic, rather than crisp and clear). The same computer can render stereo audio in a quality that approximates FM radio reception (fairly good).

If you have a computer connected to a corporate LAN, it is not uncommon to be able to stream video at 750 kilobits per second, giving full screen pictures at near DVD quality. Audio can be rendered at a quality indistinguishable from that of a compact disk.

In the future, there is no technical barrier to receiving multiple simultaneous video images at better than HDTV quality, with full-quality digital surround sound on each. Added to that could be synchronized text and graphics, perhaps even overlaid on the moving pictures. In fact, there is nothing about the look of television that cannot be emulated precisely, given sufficient bandwidth from the host to the player and sufficient processing power to allow television's visual gimmicks, like lower-third straps, fades, alpha blends, page turns, fly-ins, and other digital video effects to be rendered. Many of the latest streaming media players offer some of those capabilities already.

Web Radio

Many people have encountered streaming media as Internet radio. Audio can, in principle, be streamed constantly as a multicast stream to receivers, once again usually on a computer desktop. A multicast stream is a bit stream that is sent once, but can be picked up by multiple computers at the same time. Unlike the bulk of the Internet's traffic, it isn't a point-to-point transfer between two parties.

The important aspect of streaming Web radio is that it can theoretically reach a global audience, for no more than the cost of reaching a local one. Suddenly a vast array of choices and specialized niche programming can be made available to anyone with the means to listen in. With sophisticated jukebox software and scheduling programs, it isn't even necessary for the broadcaster to have a disk jockey. In fact, multicast protocols are generally unavailable on the public Internet, due to router incompatibilities. Today, the broadcast is simulated using a single unicast stream per user, with the content delivered from a continuous, non-stop, live audio program. In this case, each additional user costs an additional amount of bandwidth to supply.

Another thing that is easy to do with streaming media, but not so easy with broadcast radio, is to host interactive play lists, where the listeners choose what gets played next. Unfortunately, the performing rights societies, such as the RIAA (Recording Industry Association of America) that police the playing of copyright material to public audiences have rules that restrict how often you can play a particular artist, whether or not you can play adjacent tracks by the same artist, and so on. An interactive radio station that ignores these rules does so at some peril. If it abides by them, it limits choice and appears less than truly interactive to listeners. Under the current US copyright laws, a radio station cannot allow interactivity, such as skipping songs or rating artists so that they are played more frequently, unless the record companies and copyright holders specifically give direct permission. That means each and every party has to consent. To a Webcaster, this is an onerous restriction. The US Copyright Office has, so far, declined to revise the law.

In the earliest days of streaming radio, it wasn't clear what rate the artists and publishers ought to have been paid for each public performance over the Internet. The outcome of this wrangle was that some Internet stations became uneconomic and withdrew service, citing as reasons the rates the performing rights agencies wanted to levy, plus the fact that advertisers wanted to pay no more to stream their ads. Advertisers were also upset because some of their commercials were being rebroadcast in Web streams without authorization. Advertisers had paid actors in some commercials a higher rate if their ads showed up on the Web. Advertisers didn't want to be paying the talent unless they were buying the time.

Some major industry players, such as Clear Channel Communications, have recently resumed streaming. Now they create Web-only advertising for their streams. The issue of what Web-streaming companies should pay for copyrighted music on the Web was the subject of a ruling, in February 2002, by the US Copyright Office. At the time of writing, the decision has yet to be endorsed by the Library of Congress. Under the terms of the ruling, which is thought to levy fees retroactively to October 1998, commercial Webcasters are required to pay 14 cents per Internet-transmitted performance of a sound recording, per Internet listener. Commercial radio broadcasters that simultaneously stream their program output to the Web are required to pay only 7c for the same mechanical performance. Non-profit Webcasters are required to pay 5c per performance and 2c if the same recording is also broadcast via the airwaves.*

Today, many office employees leave streaming radio on all day as they work. News feeds could also be streamed in this way, though most Web streaming is predominantly music. When wireless broadband networks deploy, it will be possible to stream music to cars. Then it will theoretically be possible to listen to your favorite radio station in a rental car, no matter where in the world you happen to be. You can catch up on local news even when you are abroad. Indeed, with wireless networks and appropriate devices, you could listen to any station in the world, wherever you are, any time you like. As a competitor to broadcast radio, streaming radio is formidable.

Another future use of Web radio is to allow listeners, for an appropriate fee to, capture audio files as they stream. This would let listeners fill their personal jukeboxes with music they like, so that they could listen to it again, when they aren't connected. This would permit great flexibility. Users would have to pay to store the music, but would avoid the connection charges. In this world, CDs become a less-attractive proposition, since listeners could accumulate the music they like, track by track, discarding tracks they don't like. It would be possible to include all of the CD's artwork and notes along with the stream, for display on the device that stores the music. Indeed, with streaming audio, you can store music in less space than would be used on a compact disk, because of compression algorithms commonly used with streaming audio, that didn't even exist when the CD standard was laid down.

With Napster, a file-sharing utility that allowed people to swap music online (in flagrant contravention of copyright laws, more often than not), it was possible to listen to a piece of music while it was downloading, but the limitations of the bandwidth between peers connected via Napster's

---

*Analysts are predicting that this ruling will tend to eliminate non-profit Webcasters and favor existing commercial radio network owners, at the expense of consumer choice and programming varierty. It is obviously too early to predict the eventual effect of this ruling. For some comment and analysis, see: http://www.thestandard.com/article/ 0,1902,28450,00.html or http://www.newsbytes.com/news/02/174774.html.

peer-to-peer networking software meant that the music almost invariably stopped part way through, since the speed of the download was not as fast as the speed of playback. When broadband connectivity becomes ubiquitous, this problem will vanish. The copyright issues, however, are likely to be around a while longer.

Web radio would be alive and thriving today if the regulators and industry bodies could work out their disputes, and if broadband networks were widely deployed. However, today the industry waits with bated breath. One day, Web radio is going to be great.

Video on Demand

Another killer application enabled by streaming media technology is video on demand. Streaming media technology allows you to see any program you want, whenever you want. When video is made available for streaming, it is loaded as a file onto a streaming media server. The server then handles individual connections from machines that connect to it to request a video stream, providing a bit stream containing the video payload to the streaming media player at the other end of the network connection (Figure 2.7).

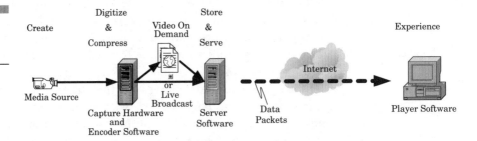

Figure 2.7
Streaming media model.

The beauty of streaming video on demand is that video material can lie dormant on a server indefinitely, until somebody comes along to play it. There is no need to schedule it for airing. There is no need to attract a certain-sized audience to justify the decision to screen it. This means it is possible to make video on even very specialist subjects available all the time, without having to worry about the number of people viewing it at any one time. Not everything has to be a blockbuster.

With digital rights management, it is possible to control access to and collect payment for any video delivered on demand. A video copyright owner can license a viewer to play the video a set number of times, or even to store it. In fact, if the cost of the bandwidth needed to stream a video drops dramatically, there will be little incentive ever to store it at home, since if you want to view it, it will be cheap enough to stream it once more.

Today, streaming video can be viewed with a PC. There aren't any television sets or game consoles that can stream video from the Internet. That won't always be the case, however. When broadband connectivity takes hold, there will be a plethora of devices to receive streaming media. Some will be in your living room; some will be in your kitchen and study. Others will connect wirelessly and deliver video on demand to your car, or perhaps to a handheld device while you are on a train or plane.

If in-flight entertainment systems were actually closed streaming video networks, with a video-on-demand server serving near DVD quality compressed video, they would be lighter in weight and less susceptible to dirt, age, and vibration than today's tape-based solutions. Passengers could opt to view the streaming media on their laptop computer or handheld device, if they wished, or else the airline could stream to seat-back screens as they do now. It would even be possible to provide personal video glasses, in much the same way as airlines issue headsets for audio, which would screen the video onto tiny personal screens set into the glasses frame. A streaming in-flight entertainment system could easily provide audio, video, games, text, and so on. As for updating the on-board streaming media server with the latest news and releases, this could be done via satellite while the plane was en route. It would also be simple enough to change a disk pack containing DVDs with all the new entertainment.

When broadband networks with guaranteed quality of service are ubiquitous, when the players are everywhere, not just on the computer desktop, and when the regulatory and copyright issues have been ironed out, streaming media on demand is the application of streaming media that could have the largest impact. It is a compelling application and all the technology already exists.

Distance Learning

The fastest-growing education movement in the US is home schooling. Today, roughly 1.5 million children learn at home. The trend is a backlash against a school system that many believe asks too little of students. According to John Taylor Gatto, author of *The Underground History of American Education*, schools are "irremediably broken. Built to supply a mass production economy with a docile workforce, they ask too little of children and thereby drain youngsters of curiosity and autonomy."

Distance learning using streaming media could be the new way people are educated in the future. The economy demands brainpower. Continuous learning throughout our entire lives, from childhood into adulthood and even on into retirement, will become an essential life skill. Education needs to push human beings to become big, self-directing, independent, and able to write their own life scripts. In the opinion of many, the current schooling system does not and cannot fulfill this role. Distance learning with streaming media content can potentially get people to learn more effectively because learning can be made thrilling.

The major access road to self-development is raw experience. Memorizing notes from the blackboard is not real work. Interacting with the finest instructors available and being challenged to explore knowledge in your own unique way and at your own pace is what streaming distance learning promises.

Today, numerous schools and colleges offer distance-learning courses on the Web, using streaming media to deliver some of the materials. In the UK, the Open University has used media technology effectively for decades, helping thousands of people obtain qualifications they might not otherwise have obtained. Streaming media offers more flexible and more compelling content creation options. Of course, the courses that have begun to exploit streaming media technology have only scratched the surface of what is possible. With Web elements and graphics synchronized and blended with moving video, first-class learning materials drawn from the finest minds can be developed. More importantly and uniquely, distance learning with streaming media allows rich collaboration and interaction between students and their peers, or with tutors, in real time. *Telepresence* is the experience of being present in an environment by means of a communication medium. For the first time, a cost-effective technology has become available that supports telepresence; this is the significant feature of streaming media that will take distance learning beyond what is possible with older audiovisual media.

When the great and the good begin to make distance-learning materials with high production values, the question becomes "what will happen to mediocre educators and schools?" The answer is that they will be swept away. However, before that can happen, production tools and techniques will have to become simpler and cost less.

There is undoubtedly money to be made serving niche audiences worldwide with on-demand, high-quality learning materials. One of the first niche applications that will influence the uptake of streaming media technology by the mass market is likely to be distance learning.

Synchronized Multimedia

I recall being taken in the early 1970s to our local town hall by my father to hear a lecture on science. A popular scientific writer of the day, whose name now escapes me, was on tour lecturing, to workingmen and their families who lived in the heavy industrial town of my childhood, about the wonders and sheer unimaginable scale of the universe we inhabit. Armed with mere lecture notes, a lectern, a modest public address system, some spotlights, and a bank of 35mm slide projectors, he proceeded to enthrall us with dissolves and wipes, as one image melted into another, aided by a dramatic narrative, tasteful lighting effects, and the unfathomable darkness of the auditorium itself. I was instantly transported across the vastness of time and space in my imagination, with commentary seemingly from the voice of God himself. It awakened in me a lifelong passion for cosmology and astronomy.

I recount this incident only to illustrate the profound and lasting impact that can be made with simple synchronized multimedia (provided that the subject material is compelling and the presentation authoritative). With streaming media, it is possible to create virtual slide shows more breathtaking than the one that changed my life as a child. Text and graphic elements can be synchronized and timed against video and audio elements, with active links to sidebars that contain more detail about particular elements. Those sidebars may even contain more video. In-picture elements, tied to the field, can be embedded into the video, allowing sophisticated overlays, and perhaps even advertisements and product placements, to be included, customized for each particular viewer. Indeed, the technology exists to present those synchronized multimedia presentations not just to a PC desktop, but also to an entire auditorium. Using modern digital light projection systems and super-high-resolution streaming, cinemalike multimedia presentations can be delivered from nothing more powerful than a laptop computer to a town hall-sized audience!

The wonderful sound and light show I experienced as a child earned money for its presenter. He sold tickets, people came, and he put a monetary value on the information he had to present. Of course, he had to subtract production, travel, equipment, and advertising costs, but he made a living at it. There was no other way he could present what he knew to an audience as specialized as the one he attracted at a time when there were only five television channels available to us and where only the popularity of Hollywood blockbuster movies helped local cinemas do a little better than break even.

Today, that same presentation I enjoyed in my youth could be streamed to a worldwide audience, on demand, in a more compelling fashion than ever before. Indeed, the content could be changed so that every single viewer experienced a different path through the presentation. For the presenter, the benefits are obvious. No longer must he endure the rigors and privations of life on the road, nor does he have to pay quite so much for the equipment to mount the presentation. However, until pay-per-view solutions and broadband networks evolve, it is difficult to charge money for the content, as was possible at the town hall. Nevertheless, the technology will undoubtedly arrive soon, and then we should expect to see a flood of life-changing synchronized multimedia suddenly available.

The open standards that make synchronized multimedia possible on the Web include Dynamic Hyper Text Markup Language (DHTML) and Synchronized Multimedia Integration Language (SMIL). However, there are many others, including proprietary quasi-open standards like Macromedia's Shockwave. On television set-top boxes, interactivity of this type is carried using a number of proprietary and open protocols. There are many ways to produce and deliver synchronized multimedia. Indeed, the problem for content producers is the sheer number of incompatible standards for authoring such presentations and the need to use particular tools to cater to each. Content producers currently face the daunting task of recreating their material for each distribution method and standard, or else of limiting their audience by choosing just one or two to support. This acts as a significant barrier to the wider adoption of the technology. There are just too many incompatible ways to do what, to the audience, looks the same. The pressure for standardization is great and will continue to increase, until the problem is solved for content producers.

Simulcast

Part of the content authoring problem is that when you originate a streaming media presentation, you typically author for a specific platform. We have already noted that in the future, the likelihood is that different streaming media receiving devices will render different media types, depending on their capabilities and that this will continue to be the case for quite some time. Imagine now that you have two media receiving devices side by side—one that can render audio only and another that can render video only. The rational expectation is that the video and audio would play back in sync, even though they might be coming down different distribution systems and even though they are rendered by different, unlinked devices. A further implication is that the content producer authored twice, once for each platform.

Unfortunately, there is currently nothing that guarantees that those two devices will render in sync with each other, but it is technically feasible to arrange things that way. With streaming media and appropriate standards, content authors can create and send packages of digital media, of varying types, together as a single bundle, with synchronization information embedded in that bundle. Receivers could choose the media types from the bundle to render, according to their capabilities, and render those in time with other devices. For authors of streaming content, the ability to create one bundle of media for all possible players is clearly attractive.

Companies like Spotmagic are developing simulcast solutions and protocols. The logic behind simulcast authoring is compelling, since it enables streaming media producers to embrace the medium cost-effectively, thereby driving its growth. Unfortunately, the problem is not well enough understood for mass demand to compel the creation of simulcast solutions. This ought to change in the future. The need for simulcast exists, even if it isn't yet widely recognized.

Mobile Streaming Media

Coming to a phone near you very soon will be full motion video. Unfortunately, today's cell phone networks don't have the bandwidth to do a good job of transmitting streaming video, and the screens are tiny, low-resolution affairs. Streaming video to mobile devices is not considered very credible, even as a future technology, because cellular networks are still prone to quality-of-service problems, such as signal drop-out and bad audio quality. These problems will need to be resolved before video can reliably stream from handset to handset. However, rest assured that these are mere temporary technical obstacles.

The range of mobile devices that already use streaming media is growing and will continue to grow. You can already buy a wristwatch that includes an MP3 (MPEG-2 Layer 3 audio compression) player, as well as a device that looks like a portable CD player, but can in reality store your entire music collection. MP3 players are also finding their way into mobile phone handsets. Just because a device is mobile, that does not necessarily imply a wireless network connection is present!

There is evidence that the industry has a commitment to wireless mobile streaming. We have already seen prototypes of streaming media cameras, which can create compressed video streams and send them to the Web directly. The mobile operating system vendor Symbian has ported the RealNetworks RealPlayer to its EPOC platform. Windows Media Player is available on the latest Windows CE devices, including palmtop computers and Web tablets. The foundations are being laid.

The Japanese have pioneered the use of streaming video on mobile devices. The Japanese telecommunications company DoCoMo has created services like video dating, where people can see each other on their handsets before they decide whether or not to make a date. No doubt, many more applications will spring up.

Device designers will create imaginative new mobile streaming products. For example, today you can buy portable DVD players, which can play a movie from a DVD disk on a small portable screen. With the addition of a third-generation cellular network connection, that same device could display streamed video. Indeed, if the DVD drive had the capacity to record as well as play back, the stream could be recorded onto a disk for later viewing. All the convenience of a VTR would be available in a portable device. The technology to make this feasible already exists.

Full-duplex mobile streaming devices, which display an incoming video stream and simultaneously transmit from a camera, will have numerous important applications. For example, an ambulance could be equipped with such a device and the patient attended to remotely by a doctor, while the patient was en route to hospital. The doctor could examine the patient with the video camera and the vital signs could be streamed back the hospital at the same time. The doctor could then advise the ambulance crew on the best course of treatment. Just as two-way radio is used today, tomorrow, richer information will enable better diagnoses during those vital minutes on the way to hospital.

Streaming Chat

One of the most popular applications on the Web, next to e-mail, is instant messaging and chat. Millions of people talk to friends, colleagues, and family every day using this technology. Yahoo!, CompuServe, AOL, and MSN attract millions of people everyday in their text chat rooms. Recently, many of these services have begun to include voice and video chat features.

PalTalk is one of the more popular on-line destinations for those seeking a live video and audio chatting experience. The service is free and includes text, audio, and video chatting either in chat rooms, or one to one, using a local ISP for global connectivity. On PalTalk, you can make an audio connection with somebody internationally, at local phone call rates. The service has its limitations. Audio quality is not high, there are perceptible time lags involved, the call is not truly duplex, in that you cannot talk while listening, and there are times when the service is unavailable. On video, the update rate is very low, with perhaps only one frame per second being the best performance achievable. However, the amazing thing is that you can do it at all! In the future, as bandwidth becomes more abundant and these types of service providers deploy better video and audio compression techniques, it is highly likely that they will make a significant dent in the revenues of long distance phone companies.

Many of these online audio and video messaging services can also communicate with regular telephone handsets (streaming to phone). Indeed, Voice over IP, a technology that is creating interest in traditional telecommunications circles, can actually be thought of as streaming audio, on a person-to-person basis.

Another technology that holds some promise in the area of streaming chat is Voice XML. VXML technology allows a user to interact with the Internet through voice recognition technology by using a voice browser and/or a telephone. The precursor to VXML was the IVR (Interactive Voice Response) system. The user interacts with the voice browser, such as the one made by Conversa, by listening to audio output that is either prerecorded, computer synthesized, or instantaneously digitized from a live audio source. The user submits audio input by speaking into a microphone attached to the computer's soundcard, through a telephone keypad or by talking into a telephone. Audio streaming technology is at the heart of these applications.

In principle, there is nothing to prevent the creation of a Web site that allows real humans to converse with each other, using VXML technology to loop one person's audio input back as output to another person. Indeed, some of the more compelling applications of VXML may be e-commerce sites you can talk to which answer back. The answer may come from a voice synthesizer or from prerecorded information messages, or else there may be a real human being at the other end, answering the query. The point is that users will not need to dial a number or do anything special to start the dialog with the e-commerce vendor. All they will need to do will be talk to their computers while on the site of interest. Mobile applications ("M-commerce") will also greatly benefit from VXML technology, since telephone handsets are optimized for audio communications.

Corporate Communications

Companies exist for the sole reason that information flows more freely and more richly within them than in the open market of competitors, customers, suppliers, and partners, according to Nobel Prize-winning economist Ronald Coase.* There are "transaction cost" advantages associated with having a company, predicated on the company's ability to coordinate activities, as a result of efficient internal communications. However, streaming media and other forms of electronic communication change the economics of using the open market, in comparison to using an internal department. If a corporation is to maintain any competitive advantage, against the open market, for a range of services it currently sources internally, it will need to lower the costs of internal communications and make them much richer.

Whether a company chooses to outsource or insource, streaming media provides richer and more cost-effective communications, over a wider geography, than any previous media technology. Companies like VideoShare are providing streaming technologies to serve this application.

Corporations can narrowcast their annual general meeting to shareholders across the globe, with no more than an eight (or so)-second delay. This allows market-sensitive information to be delivered in a fair and managed way, without giving some viewers a benefit over others.

From a human resources perspective, streaming media has many applications. Staff alignment, training, internal company news, procedures, and new recruit induction can all be delivered using streaming media, both on demand and live. Indeed, many companies may encourage outside-hours access to employees, secured using Virtual Private Networks (VPNs), to allow skills improvement and training, using the company's own internal training materials.

Streaming media employed to enrich extranet content between a company and its key partners is another obvious use of the technology. To see regular, timely updates of information, presented in accessible and high-impact ways, provides added value for both parties.

As an adjunct to a sales force in the field, streaming media is possibly unparalleled. Up-to-the-minute competitive information, sales training materials, marketing communications, and tactical communications col-

---

*This proposition was first put forward in Coase's ground-breaking 1937 book, *The Nature of the Firm*. Coase won the Nobel Prize for Economics in 1991. A reprinting of many of his seminal works, including a chapter on the nature of the firm, is currently available in the book, *The Firm, The Market and The Law*, by Ronald Coase, published by the University of Chicago Press, 1990 (ISBN 0226111016).

lateral would all be more accessible and more effective if presented as streaming video with synchronized rich media. The costs of printing and distributing hard copy would be mitigated. The sales force could be trained directly by engineers in the lab, without the need to travel. For certain categories of product, sales people often encounter questions that they must refer back to engineering when trying to make a sale. The ability to ask the question while still at the customer site and have an authoritative answer instantly could be a positive factor in making the sale. The sales staff may even send back video footage of the environment in which a piece of equipment will need to be installed to prepare the installation engineers and get an assessment of whether or not the environment needs modification, prior to making the sale.

For product launches, streaming media also provides many attractive advantages. Consumers can watch as the product is unveiled and footage of the product launch itself repurposed in other marketing communications packages thereafter. The best product demonstrator or salesman that a company has can do his or her pitch on streaming media and this can be indefinitely accessed directly by consumers, or else as part of a sales presentation.

Companies are under closer consumer scrutiny today. If a company has factories in Thailand, for example, it may be subject to accusations of using child labor or of otherwise mistreating employees. With streaming media, such companies could stream pictures from their factories live, for the entire world to see. This would provide proof of their ethics and standards and reduce consumer skepticism. To take this a level further, what would be more reassuring than seeing a video of your built-to-order car or PC at every stage of its manufacture? Streaming media can make applications like that possible, because it is digital and because it drastically changes the economics of making and delivering such a video.

The video technology for corporate communications has existed for a long time, yet isn't used as much as it could be, for three reasons. First, distribution of video was costly and cumbersome in the past. With streaming media technology, those costs fall dramatically. Compared to renting satellite time or duplicating VHS tapes, as was once the norm, the cost of deploying a video stream is much lower.

The second reason has to do with the costs of producing video material in the first place. Digital video cameras and PC-based nonlinear editing software now cost a mere fraction of what a professional quality camera and tape-based editing system once cost. Since the equipment is digital, there is less need to worry about video quality loss throughout the distribution chain, as was once the case with analog video. This also means that the source material need not be as high in quality as was once necessary, since less is lost. What this means is that the video that the end-user sees looks better, but costs less to originate.

Finally, video has not been as widely used as it could be, not just because of lack of production facilities, but, more importantly because of a lack of in-house production skills. What companies require are freelance production companies that can script and storyboard the presentation to be made, arrive on the premises, shoot the video unobtrusively, and then post-produce the package with high production values. What they do not want is production staff that acts precious, wastes time, throws tantrums, or does the job inefficiently. Unfortunately, video production, as an industry, has historically had a tendency to attract people of that nature. What companies who wish to produce video for their corporate communications need are production companies that have a "meat and potatoes" attitude to what they are doing, who do not consider their output to be high art, but believe in doing their job with integrity and to a high standard. Fortunately, those companies are becoming more prevalent, as more and more people opt for media careers and media studies training.

Streaming Cameras

Telepresence applications, where the viewer experiences being present in a distant environment by means of a communication technology, are abundant. Streaming media makes it possible for viewers to be virtually everywhere.

Cameras that incorporate video and audio compression hardware and can serve streams directly to the Internet are becoming available. Today, most Web cams actually send a series of still pictures progressively, but with the compression techniques commonly used with streaming media, full-motion video can be transmitted directly from the camera, using very little bandwidth.

During the Afghan war against Al-Qaida and the Taliban, viewers of many of the major broadcast television networks watched war correspondents filing live reports using satellite phones to transport compressed video and audio to the world. Admittedly the quality was not award winning, but the fact that this could be done at all, with equipment costing no more than a few thousand dollars, is truly remarkable.

Many applications for streaming cameras are undoubtedly surveillance applications. However, some applications encourage tourists to visit locations in far-flung places. For example, Africam.com transmits pictures of African wildlife, via the Web, all day, every day. People accessing the Web site can see what is happening on the other side of the world any time they care to watch. What better way is there to attract tourists than a free sample of what they would see if they visited? Discovery.com allows children to view sharks swimming menacingly in an aquarium in the US. Rocketry enthusiasts send wireless, streaming cameras up with their model rockets to video the flight. All these applications allow people to experience things they may never have had the opportunity to experience in real life.

Companies like National Semiconductor are working on integrating streaming video compression hardware on a single silicon substrate, shrinking the camera and its processing electronics to the size of a large die. These components will one day find application in handheld devices that use third-generation (3G) cellular telephone networks or 802.11 (WiFi) wireless networks to send instantaneous live streaming video straight from image sensor to the World Wide Web or another hand-held device.

Integrated streaming camera chips would be incredibly easy to install and conceal, which presents both opportunities and threats. On the one hand, it will be possible to get hitherto impossible "eye views" of places and events. It would, in principle, be possible to have a small streaming camera attached to your luggage, so that you could receive a real-time view of what your luggage could see, if it ever became lost. Similarly, a thief making off with your car could be caught in the act and video of his face transmitted from the stolen car in time to identify and capture the culprit. However, the ability for nearly anybody to create private spy networks has implications for privacy, national security, and civil liberties that must be sensitively and intelligently handled by legislators and governments.

Special Interest TV

Just as desktop publishing software made it economically feasible to create a vast array of special-interest magazines and publications to appeal to narrower and narrower niche interest groups, so desktop video production and desktop media streaming change the economics of creating video content for much smaller special-interest groups than can be addressed profitably using traditional television production and distribution technology.

The underpinning of the economics of streaming to niche audiences, just as with specialist magazines, is the fact that the audience is pre-qualified to advertisers that sell products and services relevant to that niche. If you sell auto parts, a good place to advertise is alongside specialist content, which is designed to appeal to an audience of guaranteed auto enthusiasts.

From the point of view of the audience, the attraction of specialist niche programming is that the content does not have to be "dumbed down" to appeal to a mass audience. Instead, programs can assume the audience has a certain degree of competence and knowledge in the subject matter. The audience, therefore, gets the feeling of being treated as colleagues, not of being talked down to, like novices.

Even cable channels fail to create this collegiate audience response. For example, in the UK, popular science cable TV channels, such as Einstein TV, still present material with narrative content that is somewhat insulting to the most scientifically literate in the audience, mainly because of a need to appeal to nonscientists, in order to keep audience viewing figures high, so that advertisers will be attracted. A streaming media presentation is still economical even with lower audience figures, so the need to appeal to the least educated in the subject matter, at the expense of annoying the most-educated enthusiast, is not there. Niche streaming media programming can segment audiences into novices and experts and create programming to suit those narrower audience segments. It also allows advertisers to segment their market more narrowly, with appeals to various levels of viewer expertise. When access to distribution channels is not limited, as it is with broadcast, satellite, and cable television, there is no need to make every program with mass appeal in mind.

Of course, the grammar of the visual editing used in mass-market documentary and factual programming must change for niche appeal streaming. Whereas with broadcast television the narrative can spin out the story to fit the available program time, specialist audiences, with the ability to skip through the boring bits, are far less tolerant of being teased and of listening to a story being told to them more slowly than they want. Program makers need to recognize and embrace this difference between streaming media programming and broadcast television's dictation model of information delivery.

The final reason why niche streaming will take off is found in the medium's ability to aggregate an audience from a wider geography. For example, people interested in junior swimming meets may live all over the world. With broadcast television, the geographic reach limitations that apply mean that no single channel can ever amass a sufficient viewer base to make a program on those swim meets pay. However, if you can draw an audience from the entire world, irrespective of time zones, it becomes viable to produce and distribute programs that appeal to that specialized audience. Streaming media technology makes it possible.

Streaming Media and e-Commerce

Streaming media content can enliven the e-commerce shopping experience. By presenting richer information, over a global reach, online shoppers can get a better feel for what they are buying. In some cases, they can even use streaming media to inspect the actual object they wish to purchase remotely. This is almost, but not quite, as good as being there in person. It is far better, of course, than not being there at all. In some cases, because of issues of consumer trust and need to examine the goods before purchase, some e-commerce applications are not viable at all without streaming media.

Already, streaming media is being used to create virtual-reality experiences. People can walk through a home they may be interested in buying, using three-dimensional camera technology to give a 360° $\times$ 360° degree view of every room in the house. Clothing can be modeled on virtual catwalks, with professional models demonstrating how the clothing will move and flow when worn; something the average clothing store cannot routinely do.

Of course, with items like clothing, people still like to try on their garments before they buy, but streaming media can help the onlines hopper eliminate garments that definitely don't feel right, before completing the purchase in a physical store. The added richness and reach provided by streaming media short circuits the consumer's process of navigation from all the possible products to the one chosen. A smart clothing retailer or manufacturer seeking to build a brand will embrace streaming media as a way of drawing attention to its particular offerings and away from the chance encounters typical in a mall. This process of helping the consumer choose a particular offering is also attractive to the consumer, because it lessens the search costs associated with looking for something, in the absence of those navigational aids. For example, if you want to buy a pair of jeans, knowing the location of a pair that looks just like the ones you wanted, viewed online via streaming media, is much simpler and takes less time than visiting every store in a mall (or in several malls) looking for a similar pair of jeans.

Those with strong brands, which already serve as short circuits in the consumer's purchase navigation process, will use streaming media to further strengthen consumer affiliation with their offerings and make purchases of their branded articles more likely.

On-line car dealers can entice you with photographs of cars for sale. Some even do a three-dimensional model of the car so that it can be viewed from all angles and in several color options. Streaming media can enhance this by adding a driver's-eye-view video of the car being driven on a twisting country road, for example, complete with engine noise! The greater the richness of information that can be provided, using streaming media, the more likely it is that the sale will be made.

Many products use live demonstrations in order to sell. For example, power-tool makers often have in-store demonstrations of the virtues of their latest models to shoppers in larger home decorating stores, such as Home Depot. The reach afforded by these in-store demonstrations is limited to the number of people that will pass the demonstrator for the duration of the demonstration. Very often, passersby will miss crucial parts of the demonstration through arriving at the wrong moment. Those product demonstrations can be made available online, on demand, using streaming media technology. People interested in buying a power tool could access the demonstration at their leisure, skipping forward and backward, to see the unique features of that tool. The presentation could include synchronized multimedia to present text sidebars explaining the details of the product, feature by feature, while the demonstrator presents. Indeed, consumers may even interact live with a product demonstrator streaming his presentation from a store in another city. Then, the consumer has the option of either buying online, or else visiting a physical bricks-and-mortar store to get a closer look. Either way, the consumer is better informed, more engaged, and able to compare models and prices before making a purchase.

Broadband is a prerequisite for e-commerce streaming, since the video has to look as good as television to attract viewers. That means that streaming e-commerce applications, from business to consumer, are in the future. However, there is already a compelling case for the business-to-business streaming of products, especially where technical three-dimensional models can be exchanged over a network. The use of streaming media to add richness to business-to-business sales is a likely first step, since the bandwidth already exists. Business LANs and interconnections are already, for the most part, broadband.

Vendors could save vast amounts of money making product demonstrations available online, rather than sending a mobile sales force on the road to do the same job at each potential customer's site, customer by customer. With a streaming product presentation online, the account manager would only need to show up at the customer's premises to forge the relationship and obtain the order, if even then. There would be far fewer wasted sales calls, and sales staff could focus their efforts on answering specific customer questions and creating new business, rather than on doing repetitive demonstrations.

Independent Film Making

The vast majority of feature films made never get distributed and hence never find an audience. It is staggering to think that more films, by far, are made than the tiny number that are released and watched by paying customers. The reason this is the case is that the channels of distribution are so limited. The exhibitors, the people responsible for deciding what to screen, who depend on "bottoms on seats" for their revenues and profits, want to minimize their exposure to risk. What they ideally want is a name director and an all-star cast, telling a great story (preferably a sequel of a tried and tested box office smash hit), using lots of stunning special effects. Blockbusters are the staple diet.

What chance does an independent filmmaker have of reaching an audience, particularly if the subject matter of the film is challenging or controversial? Not even videocassette distribution guarantees an audience. It is a little-known fact that many of the larger video rental chains do not actually buy their stock. Rather, Hollywood studios place product in those stores, sharing the rental revenues with the video rental chain. Independent video rental chains cannot afford to waste shelf space on quirky, challenging films either. In the first place, they could fill the same shelf space with more well known or studio subsidized cassettes; in the second, they cannot afford to own a large stock of films which go in and out of fashion so frequently; finally, they have a greater chance of making a return from renting blockbusters supported by multimillion dollar marketing campaigns than they do from unknown filmmakers working on a shoestring. People like what they know. Marketing campaigns help people know particular films.

With the advent of streaming media, sites like Atom Films have been established to create a portal for independent filmmakers to display their talents. Granted, the same problem of getting the audience to know about a film is ever present, but at least it is available to watch as an on-demand stream, should a viewer ever discover that the film even exists. The traditional film distribution bottleneck is bypassed.

Today, with broadband penetration still relatively limited, the market for independent films is still not strong. However, as the quality of streaming improves, so the exposure of independent films to audiences will no doubt increase. Indeed, the most pioneering uses of the new streaming medium are likely to be found among independent film makers, who tend to push the envelope of any new medium's capabilities, taking advantage of its unique characteristics and properties. Already some filmmakers have dispensed with film altogether and are shooting, editing, and post-producing their creations using digital cameras and editing systems. Digital distribution via streaming is a natural extension of this burgeoning digital filmmaking workflow.

It must be remembered that streaming media allows immersive interactivity, as in three-dimensional computer game play and the use of some media types to support story telling with another. One of the cleverest uses of digital multimedia in recent times was the marketing hype created on the Web for *The Blair Witch Project* film. Here, the filmmakers used the Web to create an illusion about the making of the film, which tended to make the audience see it as more documentary than fiction, thus enhancing the film's ability to affect the audience's emotional response to the film. The Web hype made the film seem more frightening and horrifying. As a consequence, the film did very well and made much more money than it cost to produce. Hollywood got a blockbuster for a bargain price. Although streaming played very little part in this film's success, expect to see novel uses of multiple streaming media types in the promotion of future films and to enhance the audience's emotional response. Filmmaking will become more of a multimedia authoring craft, rather than limiting itself to traditional filmmaking techniques and visual grammar. Streaming media changes all the rules.

D-Cinema

As we noted in the previous section, video compression and the relentless progress of Moore's law has made film stock redundant. It is now possible to create a high-quality feature film, from camera lens to projector, entirely digitally. Streaming underpins this "glass to glass" process. D-Cinema (D for "Digital") is the name given to the digital distribution and presentation of feature films, in place of prints made on traditional film stock.

In D-Cinema, films are delivered as digital media files, often on DVD disks. These are uploaded onto a high-resolution video server, using hard disk arrays for storage. A decompressed stream of digits is fed from the video server to the digital projector, on playback, and the digits are used to switch tiny micromirrors on a silicon die, one per pixel, to project an image onto a screen. Projectors based on micromirror technology have 750,000 micromirrors to work with. Each mirror can cycle 1.7 trillion times, at rates of up to 50,000 cycles per second, before failing. That equates to nearly 95 years. The brightest projectors currently use 13,000 ANSI lumen lamps. These projectors present images superior to those obtained with film projectors, with contrast ratios of 1000:1 typical of the new digital projectors. Other digital projector technologies exist. Some of these can present QXGA (Quadruple extended Graphics Array) images of 2048 × 1536 pixels.

The advantages of D-Cinema distribution and exhibition over traditional film-based methods are manifold. Streaming adds unique capabilities to D-Cinema, although D-Cinema distribution is possible without streaming. Streaming always takes place on projection, however.

Creating film stock prints of a film for general release imposes significant costs on the distributor of the film. Prints cost over a thousand dollars each. At the most competitive rates today, a print can be made, mounted, and cased for approximately $800. Consider that the average release requires 1000 to 8000 prints, depending on the distribution schedule, and that a single print of a five-reel film, in two cases, can weigh approximately 35kg. Making a few thousand copies and sending them around the world under secure conditions costs a lot of money. It is one of the reasons why some films are released in the US ahead of the UK. Besides making it possible to market the film more intensively, by concentrating on individual geographies at a time, the release prints made for US audiences can be reused in the UK.

Release prints made on film stock degrade. They pick up scratches and dirt and sometimes tear and must be respliced. With a digital "film," however, every screening is made with what appears to be a pristine, perfect print. Because it is just files consisting of digits, played back from highly redundant arrays of hard drives, digital film never degrades with use. Also, there is no need for a projectionist to carefully time the changeover from one reel to another, since the entire film can be played back as an unbroken stream. Sound and picture quality is extremely good, because enough bits are allocated to delivering a detailed payload. There is also no image jitter or weave, which is caused by mechanical variations in how the film travels past the projection lens. Because the only moving parts of the digital projector involve micromirrors and cooling the system, digital projectors are potentially more reliable and produce very stable images.

Streaming technology gives theatrical distributors the ability to release a film simultaneously to the entire world, with strong enough encryption and security to make digital piracy prohibitively expensive. This could severely curtail the activities of video pirates who steal first-release prints and make thousands of illegal copies almost overnight. Hollywood is still nervous about the ability of encryption methods to guarantee controlled access to multimillion-dollar assets, but there is increasing evidence that digital rights management technology will be sufficiently strong to make illegal copies too expensive to attempt.

The biggest problem that adoption of D-Cinema faces is the question of who pays for the entire digital infrastructure. Exhibitors are not keen to invest in expensive video servers and digital projection systems, if the film distributors are the ones making all the savings. There is a satisfaction with the status quo in individual cinemas. In many cases, the equipment has already paid for itself and each additional seat sold contributes to profits. If all the old projection equipment is suddenly obsolete, the revenue generated from individual ticket sales will most likely go to pay for the new equipment. This instantly erodes the operating margins of the cinema business, since consumers are not willing to pay a high premium for digital screening.

Because of fears about digital encryption and because transport of streams to cinemas is still relatively expensive and difficult, Hollywood is still opting for armed-guard delivery of digital presentations to individual theatres. Streams can be delivered via satellite, but current technology requires on the order of 45 Mb/s to stream the presentation in real time (where it takes one hour to deliver an hour's worth of media). This is a problem, since satellite bandwidth is still relatively expensive and 45 Mb/s exceeds the capacity of single-satellite transponders, meaning multiple transponders must be used, or else the payload is delivered slower than real time. The wider the release, the better the economics of multicasting to cinemas worldwide.

There is some dissent over whether or not 45 Mb/s is really necessary to render a high-resolution film image, or indeed if such lavish bandwidth usage is justified for all films. With compression technology improving, better results can be obtained every six months or so with less bandwidth. Meanwhile, metropolitan area networks with fiber optic connections are becoming capable of sustaining such data rates to individual points of presence. What is certain is that the delivery of high-bandwidth streams to many places on the globe simultaneously will one day become cheaper and relatively routine.

When streaming D-Cinema material becomes cheaper, due to Gilder's Law* (roughly stated as "bandwidth per unit cost triples every twelve months") and the relentless improvements in the hard-drive storage capacity of video servers, the necessity of playing to mass audiences is somewhat lessened. Exhibitors may find it cost effective to give films shorter runs and to partition their cinemas into smaller multiplex units, with staggered starting times and a greater choice of titles. Indeed, exhibitors may spring up to serve smaller niches, appealing to cult film audiences, for example.

The cost of storing outtakes or alternative scenes on hard disks is not as high as it once was. In 1994, it cost over $200,000 to store all the footage shot for a Hollywood feature film, in low resolution, for editing purposes. Today, the same amount of storage costs less than $2000. In these "micro-cinemas," the director's cut, consisting of no more than an edit decision list to instruct the video server to play different stored material at different times, would be very easy to distribute and present. Individual cinemas could even choose the ending of the film that seems to appeal most to their local audiences.

The existing film distribution system is not close to realizing these innovations yet, and may not be for some time to come. However, digital streaming technology makes possible many new ways of presenting feature entertainment. It is a matter of time before entrepreneurs realize the potential to use streaming digital media to make more money, while simultaneously addressing new audiences. Innovations will follow.

High-Definition Streaming

Video compression technology, developed to enable streaming media, makes other interesting applications possible. With some encoding techniques, it is now possible to render high-definition television pictures using bandwidth on the order of only 3Mb/s. Computers and mass-storage devices routinely handle data flows at this rate. Compare this to the 270Mb/s normally required for raw, uncompressed, high-definition television pictures. The cost of display technology for high-resolution images

*George Gilder, Senior Fellow at the Discovery Institute, first made this obseravation in his controversial bookm *Telecosm*, published in 2000 and revised in 2002. Gilder's Law is analogous to Moore's Law in predicting the future technological capabilities of optical networks, as compared to silicon chips. He studied politics at Harvard under Kissinger and is a regular writer for *Forbes*, *The Economist*, *The Wall Street Journal*, and *The Harvard Business Review*.

has also decreased, because of advances in micromirror device fabrication (these are used in digital projection systems, as noted earlier) and because of advances in solid-state display technologies, including light-emitting diodes bright enough to be visible outdoors.

These technologies, many developed as offshoots of streaming media, make a lot of interesting, cost-effective applications possible. For example, art galleries could feature virtual galleries or even installation art using high-definition streaming technology. Live auctions with accompanying live-streamed, high-definition images of the goods on sale are possible. Remote collaborative CAD design using high-definition streaming to exchange design data is a technical reality. Cinema-quality advertising can be displayed in shopping malls and at points of sale, with the content streamed to the display device from a central server. Live concerts or off-Broadway theatre can be broadcast worldwide, in high enough resolution to recreate the feeling of actually being present. Outdoor advertising signs and electronic billboards are now available at new, lower prices.

Film crews shooting on location can now relay the day's rushes to the producers, who may be in Hollywood. This gives producers the option of leaving the crew on location for another day, if more footage is required, rather than what happens now. Today, if there is a need to reshoot on location, the entire crew and all sets and props must be brought back to the location, at great expense, after the bulk of filming has completed. Often, the talent has already moved on to other projects, or in extreme cases, died, making reshooting even more difficult. With high-definition screening, informed decisions about whether or not to strike a set can be made overnight by decision makers located far away from the location set. Indeed, non-linear editing technology makes it possible to see the day's rushes in the context of other material already shot, with rapid turnaround.

High-definition streaming over broadband connections could allow better remote medical diagnosis, with world experts able·to view patients in great detail, wherever they happen to be located. There are some legal issues relating to the use of compression technology in medical imaging, but if live high-resolution video were to be accompanied by super-high-resolution, uncompressed still images, such as those that can be obtained with the latest 16-mega-pixel imaging devices, these problems ought to be obviated, in principle at least.

Instantaneous streaming of high-resolution video could be invaluable evidence in convicting felons, caught in the act, by surveillance cameras. Today, images of crime scenes like those issued on crime television programs, which ask for the public's help in identifying offenders, suffer from fuzzy, blurred, indistinct image quality. It is almost impossible to zoom in on the face of the offender, caught on video, without severe image degradation. High-definition imaging and relay back to a monitoring facility becomes cost effective, because of advances in streaming media technology, broadband networks, imaging devices, and video compression technology. Indeed, with detailed facial images, computer recognition of offenders could be a realistic option, provided some caution against naively interpreting the data is applied.

High-definition streaming technology can actually be powered from batteries or small electric generators. This makes it possible to take portable screening equipment to the most remote locations and most isolated populations on earth, bringing them high-quality information, education, and entertainment for the first time. While the data may not stream live to the display equipment in the jungles of Borneo (it could, if a satellite dish could sit within an appropriate satellite's footprint), playback of highly compressed high-definition streams, from hard disk or optical device, is still possible.

Other exciting applications of high-definition streaming include large displays at sporting events, relaying the action to the cheaper seats or allowing spectators to see instant live replays. Interplanetary probes could potentially transmit live, high-definition images, rather than still snapshots for detailed analysis and low-resolution video for navigational purposes. Motorway and freeway signs could include high-resolution views of the traffic or weather conditions up ahead on the road, allowing motorists either to divert or take extra care, when conditions are adverse. Arcade games and simulators of all kinds could use high-definition streaming to create more immersive experiences. All these applications become possible thanks to advances in the video compression and decompression techniques that accompanied the development of desktop streaming media.

Time Shifting and Live Pause

An obvious thing to do, when viewing an on-demand or live stream, is to store it away for later viewing. Indeed, you can imagine recording the stream without watching it at the time. You may pause a stream you are watching, shuttle back over something you want to see again, and then jump back to the live action or to the point where you paused, without stopping the stream to your hard drive. With hard drive sizes now in the hundreds of gigabytes and with video compression making good use of that space, streaming media is well suited to these applications.

Today, most desktop streaming media players do not allow you to record a stream to a storage device. Many digital media devices that use streaming discourage recording. There are three reasons for this.

First, copyright owners fear widespread piracy. If a stream can be recorded, it can be duplicated and distributed widely in seconds. For this reason, digital rights management systems are being developed to allow recording and viewing, but only under the terms the copyright owner allows, whether or not you pay per view. These systems are in their infancy, but work by encrypting the material streamed, which can only be unlocked by obtaining a digital key from an issuing authority. Keys are designed to allow playback only on an authorized device and only a specified number of times.

Second, many media presentations are subsidized by the inclusion of advertising. When you have no control over the playback of the media, as with broadcast television, if you intend to watch the entire program, you must sit through the ads, whether you want to or not. The fear advertisers and media distributors have is that with time shifting and live pausing technology, the viewer has the ability to skip all the ads. The long-term answer to these concerns is for the advertising industry to understand that in the digital media world we inhabit, the only advertising that will work is "permission advertising," whereby viewers actively seek out the information the advertiser wants to promote, or otherwise give explicit permission to trade their time, dutifully watching the ads, in exchange for obtaining cheaper access to the programs they wish to view. Ensuring that the viewer actually watches the ads is still problematic, but can be partly solved by disallowing pausing or skipping over the ads, for example. In the longer term, it won't be easy to impose advertising on viewers and force them to watch. In the shorter term, some solutions have included banner advertising embedded in the frame of the streaming media player, so that no matter where the viewer navigates within the program, the advertising is still visible. More subtly, product placement in the program itself permanently embeds the advertising in the program material. As surely as viewers will seek ways to avoid advertising, advertisers will find ways of getting their message to an audience.

Third, there is a fear of virus infection from downloaded media files. This is a major problem for the computer industry in general. There are ways to spread viruses by merely viewing a Web page! There are seemingly infinite ways to spread viruses, and detection methods rely on knowing specific signatures. Clearly, the virus authors have the upper hand! Adding to these fears is the damage that can be inflicted by high-bandwidth streams in terms of denial of service, whereby high-bandwidth streams of garbage going nowhere swamp a network, denying service to ordinary users. This is analogous to the ability of junk mail to slow down the delivery of mail and packages to regular users of the postal network. To date, effective solutions to these problems have not been found. In the long term, the answer is to build networks of such high peak capacity that the effect of rogue streams is negligible. Beyond that, the use of paid delivery networks, with guaranteed quality of service, which strictly police and remove rogue media and viruses as they are found based on viewing network traffic behavior and changes in loading, may be the way forward.

Some time shifting and live pausing systems such as TiVo players or Sky+ set-top boxes in the UK have become available. Many of these systems try to answer the three concerns raised above. Consumer response has, so far, been muted.

Streaming and Advertising

As noted in the previous section, streaming presents unique difficulties for advertisers and for advertising-subsidized programming, but it also presents unique opportunities. In the end, the power of these new features for advertisers will outweigh the perceived problems, making advertising more cost effective and able to deliver better returns than is possible with current mass-media advertising models.

When an advertiser places an ad in a mass-market media publication today, there is an acceptance that most of the impressions (i.e., the number of times the ad is reproduced) will be ineffective at causing a sale. In other words, the advertiser accepts waste. Most of the times the ad is seen, it will be ignored. The cost of those copies will be written off against increases in sales due to ad impressions that actually do cause viewers to go out and buy something. The advertising industry can only estimate the effectiveness of a particular ad. They cannot tell an advertiser, with any certainty, that for every thousand impressions of the ad, five people will buy the product, for example. All they can provide is aggregate data and averages. Analysis of why some people bought, after seeing the ad, while others didn't, is an even more elusive goal.

Streaming media technology answers some of these problems. Because streaming media can use the broadband Internet as a delivery medium, communication is two way. Unlike television, radio, or any other mass medium, it is possible to know who is watching and when. Indeed, if the viewer volunteers the necessary information, it is possible to present targeted advertising, with each viewer receiving different ads, depending on preferences, interests, location, and demographic profile. It is also possible to directly correlate this advertising to online sales, thereby providing an exact measurement of the effectiveness of any given ad campaign, not an extrapolation from a possibly representative audience sample, which is the best that can be offered with broadcast television today.

Some companies have sought to infer the data and demographics by correlating information from different databases or through the use of browser cookies, small files that Web site owners can access on your hard disk, which record your preferences. There has been a consumer backlash against this invasion of privacy. Advertisers are beginning to realize that they can't be sneaky in their data gathering and analysis. The data has value and consumers know this. Not only do they quite rightly want to maintain control over the use of their personal data, they also want some tangible reward for making it available to commercial enterprises.

Some of the ways consumers might put a value on their personal information is to provide data in exchange for cheaper access to content supported by advertising. Alternatively, they may get discounts, if they opt to buy the advertised product, in exchange for information about why they bought. A class of wealthy viewers may be given free access to content, provided they watch the ads, while poorer viewers, who are less interesting to advertisers, may have to pay per view or via subscription charges. Those wishing to preserve their privacy may also need to pay for access to content. Anybody expecting to get all media access for free will necessarily be relegated to mass media, where wealthy viewers' buying habits effectively cross-subsidize those who are only watching in order to see the show. In commercial terrestrial broadcast television, it isn't the advertisers or even the networks that are delivering the entertainment for all; it is the people who buy the products advertised, paying a proportion of the purchase price of those products to support the programs that others watch free.

Intriguing new advertising devices are made possible by streaming media technology. With streaming, it is possible to create links in the video, which, when clicked on, launch sidebar video ads, either as picture-in-picture, in another viewer, or by splitting the screen. For example, you could click on a can of cola, held by an actor, and get the latest video ad for that soft drink, along with information on where it is being sold on special in your local area. With an appropriately designed receiver, the family could be watching something when dad decides to find out more about one of the products on screen. Rather than interrupt the program for everyone watching, the information could be presented to dad as an audio voiceover in his headphones.

Advertising could also be designed to allow the viewer to ask to "tell me more." For example, the normal running time of a streaming ad might be fifteen seconds. However, if you click on options presented on screen during the ad, it may add another fifteen seconds of additional information, but only if you ask for it. Indeed, the viewer might ask for more information again and be directed to a large and detailed Web site, rich in video of other aspects of the product of interest, or else be connected to a live sales representative. In this way, advertising becomes entertainment and information, allowing the motivated consumer to explore the offering advertised in his or her own way and at his or her own pace. The more entertaining the advertising material, the more likely the advertiser is to hijack viewers from the content they were watching into watching their particular sales pitch.

Traditional broadcast television is experimenting with these new modes of advertising. The Advanced Television Enhancement Forum's (ATVEF) technology, for example, allows many of these types of advertising to be realized, with the data needed to render the ads on the set-top box carried in the vertical interval of the television signal and with the back channel consisting of a modem connected to a phone line. These solutions (there are many proprietary technologies for interactive television) suffer from severe technical limitations imposed by the bandwidth made available for enhanced program features in the standardized digital television stream, the multicast nature of DTV, the limited bandwidth of the back channel, and the lack of processing power and memory afforded by the current crop of set-top boxes. In the long run, these television enhancement technologies are merely transitional. With a broadband Internet connection to transport the same media types, much richer and more personalized enhancement data can be delivered along with the video. The future for enhanced video viewing is not in continuing to extend and milk digital television transmission standards (DTV) for additional bandwidth, it is in embracing broadband Internet delivery.

Unfortunately, at the time this was written, desktop streaming solutions were not satisfactory to advertisers either. Leaving aside the fact that most Internet-delivered streaming is viewed in the less-than-satisfactory presentation environment of the desktop PC monitor, current media players are almost incapable of butting together two media streams and presenting the join to the viewer seamlessly. This is a prerequisite of advertisers. Today, most media players switching streams stall and rebuffer, or else display a few unnecessary frames of black and silence. Not a single one can do a smooth crossfade between one stream and another. This is both because the quality of service of the Internet connection is not high, but also because the authors of media players haven't seen the need to add this attribute to their offerings. These problems will be solved by improved Internet quality of service technologies and when the authors of media players realize what broadcast television has known for decades: presentation continuity matters.

Interactive Tutorials

Streaming media, as we have previously noted, is a great aid to distance learning. However, it has certain characteristics which make it especially good for interactive training. Tutorials that help you study are easily realized with streaming media technology. Writing effective interactive tutorials with streaming media places new demands on program producers.

The following scenario might be a typical application. A world expert gives a lecture on a topic of interest. You, the student, are tested throughout the lecture, to aid in retention of information. At the end of the lecture, or at a later date, the same media package can ask review questions, automatically reiterating the piece of information that you didn't retain, should your answers to the review questions reveal a gap in your understanding. During the lecture, you can pause and ask for more detail on particular issues, or skip over parts of the lecture in which you have already established confidence. Indeed, if the sidebar information appeals, you may explore that information for a while, following links to more detailed information, before returning to the main lecture. In this way, the lecture can also be a tutorial and examination in one, with the feel of being more like a discussion with a world expert, than a dryly delivered, dogmatic, didactic lecture.

Parts of the streaming tutorial can be synchronized with other applications or media, to illustrate the concepts under discussion more richly and thus assist the student in assimilating the knowledge. If there is a live lecture in progress, individual students can pose questions directly to the expert, redirecting the flow of the lesson in order to get the information required. Live lectures and tutorials can, of course, be recorded for later on-demand viewing, or even appended over time, as interesting questions are posed by successive students.

The interactive tutorial realized with streaming media technology can grade students as they go. It can allow students to bookmark topics for later review, perhaps using an interactive digital highlighter to mark sections of the synchronized lecture transcript the student feels are especially insightful or relevant. Video clips representing those highlighted sections can be assembled into a single, individualized edit, to aid student review later.

When mobile broadband streaming becomes ubiquitous, the potential for learning wherever you happen to be becomes realistic. Students on their morning commute can participate in coursework, rather than sitting blankly waiting for the journey to be over. With mobile streaming tutorials, the wait by your broken-down car until the repairman arrives could be made productive. A virtual mechanic might lead you through all the things to check, via your mobile streaming media receiver, before the roadside assistance mechanic arrives. Any situation where the advice of an expert would be useful could be helped if streaming tutorials, were available on demand.

Information Blitzes and Search Randomizers

An interesting application of streaming media exploits the eye's ability to take in visual information rapidly. When we read text, we assimilate information at about 55 baud, according to some researchers. When we view images, the uploading of information is somewhat higher. If a search engine were to return a video clip, consisting of five frames of video for each search hit, rather than the list of text entries that search engines return today, searchers would be able to view their search results as a video montage of rapid cuts, in very little time. I call these little search result videos "Information Blitzes."

For these to work, search engines would need to catalog not only text information about what is available on the Web, but also small amounts of motion video representative of the longer streaming media clip found by the search engine's Web crawler at the site hosting the video. Indeed, for streaming media clips to be indexed and found, we need more than just text descriptions anyway. Thumbnail images are a minimum requirement. There are some copyright issues associated with cataloguing other people's content by means of samples of that content, but perhaps fair use laws would allow such samples. If not, then it is actually in the content owners' interests to have material cataloged and locatable by search engines. It would be hard to imagine an organization not wanting its media cataloged in this way.

Content owners could also create information blitzes to give a visual précis of their longer video content, so that people could view the blitz before deciding to spend their time watching the whole thing. These are the equivalent of textual metadata; in fact, they are visual metadata.

Search randomizers are just information blitzes where the clips edited into the composite blitz give a random selection of what is available, or just a random smattering of "cool clips," or even clips relevant to stated interests. Navigating the universe of streaming media, where there could possibly be millions of clips available for on-demand viewing, will not be possible using traditional program guides. The sheer number of available video clips will make electronic program guides, such as those used in digital television systems, unwieldy and effectively useless. The information blitz might be a way to present vast amounts of visual data rapidly, allowing viewers at least to make an informed choice before investing time in watching a full clip (Figure 2.8).

Figure 2.8
Information blitz.

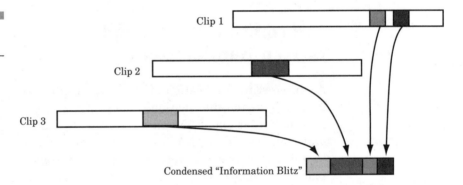

Streaming from DVD (WebDVD)

Another slightly tangential use of streaming media is "WebDVD." With this invention, high-quality video content is delivered on a DVD disk, bypassing all the quality-of-service problems and bandwidth limitation of the Internet. However, embedded in the DVD are links to the Web, which allow synchronized elements and even updated video to be delivered live, in tandem with the DVD video content on playback, giving the best of both worlds. You get the high-quality experience of a DVD, with the freshness of content that comes from the Web (Figure 2.9). E-commerce applications are making good use of this technology to deliver a product catalog showing off the products in high production-value video clips,

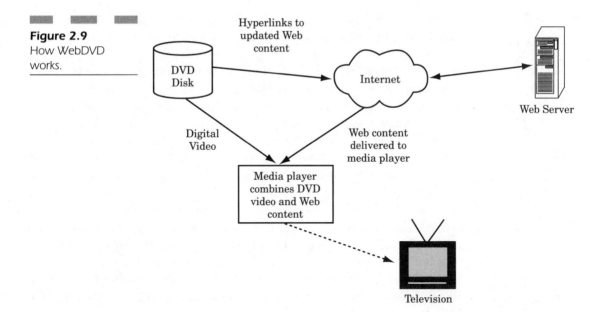

Figure 2.9
How WebDVD
works.

with online ordering, pricing, and stock availability information beamed into the electronic catalog, without the viewer's being aware that it is not coming from the DVD disk. What the customer sees is a seamless experience, which is both attractive and up to date. Whether or not this is a transitional technology while coverage for broadband connectivity is patchy remains to be seen.

How Does Streaming Media Work?

There is more than one way to skin a cat, as the old adage goes. This is as true for streaming media as for anything else. There are many technologies that can be used to complete a streaming media chain, from producer through distributor to end-user. Many of the technologies are proprietary, though some are open standards. This section will discuss the basic elements of streaming media, in general terms, then make reference to specific implementations and technologies. The aim is to understand the process, rather than implementation specifics, in the first instance, then to illustrate the different approaches various implementers have taken to solve the same set of problems.

The main building blocks of streaming media are discussed in the following sections, which explain compression technology, packaging of the streaming data, and distribution over a network. We explain how the media is played and then look at some extras that can make streaming media give a better quality of experience to the viewer. For illustration, we'll talk about technologies that stream across the Internet, since these are well known and more easily understood. However, the reader should remember that this is not the only way to stream media.

Before we can describe how streaming media works, we need to spend a little time on some digital media fundamentals. It is important to understand how video and audio get to be digital in the first place and how sampling works. We also need to explain how the digits become sound and pictures once more. If you already understand analog-to-digital and digital-to-analog conversion, just skip ahead to the next section.

What we perceive as sights and sounds, physicists understand as variations in light intensity and vibrations of air molecules. We see things because our retinas respond to variations in light intensity. We hear because we have ears sensitive to minute variations in air pressure. Brains attached to those sensors (the eyes and the ears) continuously take measurements of light intensity and air pressure and turn the data into visual and audible experiences. So, to recreate a picture or sound at a distance, all you have to do is take measurements of light intensity and air pressure rapidly enough, transmit that data through some transmission path and then recreate the light intensity and air pressure according to the data you received. Cathode ray tubes can recreate varying light intensities. Loudspeakers can recreate air pressure variations.

With analog reproduction systems, there is a continuous stream of information from the sensor to the reproducer. For example, when Edison invented his phonograph, the sensor was a large acoustic horn that wiggled a needle in response to variations in air pressure gathered by the horn. The continuous wiggling could be recorded on a wax cylinder. The groove in the wax would be *analogous* to the variations of air pressure detected by the horn.

If the same needle were then excited by the recorded wiggles in the wax, the horn would vibrate in response to the movement of the needle as it followed the groove in the wax. Air pressure variations, the same as those that were recorded, would be recreated, and ears could hear these as sound. If the recording process were perfect, there would be no way that the ear could tell the difference between a real sound and a recorded and reproduced sound. But the recording process isn't perfect.

The needle can only move so fast, the horn is only sampling the air pressure variations present in the room, and the movement of the needle along the surface of the wax produces its own noise. The important thing to notice is that the groove in the wax is actually a continuous measurement of air pressure variations received by the recording horn. It reproduces an *analog* of the original sound.

With digital reproduction systems, the aim is to make those measurements discrete values. If you measure often enough and with enough precision, you can effectively reproduce the continuous variations in light intensity or air pressure that you need to create a convincing picture or sound. To take an absurd example, if you took the wax groove made by Edison's phonograph and measured the depth of the groove every hundredth of a millimeter along the length of the groove using a measuring device that measured with fine enough gradations, then wrote the measurements down in a book, you would have performed an analog to digital conversion! The list of measurements in your book would be a digital representation of the sound you recorded. You could photocopy the pages and the copy would be an exact representation of the sound you recorded. You could substitute every occurrence of the digit 2 with the digits 22. When you sent this to your friend, if he knew to substitute every pair of digits 22 with a single 2, he could exactly recover the digital representation of the sound you recorded, whereas somebody who got your page of measurements, who didn't know about the substitution, would have little chance of recovering your original measurements. You would effectively have encrypted the information and managed the rights to reproduce the original by the act of telling only your friend how to decrypt the numbers.

Now, if you had a machine that could wiggle a needle very precisely, according to a series of numerical measurements you punched into it, such that the excursion of the needle was proportional to the number you punched in, and if this machine's needle was connected to a big acoustic horn, then if the wiggles were reproduced, number by number, at the right rate, the horn would vibrate and reproduce a sound indistinguishable from the one originally recorded in the wax groove. The machine would take a list of numbers and produce air pressure variations proportional to those numbers. It would be a digital-to-analog converter.

All digital media work this way. Measurements of light intensity or air pressure are taken fast enough and with sufficient precision not to miss any important detail and "written down" somehow. These numbers are then sent somewhere else. A machine on the receiving end converts the list of numbers, at the right rate, into variations of light intensity and air pressure identical to those originally recorded.

In an actual digital audio or video device, the analog to digital converter takes an electrical analog of the air pressure variations obtained by an electrical transducer and converts those analog signals into a series of binary digits. It does this using a sample-and-hold circuit and an analog-to-digital conversion circuit, which is often a series of comparators that change state if the voltage seen on their input pin exceeds a reference threshold. If you stack up enough comparators, you effectively divide the allowable peak signal voltage excursion into quanta (Figure 2.10). To get a sixteen-bit conversion, so that each measurable voltage level is represented by a unique sixteen-bit binary number, you need to be able to discriminate 65536 (2 to the power 16) distinct voltages and assign each a binary number. Circuits that do this are readily available.

A digital-to-analog converter takes binary numbers and, through a series of switches and a voltage divider network, presents an analog voltage (or current) according to the binary digit presented. Again, digital-to-

Figure 2.10
Analog-to-digital conversion.

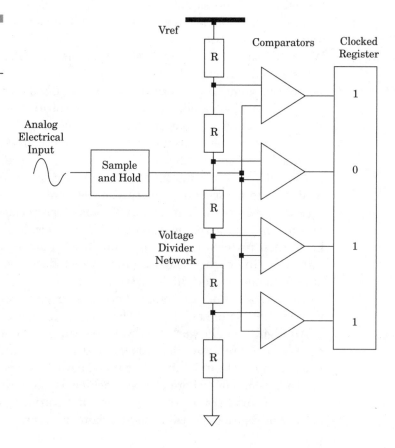

analog conversion circuits are commonly available. The analog signal can be amplified, so that it can drive a transducer, like a loudspeaker, in order to recreate air pressure variations, or sound. The same digital-to-analog converter may, instead, drive a cathode ray tube to create pixels of varying light intensity, according to the binary digit presented to the converter. When the binary digits are stored, they are kept in random access memory circuits, or else recorded as magnetic signals on a hard drive, or as optical pits on a CD-R.

The answer to the question: "How often do I need to take measurements so as not to miss anything important?" was answered over 50 years ago by Harry Nyquist and Claude Shannon. It turns out that ears are not sensitive to vibrations of air faster than about 20,000 cycles per second. Nyquist did the math to show that if you sample at twice this rate, i.e., 40,000 cycles per second (or Hz), you are able to reconstruct vibrations of half that frequency. The generalization of this, the *Nyquist theorem*, says that if you want to sample and reproduce things up to a certain frequency, you need to take measurements at a rate at least twice that frequency. Otherwise, reconstruction is uncertain. You get *aliasing*. You can fit waves of many shapes and frequencies, not just a single unambiguous one, to the sample points you have.

Claude Shannon was interested in how much data could reliably get through a noisy channel of a certain capacity. His work concerned error correction and coding schemes to minimize the effect of introduced noise in a transmission channel. His breakthrough idea was the digital representation of information, sampled at an appropriate rate, as a bit stream of the samples, coded with some redundancy to protect against corruption. These two men are the fathers of digital communications. Without their work, streaming digital media wouldn't be possible.

Compression

Compression is a big subject, about which entire books are written. My colleague, Peter Symes, wrote an excellent work entitled *Video Compression Demystified*, which is a comprehensive treatment of the subject of video compression. Audio compression technology fills other similarly sized books (for example, Markus Erne's book *Digital Audio Compression*). New compression techniques, being developed all the time, give improvements in picture or sound quality or temporal image quality, for a given amount of bandwidth. Compression is a key enabling technology for streaming media, because it makes efficient use of available bandwidth and, to date, bandwidth has been extremely limited. Without the ability to compress moving images radically, it would currently be impossible to transmit video over the narrowband Internet.

Compression is the use of coding techniques to reduce the amount of data used to convey information. Information is in the eye and ear of the beholder, so some of the techniques used in compression exploit these psycho-acoustic and psycho-visual phenomena to disguise the fact that not all the information is delivered (Figure 2.11).

Figure 2.11
Compression.

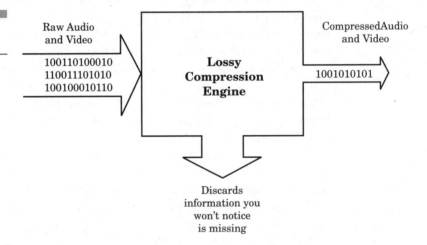

All compression exploits the fact that information has order and patterns. If you can describe the order and patterns, without explicitly transmitting all the bits required to reconstruct the data, given no a priori knowledge, you achieve a reduction in the data required to transmit the original.

With compression, it is also vitally important to start with a digital representation of only the information you want to transmit, uncorrupted by noise. If noise is present, bits will be required to represent it. Compression and noise reduction work hand in glove.

There are two kinds of compression. One is *lossless compression*, where you devise a code and a codebook that allows the receiver to exactly decode a digital transmission by looking up coded symbols. It is called lossless compression because the receiver can recover an exact replica of the original digital representation, even though a reduced data stream was actually transmitted.

Lossy compression, on the other hand, exploits perceptual anomalies of the human nervous system to send a "good-enough" digital representation of the data, with the receiver only recreating an approximation of the original digital media. When we view or hear the "good-enough" reproduction, our perceptions ignore the deficiencies. In fact, sometimes you can notice that the received digital media is quite unlike the original, because the artifacts are far too obvious to miss. However, being able to transmit an impression of the original may be sufficient, for a given application. Extremely lossy compression can be acceptable in some situations (usually when it's a choice between badly degraded digital media or nothing at all).

When a compression algorithm throws data away, it can do it in several ways. Let's consider video compression. First, a compression algorithm can reduce data by quantizing the image more crudely, either by:

- Using fewer bits to represent each color
- Reducing the number of pixels in the image
- Reducing the amount the value of a pixel can change compared to its nearest neighbors
- Reducing the number of frames per second

Many algorithms have been derived to perform these operations, ranging from crude "select-the-nearest-value" approaches, to more sophisticated processing algorithms. The best decimators use *sample lattice transforms** to deduce a lesser number of lines, pixels per line, and frames per second from the given sample data (i.e., the data obtained from the analog-to-digital conversion process). Sample lattice conversion produces fewer quantization artifacts than merely choosing some of the available values in the sample set on a nearest-neighbor basis. For a simple reduction in bits per sample value, the crudest way is simply to *truncate*, whereas the least noticeable way is to *dither* each value before truncating. Dithering is the process of adding noise to the values in frequency regions that the senses are less attuned to, in order to make the pixel truncation process appear to distort the samples less.

---

*If a video sequence is digitally sampled, you get a three dimensional lattice of samples, of dimension "pixels per line," by "lines per frame," by "frames in the sequence." In other words, all the pixel values can be represented as a cubic lattice, with axes representing the dimensions of the video sequence in both space and time. If you need to change the number of frames per unit time, or the number of lines per frame, for example, a sample lattice transform, which is a mathematical algorithm for creating the new sample lattice by three dimensionally interpolating the source lattice sample values, is used. The algorithm has the property of mathematically producing the new sample values in close agreement with what they would have been if the original source video sequence had been sampled according to the dimensions of the final lattice.

Compression algorithms can choose a smaller number of bits to represent a digital symbol denoting pixel values that occur frequently in an image and longer representations for those values that occur infrequently, so that on average you use fewer bits to represent a given moving image. This is called *entropy encoding*. Entropy encoding is lossless. *Huffman codes* are an example of entropy encoding.

We can exploit features of photographic images, for example, that make it likely that the intensity of a pixel is similar in intensity to its neighbors. Photographic images are said to be *Markov sources*.* If you know that the next pixel will be similar in intensity to the last, you can use predictive coding schemes, so that given the first pixel, you don't need information about the second, you can just predict its value from the value of the first. In fact, the more you know about the likelihood of the next pixel based on first, second, and higher orders of predictors, the less in error the predicted value will be. In other words, if you can accurately predict the next pixel, given the behavior of several near neighbors, the compression will be better.

You can break the picture up into regions and do a transform on each region to establish its frequency domain content. You perform mathematics on the block of pixels in the region to obtain a transform, which gives a spectrum of frequency values that can be represented with fewer symbols than the original block of pixels. Spectral components equate roughly to colors, and transforms establish what colors are present in the block, ignoring the unimportant colors. Further, the base functions of the transform take a given spectral component (a number representing the dominant color in the region analyzed) and fit a "pattern" to best describe all the pixels in the analysis region with the least error. The compression algorithm has several mosaic tiles representing the transform base functions, which can be thought of as textures. The encoding algorithm takes the actual pixels in a region under analysis and tries to fit the best texture to that block. When a best fit is found, the only data needed for reconstruction of the image are a color value representing the average color in the block and the base function, or texture, which was

---

*A Markov source is a source of data that has the property of allowing you to predict the next data value, given the current one, simply because some sequences of values are statistically more likely than others. In the English language, for example, *h* is more likely to follow *t* in a word, than say a *q*. This is because the *th* letter combination occurs frequently in the English language, whereas the combination *tq* is very rare. In photographic images, the next pixel is more likely to be close in color to the current one than it is to be a completely different color. Images are just like that. They behave as Markov sources.

the best fit. You don't need much information to send this, compared to the original pixel field's digital representation.*

In a decoder, the inverse transform is performed to recreate the pixel block, given information about the colors that must be present. A variety of transforms can be used, but popular ones include the Discrete Cosine Transform (DCT) and the Discrete Wavelet Transform (DWT). Both of these have the property of describing an impression of a block of pixels using a smaller number of spectral values, not because they are inherently lossy (in fact, they are lossless), but because you can ignore some of the data and get away with it. On reconstruction, what you get is a pixel field not too far from the original pixels analyzed, but effectively synthesized by the inverse transform mathematics.

Pixel fields need not be square, but for convenience of computation, they often are. Unfortunately, the result is that you get *block effects*, since the analysis performed on one region of the image bears no relation to the analysis performed on the next. If you divide the image into a mosaic of square regions and perform discrete cosine transforms on each region independently, when you reconstruct the image from the transform data, there will sometimes be noticeable and potentially sharp differences in color at the edges of the little mosaics. Ways around this problem include using non-square or overlapping regions to create some sort of average. Of course, to get bigger reductions in data, at the expense of greater error when decoding the image, you can choose bigger regions of pixels to transform.

Another solution is to use the discrete wavelet transform to avoid blocking artifacts. The wavelet filter is applied to a much wider area of pixels, so doesn't suffer from blocking artifacts, even at relatively large compression ratios. Because wavelet transforms remove high-frequency components of the image, step-by-step, wavelets are excellent at representing edges in the image. Under high compression ratios, the artifacts are localized around image edges, not spread across the whole area of the block, as they are with the discrete cosine transform method.

Yet another way to reduce data is to find only those things in the image that have changed since the last frame and encode only those. Using motion analysis, if the background, say, hasn't changed from frame to frame in a given video sequence, you need only instruct the decoder to hold on to those pixels and just change the moving ones.

---

*For a more detailed technical explanation of how actual compression algorithms work, refer to Appendix E. The purpose of the current discussion is to give simple analogies describing the effective action of the compression process.

The above is a gross oversimplification of compression techniques and serves only to refer the interested reader to more in-depth sources. I have used terminology that explains what is going on in a way I can understand it, so my explanations are more impressionistic than rigorously mathematical. What I would like the reader to take away from this discussion is that there are many techniques used to "throw data away" or to use information that the decoder doesn't need to receive (or receive often) to reconstruct an image. In fact, many proprietary compression schemes don't want you to know how they achieve their particular compromise between good image quality and heavy data reduction, as this is a trade secret and the "special sauce" in their particular compression products.

To me, it is somewhat pointless to argue that one compression scheme gives better quality than another, for a given bit rate. All lossy compression is a compromise. Every technique so far developed will do a great job with some kinds of images, yet reveal the nasty inner workings of the algorithm's throwing away the details, given other images. The art of the compromise is in hiding the fact that some of the detail is missing.

This entire catalog of techniques is ultimately used to render an image from a compressed representation of it, so that the human nervous system thinks the image is all right. There is no doubt that many new and ingenious ways of unnoticeably discarding data will be discovered yet.

At the time of writing, these compression and encoding technologies were battling for supremacy:

- Microsoft Windows Media 8
- RealNetworks Real Video 8 and Real One
- Sorenson Broadcast 3.1 in conjunction with Apple QuickTime 5
- MPEG 4 (various vendors)
- H.26L

Some of these use wavelets, some use discrete cosine transforms. This list is by no means exhaustive. For audio, there is an equally impressive array of compression techniques and vendor solutions.

Ultimately, the consumers care only that the media plays well and that the player is compatible with the media they want to watch. Several vendors have sought to create multi-codec players (that can decode anybody's compression—for example, Generic Media) and other vendors have created machines that encode into everybody's compression formats. To date, the competition between compression schemes has served mainly to create a desire, in the end-user, for a single standard. The subsequent unfortunate confusion has fragmented the market for streaming video into smaller, less-profitable slivers.

Bandwidth

Bandwidth is just another way of saying channel capacity. It describes a particular channel's capacity to deliver information. The greater the available bandwidth, the more bits you can transmit down the channel in a given time. Bandwidth is measured in bits per second, where a bit is the smallest discrete amount of information that can be transmitted. As we mentioned earlier, information theorist Claude Shannon did all the fundamental work on describing the information carrying capacity of a digital communications link in the presence of noise. When we say we have a channel that can carry 500 kilobits per second, we are saying that we can transmit 512,000 discrete bits of information every second. If a picture we wished to send were represented by 512,000 bits, after some compression had taken place, for example, the implication would be that we could only transmit one such image every second, given the constraints of our digital communication channel.

How much bandwidth do we need for streaming media? We can do some "back-of-envelope" calculations to find out. First, let's consider an extreme worst case. We'll begin with the video, which we won't compress, just for argument's sake. The best in-camera sensor available today can produce 16 mega-pixels per image (4096 × 4096 pixels). Let's say that there are three sensors, one for each of the primary colors: red, green, and blue. Now, let's imagine that we take 60 images per second (as specified in the High Definition Television (HDTV) standard, for example). That's 3,019,898,880 pixels per second. If we digitize these pixels at, say, 10 bits per pixel (this is a typical quantization choice in professional imaging applications today), that's about 30 gigabits per second. No such motion picture camera yet exists, but the technology to realize one is imminent, at the time this is being written. The pictures would be stunning!

Now, let's add five channels of surround sound, sampled at 20 bits per sample (which is the best that can be practically achieved today) at a sample rate of 96kHz per channel (the typical sampling rate used by current high-end professional audio equipment). That's 9,600,000 bits per second, or about 9.2 megabits per second for the audio alone. The grand total is still around 30 gigabits per second, since the audio bandwidth is not significant compared to the video. That far exceeds the capacity of the single OC48 fiber typically used in metropolitan area network backbones, which can only handle 2.45 gigabits per second.

Now, let's arbitrarily imagine that 100 million people in the world simultaneously wanted to watch something different from what everybody else was watching (there are about 100 million people connected to the Web in the US today, about 20 million of whom have broadband connections). That would be an aggregate load of about 3 exabits per second (an exabit is 10^{18} bits or a billion billion bits)!

As a point of comparison, today, optical fibers capable of transporting 40 gigabits per second represent the state of the art (OC768). They are not yet widely deployed. To carry 3 exabits per second, you would need 75 million of those fibers, assuming they were available. Put another way, a typical transatlantic cable consists of 850 fibers, each of which can carry 40 gigabits per second. The aggregate capacity of the cable is therefore 34 terabits per second. To carry 3 exabits per second would require on the order of 90,000 of those transatlantic cables. Yet, today, the transatlantic cables are not used to capacity, for the first time in the history of transatlantic telephony.

In March 2001, WorldCom announced it had transmitted a world-record 3.2 terabits per second down a single fiber, using 80 wavelengths of 40 gigabits per wavelength. This is the equivalent to about 41 million simultaneous telephone calls. The actual demand for bandwidth in the US is currently only a few terabits per second, with data transmissions representing the vast majority of this traffic. Voice calls are expected to account for less and less of the aggregate bandwidth demand, over time. Even with these world-record-breaking single fibers, we are still several orders of magnitude away from our target 3 exabits per second. Let's hope that 100 million people in the world never want to log on all at once and watch something different in such high resolution!

Today, a typical guideline price for bandwidth of 162 megabits per second is about $100,000 per month. That's about 4 cents to carry 162 megabits. At those rates, supplying bandwidth at 3 exabits per second would cost around $740 million per second! So, if this were what the world routinely did with its bandwidth, the industry would be worth 24 thousand trillion dollars per annum at today's prices. That's clearly absurd. At those prices, each viewer would be charged $7.40 per second! That means to watch a 100-minute feature film, on demand and in better than cinema quality, the viewer would need to shell out $44,400!!

Having established a ridiculous upper limit, let's now scale our expectations of image quality back to those typical of D-cinema presentations, where the video is compressed to 45 megabits per second. We would still require 4.5 petabits per second (a petabit is 10^{15} or a thousand trillion bits) of capacity to serve one hundred million customers with their own unique, on-demand, D-cinema-quality program. That would cost one cent per second per viewer, at currently quoted bulk bandwidth rates. A 100-minute feature film would cost around $67 to watch—still too expensive. However, if we believe that Gilder's Law will hold, where bandwidth per unit cost triples every year, it will only be three years until the cost of a D-cinema presentation is around $2.50. This is comparable to video rental prices.

If we now lower our image quality demands to match those of current digital television transmissions, where an MPEG-2 encoded stream of around 8 megabits per second would be used for a typical premium channel, at current bulk bandwidth rates, a 100-minute feature film would cost $11.90 to watch. Once again, this is too high, but applying Gilder's Law once more, it will only be two years before this same movie costs just over a dollar in raw, bulk bandwidth.

There are streaming media encoders advertised today that claim to give "near-DVD quality" images using only 750 kilobits per second. Doing the same calculation as we have been doing, this would mean we could watch a 100-minute feature film for only around a dollar today, if we bought bandwidth to our homes at the bulk rate we have been using as a guideline. In fact, with ADSL, many US customers are obtaining 768-kilobit-per-second connections for around $50 per month. At these prices, a 100-minute feature film only costs about twelve cents to deliver. Clearly, this is about one tenth of the cost of the bulk rate we have been conjuring with, but with ADSL, service-level guarantees are not as stringent as with bulk bandwidth purchases. ADSL service offerings provide for "up to 768 kilobits per second." In practice, lower bandwidths are obtained.

What the above discussion has illustrated is that for superb image quality, there isn't enough bandwidth in the world today and there won't be, anytime soon. Also current bandwidth is not priced at anything like a level that makes those super-high-quality applications feasible. However, as we lower our expectations and accept image qualities that we are already accepting with other media, streaming becomes a realistic proposition, at least in "ideal world" terms, where high bandwidth is available at your home, at a reasonable cost, and where the quality of service for the bandwidth provided is guaranteed.

Reliable delivery of the data is a serious issue, if the bandwidth available is all but completely consumed by the data stream. If some of the data packets are lost, due to errors, and have to be resent, there won't be enough channel capacity both to maintain the stream and to send the lost packets of data again. What the end-user sees, when there are packet delivery delays or retransmissions of data packets, is that the video play-back pauses and stutters, making the motion on the video look very jerky. In extremes, the need to retransmit lost or delayed packets disrupts the normal playback of all subsequent video frames. Unless there is enough capacity in the channel to allow for packet loss and recovery, the first loss of data may cripple the streaming playback process from that point onward, indefinitely.

Additional bandwidth is also useful for minimizing start-up delay and buffering. For example, if you select a particular stream for playback, you can't play a thing until there is enough data in your local streaming buffer to allow for fluctuations in the streaming data rate. If you have a channel with much more capacity than is required for your stream, the initial buffering can occur at speeds much faster than the steady state stream data rate. In other words, if I have a 500-kilobit-per-second channel and I want to watch a 50-kilobit-per-second stream, I can burst load the initial buffer, so that I could get one second's worth of my video stream into the playback buffer in only a tenth of a second. This mini-mizes the startup delay between when I select a stream for viewing and when it starts to play back.

If I wish to have smooth transitions between clips, so that switching from one to another results in a smooth crossfade, for example, I actual-ly momentarily require the sum of the bandwidth required to play the two individual streams.

What this means is that when it comes to streaming media, not only do you need enough bandwidth to sustain the stream at its particular data rate, you also need sufficient bandwidth headroom to allow for:

- Startup buffer acceleration
- Smooth transitions between two streams
- Packet loss
- Data retransmission
- Stream recovery

A rule of thumb might be that you need a channel capacity of ten times the stream bandwidth to guarantee robust playback that is resilient in all the above conditions. That would mean that reliably play-ing a 750-kilobit-per-second stream, might require on the order of 7 megabits per second of available channel capacity. I am aware of very little research that has been done to rigorously establish the peak-to-sustained bandwidth headroom needed to guarantee reliable streaming media playback in lossy, "best-effort" delivery networks.

Pipes

"Pipes" is a colloquial term used to describe the various paths through which a consumer of streaming media (or any other kind of digital data) gets delivery of bits. Some of these paths can be wireline or wireless. In wired networks, the "wire" can be copper twisted pairs, coaxial cable, regular electrical wiring, or fiber optic cabling. Wireless transmission can be signals from satellites, terrestrial radio frequency signals, point-to-point microwave signals, or infrared signals—in fact, signals from many frequencies in the electromagnetic spectrum.

All the pipes do is form a path along which the information can travel. The point about streaming media is that it can be delivered via a variety of pipes. Indeed, it can be delivered by multiple kinds of pipes at the same time, to the same receiver. This is significant because the home represents one of the last frontiers for advertisers. Currently, the gatekeepers of access to the home are the broadcast networks. They control access by dint of the fact that they have a monopoly on the terrestrial television transmission frequencies, the cable networks and the satellite dishes that deliver television. However, technology is eroding this position. One day, digital media will enter your home as data, on a multitude of pipes, offered by all kinds of companies. Streaming media technology makes it possible to get digital media through many routes.

One of the newest technologies for streaming media delivery is the emerging third-generation cellular wireless network. Called synonymously UMTS (Universal Mobile Telecommunication System) or WCDMA (Wideband Code Division Multiple Access), the system has been expressly designed to augment the voice and short messaging communications available on cellular networks today with gaming, music downloading, and video streaming. Using CDMA (Code Division Multiple Acess), users are separated by unique codes, which means that all users can employ the same frequency and transmit at the same time. A narrow band signal is multiplied by a spreading signal (which is a pseudo-noise code sequence) with a higher rate than the data rate of the message. The resultant signal appears as seemingly random, but if the intended recipient has the right code, this process is reversed and the original narrowband signal is extracted. The main benefits of using a wideband carrier, as in WCDMA, is that you can support higher bit rates, you get higher spectrum efficiency due to improved statistical averaging in the trunking system and coverage improves as the frequency diversity is improved.

The system supports four different quality classes of Radio Access Bearers (the "on ramps" to the system). These are:

- **Conversational**—Used for voice telephony with low delay and strict ordering of data traffic.
- **Streaming**—Used for streaming media with moderate delay, but strict ordering.
- **Interactive**—For Web surfing with moderate delay.
- **Background**—Used for file transfer with no delay requirement.

In a single session, a user accessing digital media remotely could potentially require all these access bearers. Initially, the transport network will be based on ATM (Asynchronous Transfer Mode), but it will migrate rapidly to IP (Internet Protocol)-based networks.

Customized "infotainment," such as interactive games, voting, chat, virtual dating, competitions, information, entertainment, sports results, betting, horoscopes, etc. is seen by the WCDMA industry as the most likely revenue leader, with other applications, such as multimedia messaging, location-based services, and full mobile Internet access appealing more to business users. This prediction is made on the basis of the share of premium-rate service minutes for UK phone users today (Figure 2.12). These services are likely to include some streaming media content, if not being streaming media services in their own right. It is predicted that by the year 2010, the average subscriber will be spending US $30 a month on such services, with consumers accounting for 65% of total revenue. The anticipated number of subscribers for third-generation network services, by that time, will be some 600 million (28% of the 2.25 billion mobile phone subscribers worldwide).

Figure 2.12
UK premium-rate services: Share of call minutes by service type.

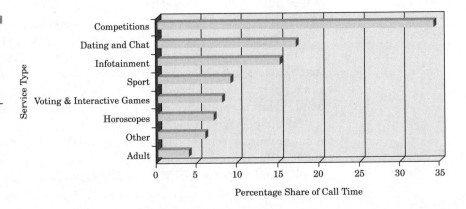

Access to these third-generation cellular networks will be priced in line with a proven preference for flat-rate services. With the declining price of bandwidth, third-generation networks will account for much high-speed Internet access, but the industry itself feels this will never be a complete substitute for wireline services. Requirements for broadband Internet access are becoming increasingly significant and cannot be entirely satisfied by third-generation mobile services. That said, mobile "per-minute" prices will approach parity with wireline pricing, for both voice and data. There is consumer and regulatory pressure on mobile Internet providers to place no commercial restrictions on users. Subscribers should not be locked into a tied portal. There should never be a "walled garden," whereby consumers can only access sites which the mobile Internet service provider wants their users to access.

Compare and contrast this with the practices of satellite digital television providers, whose business models are built around retaining eyeballs on content that they own. In a satellite television service, the service provider doesn't want viewers to have access to content from other providers. This is one of the reasons why the broadcast television industry is putting itself in peril, both from the point of view of consumer backlash, when consumers realize they can access anything they want from another bandwidth provider, and from regulators, once they realize that the distinction between broadcast television provision and mobile Internet access provision is wafer thin, in a world of digital media.

In fact, streaming media may enter the home of the future (Figure 2.13) through some or all of the following carriers. In wireline carriers, they will include:

- ISDN (Integrated Services Digital Network)
- Various flavors of DSL (Digital Subscriber Line), including:
  - ADSL (Asynchronous DSL)
  - IDSL (ISDN DSL)
  - HDSL (High-bit-rate DSL)
  - VDSL (Very-high-bit-rate DSL)
  - SDSL (Symmetric DSL)
  - RADSL (Rate-Adaptive DSL)
- FITL (Fiber in the Loop)
- CATV (Community Access TeleVision, also known as Cable TV)
- HFC (Hybrid Fiber/Coax).

Wireless carriers may in future include:

- MMDS (Multipoint Multichannel Distribution Services)
- LMDS/MVDS (Local Multipoint and Microwave Video Distribution Systems, also known as Cellular TV)

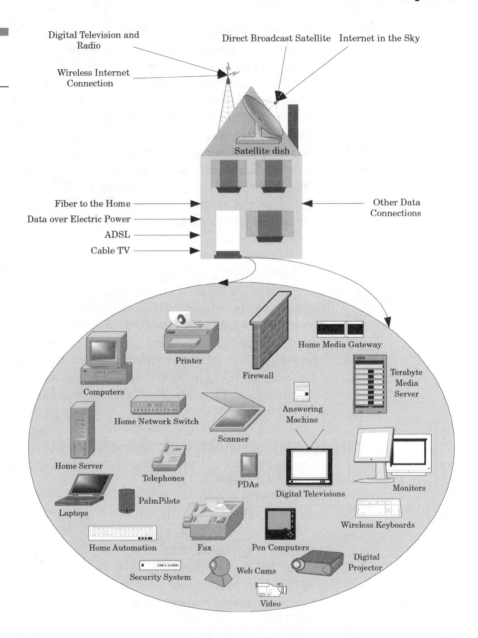

- WLL/RITL (Wireless Local Loop/Radio in the Loop)
- IVDS-DTV (Interactive Video and Data Services)
- VSAT/GEO/LEO (Very Small Aperture Terminals communicating with Geosynchronous Earth Orbit and Low Earth Orbit satellites)

The "last mile" problem, of getting a high bandwidth connection to individual homes, has many solutions. Resistance to implementation of those solutions has included regulations that maintain effective monopolies on the copper local loop infrastructure and the fact that most telecommunications companies are geared for switched-circuit technology. The provision of packet-switched network technology to individual subscribers causes significant knock-on effects to the telecommunications company, including the need to write off all investment in switched-circuit technology, the need to reorganize billing and addressing systems (what does a phone number have to do with an IP address?); the need to retrain engineers and technicians and the need to build a capability to diagnose and correct a broad variety of subscriber problems that are never encountered with the current-switched circuit phone system. These business process challenges are not trivial. Is it any wonder that the telecommunications companies have been so slow in casting off everything they know about doing business, in order to reinvent themselves as packet-switched network service providers?

Today, broadband provision to the home mostly concentrates on reusing the telephone company's existing twisted pair, copper, local-loop infrastructure, since the cost of digging up the streets and burying fiber, or of creating a network of satellites or microwave and radio masts far exceeds the cost of reusing what is already in place. ADSL is the likely current winner. However, companies are finding other ways to improve the amount of traffic that can be carried with twisted copper pairs. Companies like Actelis have developed technologies like Spatial Division Multiplexing to get 155 megabits per second over copper. VDSL will deliver up to 52 megabits per second over copper, predicated on FTTN (Fiber to the Neighborhood), whereby the fiber head end is actually positioned much closer to a local group of subscribers than is the case today. At present, the fiber head end is often in the local exchange, some miles away from the most distant subscriber, which limits the maximum bit rate of the DSL connection, due to losses over that length of copper wire. The next steps will undoubtedly be FTTC (Fiber to the Curb) and finally FTTH (Fiber to the Home), though these will take quite a while to be deployed, because of the cost involved in laying this infrastructure.

What all these technologies mean for streaming media is that the consumer will soon be presented with options, often from new entries in the bandwidth provision market, such as Digital Island. These companies will compete aggressively with television service providers, other existing Internet service providers that use the local telephone company's infrastructure, and mobile phone companies to deliver digital media to a consumer. The only technology that can deliver entertainment and information as compelling digital content over this wide variety of carrier technologies is streaming media. It is the most carrier-agnostic technology for entertainment delivery yet deployed.

With streaming media, existing protocols for the delivery of Web page content, such as HTTP (Hyper Text Transfer Protocol) can be used to deliver video and audio streams. The Web server has a piece of software on it that knows how to handle correctly formed HTTP requests, supplying the correct digital data in response to the request. Because HTTP messages are easily handled on an IP network, any symmetrical carrier that supports TCP/IP can support streaming media. Even asymmetrical carriers, which use another technology for the back channel (such as satellite IP, which uses the phone line as a channel for data requests, but the satellite signal to deliver the data requested), can support streaming media. The underlying transport protocols upon which streaming media is delivered makes streaming media the universal entertainment and information delivery technology, regardless of the physical carrier. Unlike digital television standards, which are tied to specific carrier technologies, streaming media spans all IP-based carrier technologies.

A final consideration, in discussing technologies for delivering streaming media, is the thorny question of *router hops*. When a consumer requests streaming media to be served from some streaming media server on the Internet, the data must travel over several physical networks to get to the consumer. Indeed, each packet may follow a different physical path. The way that packets of data find their way from the server to the consumer is that at every point where multiple physical networks are connected, there is a machine called a router. The router examines every physical data packet it receives and makes a decision about which network segment to pass it on to. It makes this decision on the basis of which connection has the capacity to take the next packet, so its decisions are locally optimal (or near to it). However, this does not guarantee that the best path across the entire network is being followed.

Bottlenecks and queues can form, where data packets wait in line to be redirected by a router. When this occurs, the time to deliver the packet from source to consumer increases. Worse still, successive packets may suffer different traffic conditions. So, from the point of view of steaming media, where a constant delivery delay is desirable, packets may, instead, arrive out of sequence or with highly variable delays between packets.

Some packets might never arrive at all and have to be resent. The effective constant bandwidth from source to consumer may be quite low, if the path contains many routers and firewalls, each examining the data packets before passing them on in the path. Router hops and firewalls act to degrade the effective bandwidth of the virtual connection between server and consumer and hence make it difficult to maintain smooth and unbroken streaming over the entire network. A number of traffic management techniques have been employed to solve this problem, or at least alleviate it. We will discuss these later in this chapter.

The Personal Computer and Streaming Media

Decompressing a compressed streaming media stream and then rendering it to a display device and audio device simultaneously, along with any other synchronized digital elements, is a computationally intensive task requiring significant amounts of memory. Today, the PC is a popular platform for rendering streaming media for consumption, since PCs are ubiquitous and relatively cost effective. However, other consumer devices will soon have the processing and memory bandwidth to render streaming media. A DVD player, for example, is essentially a computer that reads the data from a disk, then decompresses it and renders it for display on a regular television set. Unfortunately, the PC is not an ideal device on which to watch streaming media, since in many cases the viewer would like to be on a sofa, with the display device at some distance across the room. Today's televisions do not have sufficient processor and memory bandwidth to render streaming media. Set-top boxes have been designed to add processor and memory bandwidth to the dumb television display device. Unfortunately, the set-top box industry took the view that they could build streaming media devices specific to the MPEG-2 compression schemes employed by most terrestrial and satellite digital television systems deployed today, using embedded microprocessors, limited memory, and embedded operating systems. These devices cannot receive an IP stream and decompress the streaming media data for display, only because the makers didn't have the need or foresight to make them do that.

Indeed, the cost of the hardware and software in a set-top box is now approaching that of a regular PC with a PC operating system, as commodity PC prices have fallen. These days, a set-top box that comprises a PC motherboard and an operating system like Microsoft Windows CE costs about the same as a set-top box made from embedded microprocessors, loaded with software developed expensively for that far less-common hardware environment. Software development costs fall for operating systems that are more widely deployed. This is because more companies build software development productivity tools for widely deployed hardware and software environments, such as the Wintel platform (an Intel microprocessor, in an IBM PC-compatible architecture, running the Microsoft Windows operating system) than for more custom and exotic hardware and software combinations. Also, more programmers know how to write for that platform than for a custom embedded system. Expect the television in your living room to be a computer one day. When it is, it will be able to render any compressed data stream delivered via any technology, be that today's digital television, a DVD, or an MPEG-4 stream served from a streaming media server.

Players

A streaming media player is really a piece of application software that binds to the network interface that delivers the streaming media data packets and the display and audio devices that will show the final program. All the player does is buffer the data packets, making sure they are in the correct order and then unpack the data packets, decompressing the digital payload. Then it paints the raw video and audio data to a display buffer (a piece of memory used by display drivers to draw objects to a computer screen) and sends the data to an audio digital-to-analog converter on a sound card. The player makes sure the data continues to stream from input to rendering devices. If continuity is interrupted the player takes corrective action, like pausing, repeating frames, painting a coarser picture, rendering a lower bandwidth audio signal, or rebuffering. Players may also request data to be resent. Unfortunately, most players can only try to recover once something has gone wrong. Few have schemes that inherently avoid playback disruption ahead of the problem. Most cannot guarantee to play video and audio without interruption and with a constant frame rate, under all network and PC loading conditions. For widespread consumer acceptance, player software must evolve to the point where the quality of viewer experience is at least as high as that of television. Motion cannot be jerky. Streams cannot stop or degrade.

Most players are designed to be called from a Web page that is rendered in a Web browser (another piece of application software that can unpack Web-page data and draw it to the screen). Indeed, the difference between a Web browser and a streaming media player is only that the streaming media must keep things moving. Otherwise, both take data received from the Internet, decompress it, and render it for consumption by a user. Players often have a scripting interface that lets Web page designers control the behavior of the player from some script embedded in the Web page they serve.

It has been common industry practice to give streaming media players away. The Microsoft Windows Media Player is available for free download to desktop PCs, palm-sized PCs, handheld PCs and pocket PCs, for example. The strategy has been to capture eyeballs, so that a user group valuable to advertisers can be aggregated. For this reason, vendors of streaming media player software have sought to differentiate their products by choosing their own compression schemes and by choosing the meaning of the data payloads. They have their own file and stream formats, so that one player cannot read and render streaming media data belonging to another. Each streaming media player vendor wants to be the only streaming media technology vendor, supplying the encoding technology and serving and transport technologies to an entire industry. The world is not so simple and monopolies are resisted until one commercial product is elected the de facto standard by the overall mass of users. Hence incompatibility between media streams and players is a necessary strategic position until one standard emerges.

Microsoft Windows Media Technologies

One of the current end-to-end streaming media systems competing for dominance is the Microsoft Windows Media Technologies offering, a proprietary system in which parts of the technology are most definitely closed; but the system can be extended, through the software development kits (SDKs) that are available to developers.

There are three main components of the Windows Media Technologies—Windows Media Tools, Windows Media Services, and Windows Media Player, providing tools for creation, distribution, and playback of streaming media respectively (Figure 2.14). All of these are shipped with other Microsoft products, or available for free download. Windows Media has been shipped as part of Windows since Windows 98 SE and Internet Explorer since version 5. It is also integrated in Office 2000 and subsequent releases. Windows Media Services are built into the Windows NT/2000 Server.

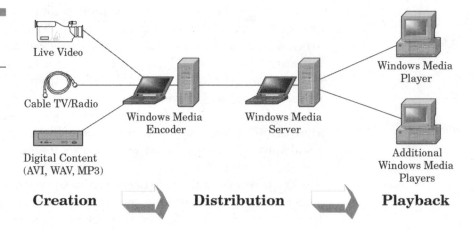

Figure 2.14
Windows media components.

Windows Media is focused on several key initiatives that have already started to broaden the impact of streaming media. It is designed to address the broadband Internet, delivering CD-quality music and near-broadcast–quality video. Coupled with rich interactivity and e-commerce capabilities, Windows Media is now delivering pay-per-use and advertising-supported content. Windows Media is also designed to enable the music industry to deliver digital music online, through digital rights management and secure distribution technologies. Microsoft is also working with the consumer electronics industry to take streaming media beyond the PC, into digital audio players, digital stereos, car stereos, and advanced television set-top boxes. Windows Media is targeted at business applications, supporting virtual company meetings, "just-in-time" learning, and instant communications with employees, partners and customers. The technology specifically addresses the needs of e-commerce sites, enabling those companies to add digital media easily, using Microsoft's Digital Broadcast Manager. This product allows content providers to manage, deliver, and sell pay-per-download and pay-per-stream content over the Internet.

To create content in Windows Media formats, you can either use several third-party tools, which have included the Windows Media Codec and file writing code, or else use Microsoft's own tool set. These tools are available for download on Microsoft's Windows Media Web site. There are three varieties of tool offered in Microsoft's Windows Media Tools suite:

- Tools for encoding media using Microsoft's codecs
- Editing and utility tools
- Tools for content creation

The encoding tools include Windows Media Encoder, which can encode files or live streams, from files and live sources. Windows Media Author enables authoring of slide-show–style content for low bandwidth presentations. Windows Media Publish to ASF (an add-on to Power-Point) converts a PowerPoint presentation into still images synchronized to an audio track, and then encodes it into a streaming file in Microsoft's Advanced Streaming Format. Windows Media Presenter (another add-on to PowerPoint) encodes a live PowerPoint presentation into a stream. Windows Media Plug-in for Adobe Premiere enables the exporting of movies from Adobe's non-linear video editing application, Premiere, to Windows Media Format. VidtoASF encodes video files that have already been captured. WAVtoASF does the same for audio files. Encoder Controller provides frame-synchronized device control, allowing the encoder to start and stop with a VTR, for real-time encoding, and allowing control of multiple encoders from a single source. Windows Media Encoder Remote Setup Utility allows the encoder to be set up under software control. Windows Media Encoder Batching Utility is used to encode multiple Windows Media files from multiple source files, with different preset profiles for each file. WMCap is a video capture application to capture a file, live preview, size the captured file, and write to AVI (Audio Video Interleave) format. Other tools available include a tool to convert MP3 (MPEG 1 audio layer 3) audio files in WMA (Windows Media Audio) format and a utility to encode video and audio using Windows Media 8 audio and video codecs.

Editing and utility tools include Windows Media ASF Indexer, which performs simple editing of files already encoded, inserting indexing, properties, markers, and scripts; ASFChop, which is a command-line–driven version of the Indexer; ASFCheck, which examines encoded files for errors; WMPcdcs, which installs the version 8 codecs into earlier players; and Stitcher, which takes in multiple files of any type supported by Windows Media Encoder and outputs a single file in Windows Media Format. The format-specific utilities include Windows Media Mobile, which verifies that the ASF header section is compatible with mobile terminals from NTT DoCoMo; WMAttr, which allows you to display and modify the metadata properties for a Windows Media File; WMAttrgui, which is the GUI (Graphical User Interface) version of the previous tool; WMProp, which shows the properties of a Windows Media File; Windows Media Metafile Creator, which is used for automatic creation of Windows Media metafiles and generation of a playlist of media content or editing an existing one; Windows Media Metafile Cleaner, which removes unsupported tags and attributes from metafiles; AudioPlayer, which plays Windows Media files; AVItoWMV, which converts files from the AVI format to the WMV (Windows Media Video) format; WMAPlay, which plays windows audio files under command line control; WMVAppend, which creates a single Windows Media Format from two of the same format butted together end-to-end; WMVNetWrite, which is used to show how a Windows Media file is streamed across the Internet by displaying each streamed sample, the time at which it starts playing, and its duration; and WMSProxy, which converts a multicast stream to a unicast stream.

Content creation tools include Movie Maker, a non-linear editing tool for editing videos shipped with Microsoft Windows XP; Producer for PowerPoint 2002, which allows users to create synchronized multimedia presentations for display in a Web browser; and Windows Media On Demand Producer, which was developed by Sonic Foundry Inc., to encode digital content, synchronize markers and script commands, and enhance video.

Although the selection of available encoding tools looks overwhelming, the actual process of content creation, including encoding, is actually relatively straightforward. The first step is to capture the audio and/or video digitally. To do this, you need a capture card, such as a Viewcast Osprey 100 for video, or a Creative Labs SoundBlaster for audio and a capture program. The usual file format for capture is AVI, but you can stream from live sources, without saving the captured material to a file at all.

It would be useful, at this juncture, to explain a file format and what's in it. A file format is just a set of rules that govern how things are written down in a file. It tells the order you will find information and what the individual parts of the file will mean. For example, there is a need to write down the sampling rate and quantization setting in a digital audio file. If you don't include those pieces of information, it is impossible to interpret the "payload," binary digits which represent the audio samples. Some files write their data in human-readable form. XML is an example of a human-readable file format. These files can be opened with a text editor, like the Notepad editor shipped with Microsoft Windows, and the person reading the data will be able to read and understand the content of the file.

Other files are purely binary, containing only ones and zeros. The only way to interpret a binary file is with a program that understands the format. Microsoft's media files are based on the company's earlier RIFF or Resource Interchange File Format. This was a general-purpose format for describing interleaved digital media data. In the one file, you could include chunks of video interspersed with chunks of audio, for example. AVI is a specific type of RIFF file, in that the header content and data payload are given specific definitions for interleaving audio and video (tagged with the suffix .avi). A WAV file is another type of RIFF file that contains only audio (tagged with a .wav suffix). The RIFF file format resembles Apple's AIFF (Apple Interchange File Format).

For streaming media use, Microsoft supports several file formats, which are optimized to deliver streaming data. The first streaming file format was ASF (Advanced Streaming Format—tagged with the suffix .asf). However, because several media players registered themselves as players for ASF files, when installed, the situation arose where audio players were being called upon to attempt to play video streams (and obviously failing). To get around this problem, Microsoft created the WMF (Windows Media Format—designated by the .wm suffix), the WMA (Windows Media Audio or .wma), and WMV (Windows Media Video or .wmv) formats, allowing streaming media players to claim to be able to play the correct media types. The ASX (.asx) file format is human readable and is used to write down information about an ASF file. Called a stream redirector file, it is actually a flavor of XML file. The equivalents to ASX files for WMF are called WAX (.wax) files.

It is interesting to understand how a media player interprets the stream of bits that it receives from a streaming media server. In the first place, the raw binary stream that enters the machine must be stripped of its transport headers, revealing the transport payload. So, for every packet of data that is received, the header information must be discarded, to get to the data of interest. Sometimes the payload is encrypted, so that as you collect the data from successive data packets into a contiguous file in memory, you have an encrypted version of the data you want. Once that payload is decrypted, you have data formatted according to a streaming file format, such as ASF. To get to the actual audio and video data, you must read and interpret the file data, separating the binary that represents information on how to decode and play the media from the binary that represents the actual video and audio. The audio and video payloads found are compressed, so they must be decompressed. The result of the decompression process, a series of mathematical operations that recreates an approximation to the original audio and video, is a data set representing the pixels comprising the moving images and another data set representing the accompanying audio samples. To get down to the data that will actually be painted on screen and converted into sound, the bit stream has had to be interpreted and decoded at various levels (Figure 2.15).

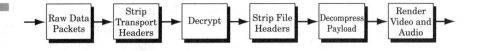

Figure 2.15
Unpacking the
stream.

Returning now to our simple content creation process, after digitization of the audio and video, the next process is encoding, using Windows Media Encoder, for example, where the captured audio and video are compressed. In fact, some encoding programs include the capture process as part of the sequence of events in encoding. Encoding usually takes care of file formatting, since most encoding programs write their output to a standard streaming format. For Windows Media, the default file format used to be ASF, but now can be a number of different formats, the usual one being WMF. After the file is encoded, it can be uploaded to a server for on-demand viewing, streamed live from the encoding application, or else processed further to add markers, scripts, closed captioning, or links, using a tool like Windows Media Advanced Script Indexer. To embed a link (i.e., a standard URL or Uniform Resource Locator) to a media stream in a Web page, you need to follow the application guides to add the right script and file calls. In most cases, Windows Media requires a metafile, which has an .asx designation. This file directs a browser to ignore the file in the tag and invoke the Windows Media Player to deal with it instead. ASX files are also useful for creating client-side playlists.

Windows Media actually supports a variety of codecs, pieces of software that convert digital media from one format to another. More specifically a codec is the piece of software responsible for compressing (and decompressing, on playback) the video and audio, so that it is compact enough to stream over the Internet. The six principal codecs shipped with Windows Media are:

- Windows Media Audio (currently at version 8)
- Sipro Labs ACELP (Algebraic Code Excited Linear Prediction), used for low bit-rate encoding of voice content
- Windows Media Video (currently at version 8)
- Microsoft MPEG-4 version 3.0 (Microsoft's own evolution of the ISO-MPEG-4 codec)
- ISO (International Organization for Standardization) MPEG-4 video codec version 1.0
- Microsoft Windows Media Screen version 7.0 (used to capture activity on a Windows desktop as a video)

Microsoft recommends that you use version 8 audio and video codecs for most purposes, because of quality and bit-rate advantages.

Optional operations in content creation include making the file downloadable and adding synchronized multimedia content, like closed captions. With Microsoft's Windows Media Technologies, you can download media files using the Windows Media Download Control. In order for you to use this control, media files must be packaged as Windows Media Download packages, (.wmd). Microsoft provides a tool to convert .wm files to .wmd. Synchronized multimedia can be authored with Microsoft's Producer for PowerPoint 2002, or by a number of other techniques. For example, Microsoft has a file format called SAMI (Synchronized Accessible Media Interchange), which is a text file, resembling HTML, used for delivering subtitles, closed captioning, and audio descriptions in a file separate to the media file. This allows closed captioning to be prepared with nothing more than a text editor, though there are some third-party SAMI authoring tools becoming available.

The second significant piece of the Windows Media Technologies suite is Windows Media Services. There are two major methods of delivering streaming audio and video content over the Internet. The first uses a standard Web server to deliver the audio and video data to a media player. The second method uses a separate streaming media server, specialized for the task. A streaming server is a more efficient and flexible solution that provides a better user experience.

When you host streaming media on a regular Web server, the activated Web page launches the client-side player and downloads the media file, just as it would if you were downloading the entire file before playing it. The difference between the download-and-play and the streaming case is entirely due to how the media player client behaves. The streaming client starts playing the audio or video while it is downloading. The Web server isn't aware that the player is doing this. With this delivery method, the client retrieves data as fast as the Web server, network, and client will allow, regardless of the bit rate of the compressed stream. Only certain media file formats support this type of "progressive playback." Microsoft's ASF is one of them.

What's wrong with streaming media like this? HTTP, the protocol used to deliver elements of a Web page to a browser, operates on top of the Transmission Control Protocol (TCP). For each HTTP request, there is actually some TCP traffic handling the data transfers. Optimized for non–real-time applications such as file transfer and remote login, TCP's goal is to maximize the data transfer while ensuring overall stability and high throughput for the entire network. To achieve these goals, TCP uses an algorithm called "slow start." By now, anybody familiar with buffering delays before playback starts ought to be hearing alarm bells. Slow start works by sending data at an initial low data rate, gradually increasing the rate until the destination reports packet loss. In other words, deliver data at a trickle until you build up to a speed that breaks the connection and interrupts the flow of packets to the player. TCP achieves reliable data transfer by retransmitting lost packets. However, it cannot ensure that all resent packets will arrive at the client in time to be played as an uninterrupted media stream.

With a streaming media server, in contrast, the media are not held on a Web server, but copied to a specialized media server (such as Microsoft's Windows Media Services). This need not be a separate piece of hardware. You can run a Web server (such as Microsoft's Internet Information Server) and a media server on the same computer, but they are often divided among different machines. In contrast to the passive burst method employed in Web-server streaming, with a media server the data is actively and intelligently sent to the client, meaning that the content is delivered at the exact data rate associated with the compressed audio and video streams. The server and the client stay in close touch during the delivery process and the streaming media server responds to feedback from the client.

While media servers can choose to use the same HTTP/TCP protocols used by Web servers, they can also use protocols like UDP (User Datagram Protocol) to improve the streaming experience greatly. Unlike TCP, UDP is a fast, lightweight protocol without any retransmission or data rate-management functionality. This is good and bad. Fast and lightweight are desirable properties. The inability to recover lost data and to deal with variable data rate channels is not so good. However, UDP is a good choice of protocol for transmitting real-time audio and video data, which can tolerate some lost packets.

When we say audio and video transmission can "tolerate packet loss," this is only if we accept throwing some video frames away or if we allow audio to stutter. If we wish to maintain smooth motion and glitch-free audio, we need to be able to recover the data exactly from a lossy stream. This requires some redundancy in the transmission and probably some forward error correction coding. In early implementations of the Windows Media Services, data loss meant lost video frames or lost sections of audio. Data loss was tolerable only if you didn't care about presentation continuity or quality of user experience.

A bonus with using UDP for streaming is that, because of the back-off policies implicit in the TCP protocol, UDP traffic gets higher priority than TCP traffic on the Internet. Instead of the blind retransmission scheme employed by TCP, streaming media servers like the Microsoft Windows Media Services use an intelligent retransmission scheme on top of UDP. Microsoft's UDP Resend feature ensures that the server only retransmits lost packets that can be sent to the client in time to get played. Those that can't, don't get resent and the player glitches.

There is really only one major advantage to using a standard Web server to stream media: utilizing existing infrastructure. However, that also means subjecting that infrastructure to the added relatively heavy burden of streaming. Web servers like to deliver small things. Media streams are large.

The advantages of a streaming media server are manifold. First, you make more efficient use of network throughput, since you know what the data rate is going to be, based on the headers of the compressed media file. Windows Media Server sends data to the client Windows Media Player only at the required bit rate and the network is never overdriven to the point of loss, at which time it becomes a traffic bottleneck. Because the server and player remain in contact throughout playback, the server can dynamically respond to client feedback. If network congestion occurs midway through playback, the server can decide to retain the audio quality, but lower the frame rate of the video stream to suit the bandwidth now available on the degraded network, for example. This is not possible with a simple Web server, since there is no feedback from the client player to the server. With a regular Web server streaming the media, if the network degrades, the client player stops and goes, causing the insidious "rebuffering" delays common to early (and let's face it, current) implementations of streaming media. Use of a specialized streaming media server also allows detailed reporting of streams played, VCR controls (seek, fast forward, rewind), live video delivery and delivery of multiple streams to the client. Windows Media Server also greatly improves performance by optimizing how media files are read from the disk, buffered in main memory, and streamed onto the network. This improves *scalability*, the number of users that can be served at once, by a factor or two or three over a Web server. Also, with a regular Web server, there is no way to prevent end-users from copying the local cached copy of the media file being played to their hard drives for later replay. With Windows Media Server, users can only stream data and are prevented from downloading the file directly to their hard disks without explicit permission of the copyright holder. As data packets are received over the network, they are delivered directly to the client application with no easy way for the end-user to intervene and make a copy.

To summarize, then, a streaming media server has multiple delivery options. The four supported by the Windows Media Server are UDP, TCP, HTTP + TCP, and Multicast. UDP provides the most efficient network throughput and can have a positive impact on player performance. However, administrators normally close their firewalls to UDP traffic, so UDP transport is problematic on the public Internet. TCP provides adequate support for delivering streaming media, but suffers from all the same problems that a regular Web server does. TCP traffic normally permeates firewalls. Microsoft implemented its own version of HTTP to enable streaming through firewalls and proxy servers, while retaining the advantages of a media server. This allows users to fast-forward and rewind, but adds some overhead to the raw TCP stream that decreases scalability. Finally, there is IP Multicast, which allows very efficient delivery of streaming content to large numbers of users, in much the same way as television broadcasts do. Multicast is finding a home on corporate networks, but is still very rare on the public Internet. The Windows Media Server will automatically switch to the appropriate protocol so that no client-side configuration is necessary. The server will initially attempt to transmit files using the optimal UDP or multicast protocols. If this does not work, the server will then attempt to send first via the raw TCP protocol, then via TCP with HTTP-based control. Windows Media Server also supports the legacy MMS, Microsoft's proprietary Microsoft Media Streaming protocol, a derivative of the Real Time Protocol (RTP). Although still included, it is considered obsolete and is not recommended for new installations. Media servers like Windows Media Server can support live and on-demand programming, using unicast or multicast protocols. Common scenarios include live ad insertion and Web radio, using server-side playlists.

Microsoft's Windows Media Server also includes a technology called Intelligent Streaming. This combines multidata rate encoding, intelligent transmission, and a video playback filter to detect network conditions and adjust the properties of the video stream automatically to maximize playback quality. In the public Internet, connection speeds can vary by 50% or more of the maximum, depending on network and ISP (Internet Service Provider) congestion. Because Windows Media Technologies is a connected, end-to-end client/server system, the server and the client communicate with each other to establish actual network throughput and make a series of adjustments to maximize the quality of the stream. Intelligent Streaming maximizes use of the available bandwidth. Users receive content tailored to connection speed. This greatly improves user experience. Modem-connected users immediately notice the presentation is smoother, less jerky, and generally of higher quality.

Intelligent Streaming works by automatically adjusting between multiple video bit rates and by cleaning up the video streams. Buffering is the biggest problem with streaming digital media. If the bandwidth available between server and client drops below the data rate of the stream, the player will always run out of material and have to rebuffer. Just because the connection speed is fast, does not mean that the bandwidth supports the bit rate. On the public Internet, unpredictability of bandwidth is taken as a given. Actual bandwidth available is determined by network conditions. Traffic on the Internet is constantly fluctuating. If a user attempts to view video streamed at one bit rate, but the bandwidth available often plunges below that, presentation continuity will be badly and obviously affected. Intelligent Streaming solves this by sending a stream with the appropriate bandwidth when the user first connects, then by dynamically and seamlessly adjusting the bit rate as network conditions change.

The first technique used in Intelligent Streaming is Multi Data Rate Encoding. Windows Media Technology supports a technology equivalent to RealNetwork's SureStream, which allows graceful playback degradation as the bandwidth available on the streaming link degrades. Called Multi Bit Rate Encoding, it encodes up to ten discrete, user-definable video streams and one audio stream into a single Windows Media stream. The video streams are encoded from the same content, but each is encoded for a different bit rate. When a multiple-bit rate Windows Media file or live stream is played on Windows Media Player, connected to Windows Media Server, only one of the video streams is received: the one that is appropriate for the current bandwidth conditions. The process of selecting the appropriate stream is invisible to the user.

The second technique of Intelligent Streaming is Intelligent Bandwidth Control. There are a number of steps in the process. Each is a strategy to modify the bit rate so that the stream remains continuous on the client, regardless of the bandwidths currently available. As bandwidth fluctuates between server and client, the server detects changes and adopts the best strategy. When bandwidth is at its best, the server employs the first strategy. As conditions worsen, the server checks its list of options one by one until the bit rate is optimized for the current available bandwidth. The strategies are as follows: in the first place, the server and client automatically determine the available bandwidth, then the server selects and serves the video stream at the appropriate rate. If the available bandwidth decreases during transmission, the server automatically detects the change and switches to a lower bandwidth stream. If the bandwidth improves, the server selects a higher bandwidth stream, but never one higher than the original bandwidth (which is a pity, if you first connected during a period of severe network congestion). If the connection no longer has the bandwidth to support streaming video, the client and server intelligently degrade image quality. When a network is extremely congested, the server attempts at least to maintain a continuous audio stream by decreasing the video frame rate to minimize playback interruptions caused by buffering. If necessary, the server will stop sending video frames altogether, maintaining the audio. If audio quality starts to degrade, the client tries to reconstruct portions of the stream to preserve quality. At this point in proceedings, there really is no quality.

The third technique of Intelligent Streaming is Intelligent Image Processing. With this technique, the client post-processes the video stream to enhance quality even at low bit rates. The Windows Media Video 7 codec (and its successors) includes an intelligent filter to smooth image blockiness and remove ghosting artifacts, to improve the overall appearance of the video. Blockiness also occurs on high-bit rate stream, but isn't as noticeable. The video filter smoothes the edges of blocks in the image and erases ringing artifacts, so that the resulting video is more pleasing to the eye.

To encode with multiple bit rates, the content creator just uses one of the multi-bit rate profiles in the Windows Media Encoder application. Client post-processing and bit rate optimization are both automatic, on-the-fly features of the Windows Media Server and Player. They also work with live video streams. The user and the content producer do nothing whatsoever to configure these.

Windows Media Services also comes with some very useful tools. These include GetDynamicIP, which establishes a static port connection between a Windows Media unicast publishing point and a computer running the Windows Media Encoder that has a dynamically assigned IP address; SendIP, which works the other way around; Windows Media Load Simulator, which simulates real-world load on a media server; and Windows Media Monitor, which is used to monitor the connections of up to eight streams simultaneously.

The Windows Media Player is the final piece of the Windows Media Technologies suite. It is a helper application for the Internet Explorer Web browser, which performs the job of playing streaming media files. It is also a standalone desktop application. When the Web browser detects an ASX or WAX file, responsibility for playing the media files described within passes to the media player application. The player launches and begins playing back the stream. The ASX file can describe a single media file, or else a list of files to be played back in sequence, with the media picked up according to the URLs given. Hence, a playlist can play media served from a variety of media servers. The instructions embedded in the metafile also tell the player which streaming protocol is in use, so server and protocol rollover is a feature of metafiles. If the player cannot connect to the media server specified in the file, or if the protocol cannot be established, the player can skip to an alternate server and/or protocol, provided that this information is given in the redirection file. Microsoft's Windows Media Player can also play MPEG files with extensions .m3u, .mp2v, .mpg, .mpeg, .m1v, .mp2, .mp3, .mpa, .mpe, and .mpv2. Musical Instrument Digital Interface (MIDI) files (with extensions .mid, .midi, and .rmi); Apple QuickTime files (with extensions .qt, .aif, .aifc, .aiff, and .mov—versions 1 and 2 of QuickTime only), and UNIX media files (with extensions .au and .snd) can also be played by Windows Media Player, but support for RealNetworks' .ra, .rm, and .ram formats has been discontinued. Microsoft issued a press release in April 2001, announcing the results of an independent, double-blind viewing test comparing the quality of Windows Media Video 8 against RealVideo 8. The tests were conducted by eTesting Labs using ISO/MPEG-4 reference clips. The press release claimed that viewers chose Windows Media Video over RealVideo by a ratio of almost 3 to 1, claiming better motion smoothness, image sharpness, and general preference. This press release serves to illustrate how competitively the vendors of streaming media technology take their compression abilities. I would have liked to see a double-blind image preference test of Microsoft Windows Video 8 against cable television or DVD, but that will have to be done another time.

Although optional, Microsoft's Digital Rights Management is also an important, if not core, part of the Windows Media Technologies. When a consumer receives a media file, encrypted using Windows Media Rights Manager, from a Web server (or media server), a key must be obtained to unlock the content before it can be played. Content owners can easily set these licenses and keys in motion by protecting their files using the Rights Manager tool and then distributing content in the usual way. It lets content providers deliver media over the Internet in a protected, encrypted file format. Rights Manager is the tool used to package the files in this way. A packaged media file contains a version of the media file, which is encrypted and locked with a digital key. It also contains additional information from the content provider. The package can only be unlocked and played by a person who has obtained a license.

Figure 2.16
Windows media
rights manager flow.

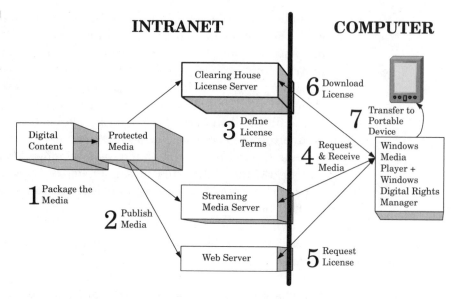

Rights Management is a five-stage process: packaging, distribution, establishing a license server, license acquisition, and playback (Figure 2.16). To package a file, the Rights Manager is used to encrypt it and lock it with a "key." The key is stored in an encrypted license, which is distributed separately. Other information is added to the media file, such as the URL where the license can be acquired. The packaged digital media file is saved in Windows Media Format. To distribute the rights managed package, it can be placed on a Web site for download, placed on a media server for streaming, distributed on a CD, or e-mailed to consumers. Consumers are free to redistribute the rights managed package to their friends. In order for consumers to be able to unlock the rights managed packages, the content owner chooses a license clearinghouse to store the specific rights of the license and implements the Windows Media Rights Manager license services. That's all they need do to establish a license server. The role of the clearinghouse is to authenticate the consumer's request for a license. Because the media files and licenses are stored and distributed separately, it is easier to manage the entire system. To play a packaged media file, the consumer must acquire a license key to unlock the file. The process of acquiring a license begins automatically when the consumer attempts to acquire the protected content, acquires a pre-delivered license, or plays the file for the first time. Windows Media Rights Manager either sends the consumer to a registration page where information is requested or payment required, or "silently" retrieves a license from a clearinghouse. To play the digital media file, the consumer needs a media player that supports Windows Media Rights Manager. The consumer can then play the media file according to the rules or rights included in the license. Licenses can have different rights, such as start times and dates, duration, and counted accesses. For instance, default rights may allow the consumer to play the digital media file on a specific computer and copy the file to a portable device. Licenses are not, however, transferable. If a consumer sends a packaged digital media file to a friend, the person receiving the file must obtain his or her own license to play the file. This PC-by-PC licensing scheme ensures that packaged digital media files can only be played by the computer that has been granted the license key for that file.

This scheme erodes consumer rights significantly. If you purchase a CD, the content owner grants you license to play that CD on any player you choose, provided you don't make public performances or illegal copies. Imagine the feeling of having purchased licenses to your entire music collection, exchanging personal data in the process, then having your PC stolen. How long would it take, in practice, to get all those licenses again on your replacement PC? How would a consumer even prove that legitimate, fully paid up rights had been stolen or even keep an inventory (not on the now-missing PC) of the licenses owned and the conditions attached to each one? What happens when a virus corrupts the hard drive on which the licenses granted are stored? Will a bad block on your hard drive prevent you from accessing music you have paid to access? If the licensing clearinghouse refuses to acknowledge your existing rights, so that the deal is effectively "off," how will you get your personal data back from them? On the one hand, streaming media makes it possible for people to save their precious time by bringing the music directly to them, but then wastes their time with the licensing issues. What the large print giveth, the fine print taketh away.

How keys work is shown in Figure 2.17. To generate a key, the license key seed and key ID are needed. The license key seed is a value that is known only to the content owner and license clearinghouse. The content owner creates the key ID for each Windows Media file. This value is included in the packaged file. When the license clearinghouse needs to issue a license for a packaged file, a key can be recreated by retrieving the key ID for the packaged file. The Windows Media License Server uses the license key seed (which the clearinghouse provides) and the key ID from the packaged file to create the key. The key is included in the license sent to the consumer's computer. Using the key included in the

Figure 2.17
How keys work.

Content Owner

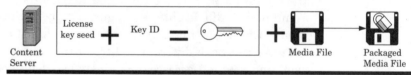

License Clearing House

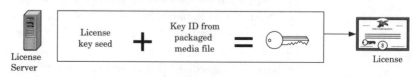

Content Consumer

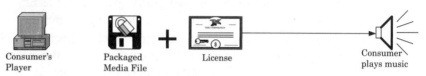

license, the player on the consumer's computer can open and play the protected file. The license also contains rules that govern the use of the digital media file; these are set by the content owner. Content owners can dictate how many times a file can be played, which devices a file can be played or transferred on, when the user can start playing the file and the expiration date of the license, if the file can be transferred to a CD burner or not, if the user can backup and restore the license, what security level is required on the client to play the file, and many other things.

License management, for consumers, is a non-trivial task and certainly not invisible and seamless, unless they always default to the content owners' interests and buy new licenses whenever they think they need them, whether or not they actually do. Licenses can be delivered in different ways and at different times, depending on the business model. The content owner might want licenses predelivered, or delivered after a consumer has downloaded and attempted to play a packaged file for the first time. Licenses can be delivered with or without the consumer's being aware of the process, using silent or non-silent license delivery.

Content owners need to be very careful about how they exercise their rights. If they prefer the more Draconian options, they are implicitly giving the technology more opportunities to say to the consumer "you are wrong," thus alienating paying customers. In a world where good business practice is all about maintaining that the consumer is always right, there is a built-in conflict in the blind, inflexible use of rights management technologies. Not only should the buyer beware, so should the content owner.

Microsoft's Software Development Kits (SDKs) allow developers to include Windows Media Technologies in their applications. Microsoft likes this, because it broadens support for proprietary formats and completes the "whole-product offering" to users of the technology. Third-party support is essential to establish a technology as a market leader. Although the SDKs are free, access to them sometimes requires registration and there are licensing conditions attached to use of Microsoft's code in commercial products.

Microsoft's Windows Media SDKs include the Format SDK, which allows application developers to read and write Windows Media Formats; the Player SDK, which allows extensions to the media player; the Encoder SDK, which allows application developers to add Windows Media encoding to their applications; and the Services SDK, which provides extensibility for Windows Media Services. Most developers also make use of Microsoft's DirectShow SDK, including the DirectShow Editing Services and Broadcast Digital Architecture, in their development projects. DirectShow is part of Microsoft's DirectX architecture. Also available are the Windows Media Embedded Product Adaptation Kit, used by hardware vendors to include Windows Media Technology in their devices, and the Rights Manager SDK, which allows developers to include Digital Rights Management in their products.

Corona is the code name for Microsoft's third-generation Windows Media Technologies, which will ship with the forthcoming .NET server. Microsoft touts the main features as faster stream starting, home-theater quality streaming, improved economics, and extensibility of the platform.

Corona's Fast Stream delivers an "instant on, always on" streaming experience for broadband users, effectively eliminating the buffer delays which are the bane of streaming media users. Fast stream also optimizes the delivery of streaming audio and video to take advantage of the full bandwidth available to the user, which vastly reduces or eliminates the impact of congestion on the Web for broadband users. Fast Switch seamlessly transitions between clips in client-side or server-side playlists, providing a smooth user experience and presentation continuity. Fast Recovery eliminates content discrepancies and interruption over high-latency networks, using forward error correction techniques to ensure an uninterrupted viewing experience.

Two new professional-level codecs are introduced along with Corona. The new WMA (Windows Media Audio) professional codec is the first to deliver 5.1 channel surround sound with full spectrum, full-resolution audio (24 bit/96 kHz sample). A new version of the WMV (Windows Media Video) codec provides a 20% boost in efficiency compared with the previous version and introduces the ability to provide HDTV-like video quality at file sizes half those of today's DVDs, for local playback on the PC. These features combine to deliver home-theater quality presentation.

New dynamic content programming capabilities, with server-side playlist support, enables real-time ad insertion, including lead-ins and interstitials. Because fast stream delivers a better user experience, ad-driven streaming media business models can now be implemented. Corona's advanced compression technology means bandwidth costs can be lower, or else quality delivered higher. With twice the scalability of the previous server platform, Corona also answers the call for a cost-effective solution on which to build a profitable streaming business, or to use streaming to reduce costs of high-quality enterprise communication. The economics of streaming are further improved by Corona's cache and proxy support, enabling operators to conserve bandwidth, decrease network-imposed latency, and decrease the load on Windows Media origin servers. The server-side playlists, which allow dynamic changes to the order of clips played, insertion of new clips, and insertion of ads, without interruption to the viewer, integrate with third-party ad servers and include advanced usage reporting.

The Corona platform is being billed as "industrial strength," because it supports twice as many concurrent users, enabling streaming for the largest enterprises and content delivery networks (more on this later). Plug-ins run in protected memory ensuring maximum system reliability, the administration tools are more flexible, and there are wizards to help systems administrators set up common streaming scenario configurations and manage activities. Secure server-to-server and server-to-client content delivery is made possible using a variety of common authentication and authorization mechanisms, including support for the new HTTP Digest. Digital rights management support is included for on-the-wire and persistent client-side security.

Application developers can deliver exciting new products and services, built on top of Corona, via a state-of-the-art plug-in model for the player, the server and the encoder. A vastly improved SDK will allow developers to easily incorporate digital media into their applications and solutions, both on the server side and the client side, using the programming languages with which they are already familiar. The plug-in architecture will allow integration with storage, billing, and logging applications. The SDK's object model and event mechanism enables developers to build custom applications for configuring and monitoring Windows Media Services, using WBEM/WMI (Web Based Enterprise Management/Windows Management Instrumentation). The object model is extensive, with over 700 server interfaces. Developers can choose to write in C++, C# (the native language of the .NET platform), VB Script, and Perl.

Corona is not thought to support variable bit rate encoding, whereby the encoder allocates more bits for more detail, in order to achieve higher quality images and sounds, regardless of the complexity of the source. I was not able to discover details about Microsoft's view on this.

It is almost ridiculous to try to list the advantages that the Microsoft Windows Media Technologies have over their nearest rival, RealNetworks' system, because both are such rapidly moving targets. However, in a comparison of Windows Media 8 (the release prior to Corona) with Real System 8 (the one before the RealOne platform), Microsoft could still claim advantages in cost effectiveness (much of the technology shipped at no charge with other products. Microsoft charges no per-stream license fees such as there are with RealNetworks' server). Windows Media 8 was arguably the most scalable streaming media platform on the market and it excelled in administration flexibility, with additional features to improve performance and usability for authors and administrators. Windows Media also offered easier configuration and management than Real-System did. However, with two vendors as fiercely competitive as these two, all of these advantages are likely to be temporary.

Another interesting technology recently included under the Windows Media Technologies banner is HDCD (High Definition Compatible Digital). This currently has nothing to do with streaming media, but may find its way into streaming applications in the future. HDCD is a patented encode/decode process for delivering the full richness and detail of the original microphone feed on Compact Disks and DVD-Audio. HDCD-encoded CDs sound better because they are encoded with 20 bits of real musical information, as compared with 16 bits for all other CDs. HDCD uses a sophisticated system to encode the additional four bits into the existing 16-bit CD format, while remaining completely compatible with the existing format. Originally developed by Keith Johnson and Pflash Pflaumer of Pacific Microsonics Inc., the technology has been licensed to Microsoft. The bits required for selection of decimation filter, peak extension, and low-level range extension are embedded in a pseudo-random noise sequence, inserted into the least significant bit of the audio. This subcode is completely inaudible, as it is inserted only a small percentage of the time. However, the decoder can recover the instructions for range extension from the audio signal and control the digital-to-analog conversion process to reproduce extended audio range signals. The same techniques could potentially be used to extend the range of encoded streaming audio, or else use fewer bits to represent audio of a given quality.

QuickTime and Sorenson

Apple invented QuickTime as a patented, extensible, track-based container format for multimedia files. Each track delivers a different element of content, such as video, audio, interactivity (such as Flash), HTML behavior, etc. QuickTime includes a player, which runs on both the Apple Macintosh and the Microsoft Windows operating system. At the time of writing, QuickTime 5 was the current version. The player allows users to play back audio and video on their computers, but Quick-Time is many things, including a file format, an environment for media authoring, and a suite of applications for playing, viewing, authoring, and serving multimedia. QuickTime includes a Streaming Server, which runs on the Apple Mac OS X operating system, for delivering streaming media files on the Internet in real time; and the Darwin Streaming Server, an open-source streaming media server for Linux, Solaris, FreeBSD, and Windows. Apple claim to have 150 million copies of QuickTime Player in distribution. Every second of every day, four people download QuickTime. The player, browser plug-ins, and servers are free, but the authoring application, QuickTime Pro is sold.

Streaming was a late feature addition to QuickTime. The QuickTime player originally required complete media download before playback, or else played media back using progressive download techniques.

The player supports a vast number of import file formats, including virtual reality formats. The QuickTime container is codec-agnostic, so several compression formats are supported by the player. The most relevant codec for video streaming is the Sorenson Video 3 codec, produced by Sorenson and first included with QuickTime 5, though earlier versions of the Sorenson codec have been shipped with QuickTime since QuickTime 3. The QuickTime player can also accomplish a number of video effects, including blur, recoloring, sharpening, and lens flare, among others. Besides playing MP3 audio content, QuickTime supports timecode tracks as well as MIDI (Musical Instrument Digital Interface) files, such as the Roland Sound Canvas and GS format extensions.

QuickTime also supports key standards for Web streaming, including HTTP, RTP, and RTSP. It supports every major file format for images, including JPEG, BMP, PICT, PNG, and GIF. QuickTime also features built-in support for digital video, including DV camcorder formats, as well as support for AVI, AVR, MPEG-1, H.263, and OpenDML.

Other vendors who supply QuickTime Streaming Servers include SGI, IBM, Sun, and Cisco. Sorenson's latest Video 3 codec includes automatic keyframes through scene-change detection, bidirectional prediction, two-pass variable-bit rate compression, block refresh for packet loss correction, and compression time packetization for error resiliency to packet loss. Sorenson and Apple are independently developing MPEG-4 codecs that conform to the ISO specification.

The key features of QuickTime 5 include support for Macromedia Flash 4 and the On2 codec, enhanced DV support, Cubic VR support (providing 360-degree immersive images), and the patented "Skip Protection" technology in the QuickTime Streaming Server, which protects streams from disruptions on the Internet, producing better playback quality. Skip Protection works by using any excess bandwidth to buffer data faster than real time on the client machine. This way, if packets are lost, the stream continues to play from this local buffer, resulting in smooth, high-quality media streaming. Like the Microsoft Windows Media Server, the QuickTime Streaming Server can deliver video on demand and live streams. It can act as a reflector for live broadcasts, fanning out the stream to many more users. Access control is built in, using authentication modules. The Server supports both unicast and multicast operation.

Apple calls their progressive downloading technique, using HTTP and FTP protocols, "Fast Start." With Fast Start, the audience downloads an entire QuickTime movie at the highest data rate their connections can support. As soon as the initial part of the movie has been downloaded, the QuickTime player plug-in begins to play it back in the browser while it continues to download the rest. This process downloads the complete movie to a user's hard drive, which allows it to be played back as often as desired. A Fast Start movie can start playing long before the whole file has downloaded, typically within a few seconds of starting the download. If the net connection is faster than the movie's data rate, the movie plays smoothly as it arrives, with no waiting. A Fast Start movie can also include pointers to local data located in other files on the Web server, a local disk or CD, or any Web URL. The benefits of using Fast Start are that no special server software is required. The movie gets through to the player eventually, no matter how slow the connection. With fast connections, the movie plays as it downloads, as if it were streaming. Fast Start delivers all types of QuickTime media, including sprites and QuickTime VR. Lost packets are retransmitted until they are received and there are no problems with firewalls or Network Address Translation. However, Fast Start cannot broadcast, multicast, or transmit live video feeds. Users cannot skip forward through the movie until the entire movie has been downloaded, and a copy of the movie remains on the local hard drive.

For true streaming, the QuickTime server supports RTP (Real time Transport Protocol). RTP is similar to HTTP and FTP, but tailored for the special needs of real-time streaming. Unlike HTTP, RTP does not download an entire movie to the client computer. Instead it siphons out a thin, one-way data stream at a constant data rate that plays the broadcast in real time, after a few initial moments of delay for hand-shaking and data buffering. A streamed one-minute movie plays in exactly one minute (plus buffering delay). As long as the connection has enough bandwidth to handle the data stream, the movie will play. Once data has been displayed by the player, it is discarded. No file on the local hard drive remains after playback. Viewers can see the broadcast again only by requesting it from the streaming server.

RTSP (Real Time Streaming Protocol) is a companion protocol to RTP, used when viewers communicate with a unicast server. RTSP allows two-way communication, so that viewers can command the server to do things like rewind the movie, skip to the next chapter, and so on. RTP, in contrast, is a one-way protocol. Once playback starts, it can only continue or be halted. The user cannot skip through the stream or rewind. Viewer interaction with the on-screen movie via the player's motion controls can only be relayed to the server using RTSP requests, which the player forms in response to user interactions. RTP uses low-level UDP (User Datagram Protocol) transport, which is faster and more efficient than the usual TCP/IP (Transport Control Protocol/Internet Protocol) used for most Web traffic. However, UDP lacks a mechanism for reporting lost packets, so that streaming over the Internet almost always involves some data loss. Also, because most corporate and personal firewalls block UDP, viewers behind a firewall may not be able to receive live streams, even though they can request them via RTSP. Apple provides proxy serv-er software to get around firewalls, but the system administrator who maintains the firewall must install the proxy software.

RTP also uses port addresses that may confuse some Network Address Translators and prevent users from receiving live streams. A final drastic solution to using RTP with firewalls is HTTP *tunneling*, where RTP packets are wrapped inside ordinary HTTP packets so that they can pass through the firewall. Unfortunately, HTTP tunneling adds significant overhead to the stream, so takes more bandwidth.

When a QuickTime Player tunes in to a live broadcast, it sends a request to the streaming server. The server looks for a Session Descriptor Protocol (SDP) file. If this is found, it begins to stream the media to the computer via RTP. An SDP file is a text file containing information about what will be streamed and telling how to tune in. SDP files are created by broadcast software on the computer that captures live media. The SDP file must be copied to the Streaming Server before the media can be broadcast. RTP is the only way to transmit live feeds with Quick-Time. Multicast also requires RTP support. Because QuickTime movies consist of many concurrent media tracks, individual tracks can be served from any streaming server simultaneously, using RTP. However, some QuickTime media types, such as QuickTime VR, Flash, and sprites don't stream. The solution is to stream the audio and video parts of the QuickTime movie using RTP and stream wired sprites or chapter lists over HTTP, using Fast Start. The movie author can organize production to enable this workaround.

You create a streaming movie for an RTP server by adding *hint tracks* to an existing movie. The hinting is performed by media packetizer components in QuickTime Pro. QuickTime selects an appropriate media packetizer for each track and routes each packetizer's output through an Apple-provided packet builder to create a hint track. One hint track is created for each streamable track in the movie. Hint tracks tell the RTP server how best to break the media down into packets for optimally smooth transmission to the QuickTime player, preserving video frame integrity where possible. Hint tracks are quite small compared with audio or video tracks. A movie that contains them can be played from a local hard disk or streamed using HTTP Fast Start, but the hinting information is only used when streaming over RTP.

In December 2001, Apple began talking about the forthcoming Quick-Time 6. The company announced that it would use MPEG-4 as the file format, adding MPEG-4 video compression and AAC (Advanced Audio Coding) to the platform. Apple also reinforced its commitment to make every piece of its architecture and infrastructure move toward open standards, while attempting to bring other developers along with them. The ISO actually selected the QuickTime format as the basis for its MPEG-4 file format, so the move is an easy one for Apple to make. The aim is to create ISO-compliant .MP4 files rather than Apple's proprietary .MOV files, so that they will play back on any ISO-compliant MP4 player. Apple also promised incredible video quality using the new MPEG-4 codecs and publicly stated that the AAC audio component for music will likely replace MP3 as the default for a new audio standard on the Web, because the audio is much better sounding, with smaller file sizes and lower data rates than MP3. Editing applications such as Final Cut Pro, iMovie, or Adobe Premiere are already based on QuickTime and should adapt to QuickTime 6 without problems.

One of Apple's claims about QuickTime is that the playback experience is the same whether the viewer owns a Mac or a PC. Microsoft and Real can't make that claim, because they don't treat both platforms the same. Content looks worse on one platform than the other. QuickTime 6 was due for release in early 2002 and was publicly previewed at the Los Angeles QuickTime Live conference held on February 12, 2002, but it has been withheld from general release pending the outcome of a licensing dispute with the MPEG-4 Licensing Authority.*

MPEG-4

In my opinion, MPEG-4 is one of the most interesting multimedia technologies ever invented. It is so powerful and many-faceted that a proper treatment of the subject could easily fill an entire book. Our discussion, therefore, will have to be a brief gallop through the core technologies and vast range of streaming applications made possible by the broad standard that is MPEG-4. We will also touch on the work relating to MPEG-7 and MPEG-21, as it pertains to streaming.

MPEG-4 is not specifically about streaming, but it accommodates streaming extremely well and provides unity of description for media objects, regardless of the distribution technology. It is the closest anybody has gotten to "author once, play anywhere" in media authoring. MPEG-4 can do much more than any of the leading streaming media systems can, in terms of media object streaming, multiplatform interoperability, and end-user interactivity in three dimensions. Figure 2.18 describes the general reference model for MPEG-4. However, from a pure audio and video compression point of view, both Real Video 8 and Microsoft Windows Media Technologies 8 do a better job, by many subjective criteria.

---

*The dispute concerns the MPEG's insistence that content creators pay a fee per hour of MPEG-4 streaming media served. Apple feels that consumer reistance to this licensing model will hinder the adoption of MPEG-4 as a streaming media standard worldwide. Their statement on the matter can be read at: http://www.apple.com/pr/library/2002/feb/12qt6.html.

Figure 2.18
General reference
model of MPEG-4.

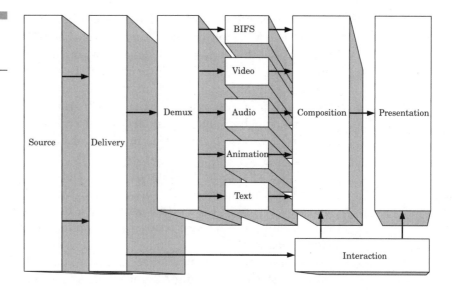

In contrast to Real System, Microsoft Windows Media Technologies and now Apple QuickTime, MPEG-4 says virtually nothing about how the streams get to the player. The point of MPEG-4 is that the streams can arrive at the player many ways, through many different networks and technologies, at the same time. MPEG-4 specifies how the scene to be displayed is reconstructed and hence how the individually arriving streams will be resynchronized. MPEG-4 also includes specifications on how to provide quality-of-service demands to the delivery networks, though the detail of how to map these quality levels to actual network quality-of-service schemes is left to industry. In fact, as with previous MPEG standards, MPEG-4 describes and specifies the decoding process. Encoding and content creation techniques are also left to industry to figure out.

Standardization efforts for MPEG-4 started in 1993, two years before Progressive Networks (which became RealNetworks) introduced their streaming audio system. The standard was not finalized until 1999, by which time Microsoft, Apple, and RealNetworks had been through multiple revisions of their systems. MPEG-7* was scheduled for finalization

---

*MPEG-7, also known as ISO/IEC 15938, consists of seven parts. As of April 2002, only part 2 concerning the Description Definition Language has reached International Standard status. Parts 1, 3, 4 and 5 are currently Final Draft of Internation Standard status. It is believed that by May 2002, parts 1–5 will be standards. Parts 6 and 7 are taking longer. It is likely that these will be completed in September 2002. In October 2001, the MPEG-7 Industry Alliance was established. Their web site: http://www.mpeg-industry.com is a useful resource on matters relating to MPEG-7.

in September 2001 and MPEG-21 won't be ratified until early 2003. MPEG is a working group of the International Organization for Standardization (ISO) based in Geneva. It works in conjunction with its sister organizations, the ITU (International Telecommunications Union), a United Nations agency that maintains standards for telecommunications and broadcast and the IEC (International Electrotechnical Commission), which produces standards on electrical and electronic matters. Some of the codec technologies in MPEG-4 have ITU designations. The MPEG body actually resulted from a 1987 decision to merge the activities of the ISO Technical Committee on Data Processing with the IEC Technical Committee on Microprocessors. One of the subcommittees of this joint technical committee, established in 1988, was the one dealing with Coding of Moving Pictures and Audio, which was subsequently nicknamed MPEG.

Two important industry associations promote MPEG-4 adoption. The Internet Streaming Media Alliance (ISMA) is a consortium dedicated to accelerating the adoption of open standards for streaming media over the Internet (see www.isma.tv). Founding members include Apple, Cisco, IBM, Kasenna, Philips, and Sun Microsystems. The ISMA aims to realize the exciting promise of streaming media by developing a single standard for consumers, service providers, network operators, equipment suppliers, and content providers. The other industry association is the MPEG-4 Industry Forum (www.m4if.org), which seeks to foster communication and cooperation between MPEG-4 technology adopters and vendors.

The aims of MPEG-4 were to:

- Maintain independence of applications from lower-level details, while maintaining technology awareness to promote scalability and error robustness
- Provide usable results over a wide bit rate, ranging from a few kilobits per second to a few megabits per second
- Reuse encoding tools and data from previous MPEG standards
- Support identification and management of digital rights
- Provide interactivity and hyperlinking with individual audiovisual objects simultaneously
- Handle natural and synthetic information, as well as real-time and non-real-time information, in an integrated fashion
- Provide the capability to composite and present information according to the user's interaction and needs, as in Virtual Reality

The main benefits of MPEG-4 are:

- **Performance**—Streaming media benefits from excellent coding efficiency and robustness in error-prone environments such as the Internet.
- **Interoperability**—A cross-platform open standard that can run on a multitude of devices and work well at all supported bit rates, over a variety of delivery networks and technologies.
- **Scalability**—Video quality can be adjusted in response to network congestion in a media format that is encoded once and played anywhere.
- **Interactivity**—Since a scene is composed of individual media objects, users can manipulate each one individually. Animated objects can be mixed with natural objects. The user can discard media objects at the player, making it possible to view any objects desired.

When people think of video, they normally think of what the camera sees. They think of a linear sequence of snapshots, played back fast enough to create the illusion of motion. Every picture is a rectangular object, containing background, foreground, etc. Similarly, the usual way of thinking about audio is as a CD track, where the sound comes out premixed and produced, as a stereo or mono sound image. Interactivity, with this kind of thinking, is limited to playing, pausing, fast-forwarding, rewinding, skipping to a location, or playing at a faster or slower rate than real time. I call this way of thinking about audio and video the "flat earth" perspective.

However, what do we really look at when we view the real world? What do we really hear when we listen? What we see are individual entities—people, trees, furniture, the room, the landscape, insects, flowers, birds, clutter. When we listen, we hear individual people speaking, background noises, the general ambiance of the room, individual musical instruments playing their parts. There are audio objects just as there are video objects. These Audio Visual Objects (which MPEG-4 calls AVOs) have a relationship in space and with each other. As the listeners and viewers of the scene, we also have a spatial relationship to the objects. What we see and hear depends crucially on that relationship. MPEG-4 allows each of these objects to be encoded in a way optimal to the nature of each object and permits each of these elemental AVOs to be streamed to end-users as elemental streams, via a variety of delivery channels. The MPEG-4 player then receives and reassembles the individual AVOs into a scene and allows end-users to manipulate the position in space of every audio and video object. What they ultimately see and hear depends on their choices.

We don't have microphones that can discriminate between individual sound sources, recording only those sounds we wish to capture. We don't have cameras that can pick out individual objects in space and record only those, discarding all other visual information surrounding the object. To encode naturally occurring scenes and soundscapes into MPEG-4, we must either accept a static flat-earth version, transmitting that to the end-user, or else we must use signal-processing techniques to analyze the data to pick out the individual objects. How this is achieved is the subject of considerable research, at present. However, we can quite easily generate computer animations and synthesized sounds. MPEG-4 allows us to mix these synthetic computer-generated objects with our natural objects, sending either a compressed representation of the objects to the end-user, or else a series of parameters that allow generation and animation of the object at the end-user's machine. Using chroma-keying techniques, actual people can be filmed against a green screen (which is relatively easy to remove with a computer) and superimposed on natural or synthetic backgrounds, or a mixture of the two. Similarly, individual sounds can be recorded in soundproof rooms and composited to create an overall soundscape. Anyone familiar with the recent BBC TV series "Walking With Dinosaurs" witnessed the mixture of natural and computer-generated images and sounds to create the convincing illusion of photo-realistic dinosaurs roaring at rivals and roaming the earth. With MPEG-4 coding, every dinosaur and background plate could be streamed and manipulated independently by the end-user, as could each sound, the closed-captioned narrative, and the on-screen titles and graphics. What MPEG-4 lacks, however, is explicit support for interactions between individual audio and video objects. For example, video objects cannot cast shadows on other objects in the scene, cannot radiate light onto other objects (radiosity and reflections) and cannot detect collisions with other objects. MPEG-4 scene-rendering processes are not as sophisticated as the three-dimensional video game-rendering architectures found in specialized graphic processors, such as the Nvidia GeForce series, though this is an obvious area for future standardization and work.

Although RealNetworks, Microsoft, and Apple can stream audiovisual information, they can only transport a fixed view of the information to the end-user. MPEG-4 goes far beyond the underlying audio and video compression technology. Each of these companies claims to have an ISO-compliant MPEG-4 codec, but what they really mean is that they have adopted the MPEG-4 compression standards, not the full-blown decomposition into individual AVOs and the user manipulation of each of them. At least not yet.

The MPEG-4 standard provides a set of technologies to support:

- The coded representation of arbitrary-shaped AVOs, whether natural or synthetic, in real time or non-real time
- The way individual AVOs are composed in a scene
- The way AVOs are multiplexed and synchronized, so that they can be transported over any network channels providing a quality of service appropriate to the specific nature of each AVO or the user's requirements
- A generic interface between the application (i.e., the player) and the transport mechanisms
- The way the user interacts with the scene (changing the viewpoint, for example) and the individual objects in a scene
- The projection of the scene so composed on the desired viewing/listening point

MPEG-4 standardizes a number of types of primitive AVOs, capable of representing both natural and synthetic objects, which can be two or three dimensional. Additionally, MPEG-4 also defines coded representations of objects such as text and graphics, talking heads, and the associated text needed to synthesize the speech and animate the talking head at the user's end, as well as animated human bodies.

MPEG-4 coding provides tools for representing natural sounds, such as speech and music, and for synthesizing sounds based on structured descriptions. The audio representations allow for text descriptions of what musical notes to play and for descriptions of instruments. MPEG-4 also provides for parameterized control of reverberation and aural spatializations. The advanced audio coding (AAC) of the specification provides stunning encoding of natural audio at much lower bit rates than the popular MP3 encoding (MPEG-1 Layer 3 audio). Synthesized sounds can be generated based on structured inputs. Text can be converted to speech, while more general sounds, including music, are synthesized in accordance with a musical score, which may be in MIDI format (Musical Instrument Digital Interface). The text-to-speech converter allows use of prosodic parameters to modify pitch, contour, phoneme duration, and so on, to provide for more natural-sounding and intelligible speech generation. It also allows facial animation control with lip shape patterns or with phoneme information, pausing, resuming, or jumping forward/backward through the text. The standard also supports international text and phonemes. Finally, MPEG-4 allows for manipulation of audio on the end-user's machine, to provide special effects such as reverberation, compression, equalization, varispeed playback, flanging, chorus effects, etc. It can also position sounds in three-dimensional surround-sound fields and allow the user to remix the audio track, emphasizing or de-emphasizing individual sonic elements.

Visual objects, as we have already mentioned, can be of either natural or synthetic origin. Natural objects can be textures, images, or video. The standard provides tools for efficient compression of images and video, treating rectangular imagery as a special case of arbitrary-shaped video objects. It also provides for efficient compression of textures for texture mapping onto two- and three-dimensional meshes. Efficient compression of implicit two-dimensional meshes and the time-varying geometry streams that animate meshes is also included. The standard allows random access to all types of visual objects, extended manipulation of images and video sequences, content-based coding of images and video, content-based scalability of textures, images, and video; spatial, temporal, and quality scalability and error robustness and resilience in error-prone environments. MPEG-4 encompasses all the video compression techniques of MPEG-1 and MPEG-2, plus new VLBV (Very Low Bit rate Video) compression techniques and new content-based encoding. For error robustness, MPEG-4 provides mechanisms for resynchronization, when data is lost. Because MPEG bit streams can be decoded forward as well as backward, once synchronization is re-established, the decoder can back up the data it received after the packet loss, using the data already received to reconstruct the video stream. This is the data recovery feature of the standard. Finally, separating motion data from the texture can conceal errors.

Whereas MPEG-2 specified seven video profiles, MPEG-4 provides for no less than 38. (See Appendix A for a detailed description of MPEG-4 profiles.) MPEG-4 also provides several techniques for adapting to the varying bandwidths of the network carrying the streams to the end-user. In stream switching, a user is switched to another stream, encoded at a lower bit rate, when network congestion is detected. With temporal scalability, some frames of the program are sent in a separate stream that can be turned off, preserving image quality, but reducing the frame rate at the player. With fine granular scalability (called FGS in MPEG-4 parlance), high-frequency detail in the images is sent in separate "enhancement streams," parts of which can be discarded as needed, maintaining the frame rate and reducing the image quality. FGS may also use temporal scalability techniques. This suite of techniques contrasts with Microsoft's Windows Media Technologies Intelligent Streaming and RealNetworks SureStream technology, which only provide for stream switching.

Synthetic visual objects are coded in a manner similar to VRML (Virtual Reality Modeling Language). MPEG-4 allows for the creation of things that are not live, like tables and chairs, etc. However, VRML is poor at representing faces and human bodies. MPEG-4 Face is an object capable of producing faces in the form of three-dimensional polygon meshes, which can be rendered. A Facial Definition Parameter stream and/or the Facial Animation Parameter sets control the shape, texture, and expression of the face. Similarly, the MPEG-4 Body is an AV object capable of producing virtual human body models and animations, controlled by the Body Definition Parameter set and Body Animation Parameter set.

MPEG-4 image compression is based on wavelets, which is a good choice, since wavelets degrade much more gracefully at high compression ratios, losing some detail but retaining good color quality and remaining artifact free. Video compression is largely based on H.263, which traces its roots to the MPEG-1 standard. Microsoft officials have claimed that noncompliant versions of the MPEG-4 codec can exploit several limitations in the standard to boost quality by over 30%. As a result, MPEG-4 is the first standards-based video codec that actually delivers less quality than its competitors, at the time of adoption. However, the MPEG-4 standard is broad enough to incorporate a range of very creative encoding techniques, so it is extremely likely that we will witness the development of higher-quality MPEG-4 codecs in the future. However, interoperability between platforms is compromised if the encoding techniques require different decoding strategies from those used by the ISO-complaint video decoder. Interestingly, MPEG-2 encoder vendors have consistently found ways to improve image quality, even though the standard was set in stone many years ago. I am unable to comment on whether or not interoperability between different manufacturer's implementations was compromised as a result, though anecdotal broadcast industry evidence suggests it may have been.

MPEG-4 provides a standardized way to compose a scene, allowing:

- AVOs to be placed anywhere in a given co-ordinate system
- Grouping of primitive AVOs in order to form compound AVOs
- Application of streamed data to AVOs to modify their attributes
- Interactive changes to the user's viewing and listening points anywhere in the scene

MPEG-4's language for describing and dynamically changing the scene is named the Binary Format for Scenes (BIFS). BIFS commands are available not only to add or delete objects from the scene, but also to change the visual or acoustic properties of an object without changing the object itself. For example, you can change the color of a three-dimensional sphere without modifying anything else about the sphere. BIFS is a very nice way to create an interactive, synchronized, multimedia experience. In principle, BIFS could place a Web page as a texture in the scene.

BIFS borrows many concepts from the Virtual Reality Modeling Language (VRML). However, in VRML, objects and their actions are described in text, whereas BIFS is coded as binary, so that it is 10 to 15 times faster to describe the same objects. BIFS can be streamed in real time, whereas VRML information must be fully downloaded before it can be played. BIFS also allows for the definition of two-dimensional objects, like lines and rectangles.

AVO data is conveyed to the end-user as one or more elementary streams (spot the similarities with QuickTime!). The streams are characterized by the quality of service they request for transmission (for example, the maximum bit rate, the bit error rate, tolerance to packet loss, etc.). Other information that travels with each stream defines the stream type and the precision of decoder timing required, which tells the decoder which resources are needed to play the stream.

Figure 2.19
MPEG-4 system layer model.

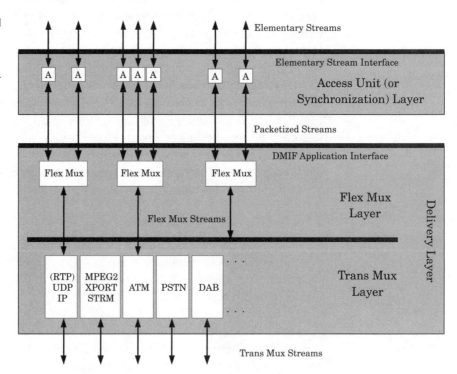

Figure 2.19 describes the MPEG-4 system layer model. Packets are synchronized at the Access Unit layer. Access units are the atomic pieces contained in an elemental stream (e.g., video frames or individual scene description commands). The Access Unit layer recovers the AVO's or scene description's time base, and enables synchronization between all AVOs and scene description commands received from all the elemental streams. The Access Unit layer interfaces to the FlexMux layer (Flexible Multiplexing). The FlexMux groups elementary streams with similar quality-of-service requirements, for example, using a low-overhead multiplexing scheme, to streamline delivery from the MPEG-4 server to the player. The FlexMux layer, in turn, interfaces to the TransMux layer (Transport Multiplexing). This layer manages the actual data transport, whether the data travels over broadcast networks, wireless cellular networks, or the Internet (or from a local hard drive!). Any suitable transport protocol stack such as RTP over UDP, ATM (Asynchronous Transfer Mode), or MPEG-2's transport stream may be a specific TransMux instance. Use of the FlexMux is optional, if the underlying TransMux provides equivalent functionality. The Access Unit layer, however, is always present. These multiplex layers provide MPEG-4 with an advantage over other synchronized multimedia schemes, such as the Synchronized Multimedia Integration Language or SMIL (discussed later in this chapter). Without these negotiation layers, SMIL (and QuickTime synchronized multimedia, for that matter) cannot be transported over media other than the Internet.

The Delivery Multimedia Integration Format (DMIF) is functionally located between the MPEG-4 application and the transport network. That means there is a DMIF layer in the player and the server. The DMIF presents applications with a transparent interface, irrespective of the actual networks conveying the streams. This allows applications to deal with generic streams with assured quality of service, relating to AVOs and scene description commands, irrespective of how they got there. In fact, the DMIF can serve multiple peer applications. This makes DMIF ideal for home media gateways, for example, since it takes all the incoming digital media streams and presents them to various players in the home as undifferentiated streams, which may or may not be played, according to the device's capabilities. For example, a high-powered computer with a DSL connection could receive all presentation elements, whereas a wireless PDA user might receive just a background image with some animated sprites and a low-bandwidth audio track. The cell phone user may receive a stock ticker text track, using the text-to-speech feature to read the content. All streams emerge from the

DMIF synchronously; so multiple devices can, in principal, tune in to the same set of streams and play individual elements (AVOs) in synchronism with each other. In other words, in a home media gateway, DMIF provides a local simulcasting capability. Control of the DMIF spans the FlexMux and TransMux and manages the quality of service requested by the application. The DMIF is a very clever piece of the MPEG-4 standard.

The MPEG-4 standard provides for both client-side and server-side interaction with the presented content. Client-side interaction involves content manipulation rendered by the client machine. AVOs can be moved, made invisible, changed in size, rotated, etc. Server-side interaction, like starting and pausing a stream for example, requires a back channel. Figure 2.20 shows the architecture of a typical MPEG-4 terminal (or player).

The MPEG-4 specification specifies a file format known as .MP4, which can be used for the exchange of content and is easily converted. MPEG-1 and MPEG-2 did not include such a specification, but the intended use of the MPEG-4 standard on the Internet and with personal computers made it a necessity. The .MP4 file format bears many similarities to Apple's QuickTime .MOV format.

We have already touched on some of the novel uses of MPEG-4, such as text-to-speech voice synthesis for a mesh-warped facial model, which creates an avatar. Mesh warping can also be used to represent simple video objects. For example, the texture of a flag could be mapped onto a mesh and the mesh manipulated to simulate waving in the breeze. Structured music synthesis allows for a variety of well-established music synthesis techniques, so that whole orchestras can be simulated. Text and graphics can be linked to the playback of a video, allowing synchronization of the media elements. A user can be walking through a three-dimensional representation of a house and the acoustics of each room can adjust according to the size of the room, so that sounds become more or less reverberant. As you walk around the house model, the virtual real estate agent's voice may appear louder in one ear than the other as he passes you in the corridor. Music tracks can be remixed, to bring up the drums and bass, if that is what is required. If the dialog is hard to hear among all the sound effects, the dialog can be equalized and compressed to make it sound louder and more intelligible. All of these end-user modifications are supported by the MPEG-4 standard, though implementations of the full standard are exceedingly rare today.

There are, however, some commercial end-to-end MPEG-4 solutions available. For example, Envivio has a range of products including a

Figure 2.20
MPEG-4 terminal
architecture.

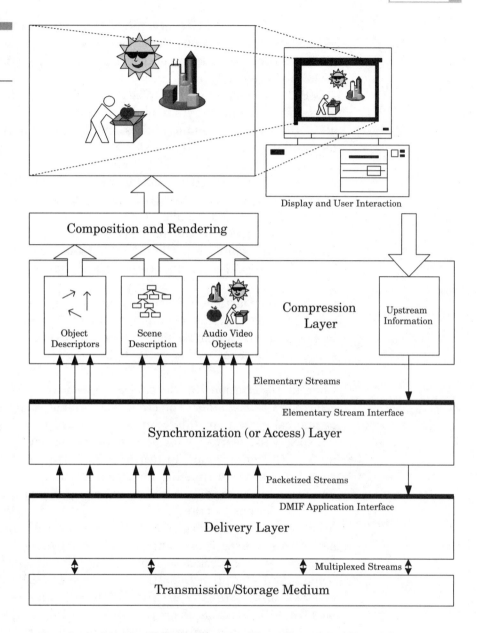

streaming server, a live encoder, and broadcaster and a multi-user content-authoring system. Envivio also produces software to add MPEG-4 capabilities to set-top boxes. The Sarnoff Corporation spin-off e-Vue has also announced an end-to-end MPEG-4 streaming solution, including content-creation tools, a content-delivery platform, and an MPEG-4 player. Another company with a solution is iVast, which produces content creation tools, an MPEG-4 server, and an MPEG-4 player. iVast also provides an SDK for its content creation tools. Philips also has its WebCine platform, which includes an encoder application, an MEPG-4 server, and WebCine Player software.

MPEG-4 compliance seems to be a term that might be open to common abuse, since manufacturers tend not to specify whether they comply with all of the video and audio profiles, whether they provide all the AVO manipulation capabilities, if there are any limitations to the scene description implementation, etc. In many cases, compliance claims may be based on nothing more than implementation of a particular subset of compression profiles and support for the .MP4 file format. Caveat emptor.

What the MPEG-4 standard did not do was specify a way for uniquely naming and describing all the AVOs and MPEG-4 composite presentations that are going to proliferate. That task was left to the MPEG-7 standard. MPEG-7 is a standardized description of various types of multimedia information. This description is associated with the content itself, to allow fast and efficient searching for material of interest to the user. The searcher, for example, could input a sketch as a wildcard image or specify a generic shape, generate a verbal description (e.g., "a sunset over a lake," or "two men fighting"). Music could be found in a "query by humming" format. MPEG-7 is formally called "Multimedia Content Description Interface." Descriptions of multimedia are often referred to as *metadata*. The standard does not comprise the automatic or manual extraction of descriptions and features or AVOs. It also does not specify the search engine or any other program that can make use of an MPEG-7 description. In fact, MPEG-7 can be used to catalog and search for any multimedia objects, including broadcast video tape, archived film, 3D models, audio recordings, still photographs, etc.

MPEG-7 aims to standardize a core set of descriptors that can be used to describe the various features of multimedia content, predefined structures of descriptors and their relationships (called *description schemes*), and a language to define Description schemes and descriptors (called the Description Definition Language or DDL). Figure 2.21 is an overview of the MPEG-7 multimedia description schemes. Yet another book could be written about the MPEG-7 standard, so once again we shall have to content ourselves with a sampling of the full menu.

MPEG-7 provides for a hierarchy of classification, allowing different granularity in its descriptions. Because the descriptive features must be meaningful in the context of the application, they will be different for different user domains and different applications. Consider the example of visual material: a lower level of abstraction might be shape, size,

Figure 2.21
MPEG-7 multimedia
description schemes
overview.

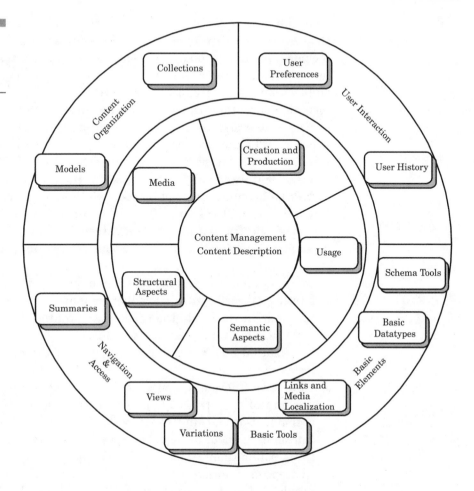

color, movement (describing trajectory), texture, and position in the scene's space. Similarly, for audio material, low-level descriptions might be key, mood, tempo, location in sound space, tempo changes, etc. However, the highest level would give semantic information, such as "the scene with the barking brown dog next to the boy in blue, with lots of traffic noise in the background." Intermediate levels of abstraction may also be used. The lower-level descriptions lend themselves to automatic extraction from the data, whereas the higher-level ones will probably require human intervention.

After giving a description of the content, MPEG-7 defines other types of information about the multimedia data. The "form" of the data is defined. An example of the form might be the coding scheme used (i.e., JPEG, MPEG-4) or the overall data size. This information helps in determining whether or not users can read this type of information with the devices they are using. Conditions for accessing the material are included, including links to a registry with intellectual property rights information and price. This is where MPEG-7 metadata links to digital rights management expressions. MPEG-7 defines a "classification," including parental ratings and division, into a number of predefined categories. There is also provision for links to other relevant material, which may help a user search more rapidly. Finally, "context information," such as the occasion of the recording (e.g., Sydney Olympic Games, 2000—women's 100 meter sprint final) is stored.

MPEG-7 content description tools will allow the creation of descriptions. These will take the form of a set of instantiated description schemes (the framework used to categorize the information) and the descriptors defining that content. Some schemes may include:

- Information describing the creation and production process of the content (e.g., director, title, short feature movie)
- Information related to the usage of the content (e.g., copyright pointers, usage log, and broadcast schedule)
- Information about the storage features of the content (e.g., storage format, encoding)
- Structural information on spatial, temporal, or spatio-temporal components of the content (e.g., scene cuts, segmentation in regions, region motion tracking, thumbnails)
- Information about low-level features in the content (color palette, textures, sound timbres, melody description)
- Conceptual information of the reality captured by the content (e.g., people and events, interactions between on- and off-screen objects or people)
- Information about how to browse through the content in an efficient way (e.g., using summaries, variations, change control logs, spatial and frequency sub-bands)
- Information about collections of objects
- Information about the interaction of the user with the content (e.g., user preferences, usage history)

All of these descriptions are coded in an efficient way for searching and filtering.

MPEG-7 data may be physically located with the associated AV material, in the same data stream or on the same storage system, but the descriptions could also live somewhere else on the globe, in some massive database, for example. When the content and descriptions are not colocated, mechanisms that link AV material and their MPEG-7 descriptions are needed and these links need to work in both directions.

The MPEG-7 standard does not define a monolithic system for content description, but rather a set of methods and tools for the different viewpoints of the description of audiovisual content. With this in mind, MPEG-7 has taken into account more application-specific viewpoints under consideration by other standards bodies, including the SMPTE (Society of Motion Picture and Television Engineers) Metadata Dictionary, Dublin Core, EBU P/Meta, and TV Anytime. MPEG-7 uses an XML (Extensible Markup Language) schema as the language of choice for the text representation of descriptions. This choice should allow interoperability with other systems in the future. There is a binary format (BiM) defined in the standard as well.

For describing audio, MPEG-7 provides five technologies:

- The audio description framework (which includes scalable series, low-level descriptors and the uniform silence segment)
- Musical instrument timbre description tools
- Sound recognition tools
- Spoken content description tools
- Melody description tools

Within the audio description framework, audio may be described using samples taken at regular intervals, or by defining segments that demark regions of similarity and dissimilarity within the sound. Samples or segments can be described by both scalar values, such as power or fundamental frequency, and vector values, such as spectra, summarizing the values for that sample or segment. "Scalable series" allows progressive downsampling of the sample or segment values contained in the series, providing various summaries along the way, like minimum fundamental frequency, maximum peak audio level, or variance of the descriptor values, as examples. Low-level descriptors include instantaneous waveform and power values and a power spectrum, with spectral centroid, spread and flatness, fundamental frequency and harmonicity, attack time, timbral spectral, and temporal characteristics and spectral basis. The silence descriptor attaches the simple semantic of "silence" to an audio segment, which may be used to segment the audio stream further, or else as a hint not to process the segment.

Musical instrument timbre-description tools aim at describing the perceivable features of instrument sounds. Timbre is defined as the features that make two sounds having the same pitch and loudness sound different. These descriptors relate to notions like attack, brightness, or richness of the sound.

Sound recognition tools are for indexing and categorizing a broad range of general sounds, with immediate application to sound effects. Spoken content description tools allow detailed the description of words spoken within an audio stream. Melody description tools allow for efficient and robust melodic similarity matching, for example, in query-by-humming. A five-step contour representing the interval between adjacent notes and some basic rhythmic information is used. Higher-precision description schemes are allowed, as well as a series of optional descriptors such as lyrics, key, meter, and starting note.

MPEG-7's visual description tools consist of basic structures and descriptors that describe the following basic visual features: color, texture, shape, motion, localization, and others. Each category consists of elementary and sophisticated descriptors.

The basic structures to describe visual data are the grid layout, the time series, multi-view, the spatial two-dimensional coordinates, and temporal interpolation. Grid layout is a splitting of the image into a set of equally sized rectangular regions, so that each region can be described separately, in terms of other descriptors, like color or texture. Each rectangle's descriptor allows the assignment of subdescriptors to all rectangular regions, as well as to an arbitrary subset of these regions. The time series provides for frame matching and image-to-frame matching descriptors. These may be regular time series, or irregular time series. Multi-view is a 2D/3D descriptor representing the visual features of 3D objects seen from different viewing angles. Spatial 2D coordinates describe the location of objects in images, using either local or integrated coordinates. In local coordinates, all images are mapped to the same origin. In integrated systems, each image or frame may be mapped to different areas, and can thus be used to represent coordinates on a mosaic of a video shot. Temporal interpolation describes a time-variant multidimensional variable using connected polynomials. This is useful because the description size is much smaller than explicit description of all real values would be. For example, a camera pan could be described as a series of sample values, or else as a series of linear interpolation functions. These basic structures are used when describing visual features, like color, texture, shape, etc.

Color descriptors include color space (RGB, HSV, or YUV descriptions, for example), color quantization, dominant colors, scalable color (used for image matching based on color features), color structure descriptor (capturing both color content and the structure of this content), color layout (the spatial distribution of colors), group of frames, and group of picture descriptors (which extend the still image-matching structures for video).

Textures are described using three texture descriptors: edge histogram, homogeneous texture, and texture browsing. Shapes are represented by four shape descriptors: region-based shape, contour-based shape, 3D shape, and 2D/3D multi-view. There are four motion descriptors: camera motion (e.g., dolly forward/backward, track left/right, boom up/down, pan, tilt, roll); object trajectory (describing the object's motion in space); parametric object motion (e.g., description of hand shapes and movements as sign language); and motion activity (e.g., "high-speed car chase," "scoring a goal in a soccer match," "news-reader head shot," two-person interview shot," or "landscape scene"). There are two localization descriptors: region locator and spatio-temporal locator. The region locator locates objects within a frame or image, using a box or a polygon representation. The spatio-temporal locator describes how regions in a video sequence, like a bouncing ball for example, are located within those frames over time. This is useful for checking whether the object has passed particular points (e.g., determining the frame when the ball bounced through the basketball hoop). Lastly, MPEG-7 provides a face recognition descriptor, which allows for image matching to a particular face. Detailed description of the above descriptors is beyond the scope of this book (and the expertise of this author).

MPEG-7's relevance to streaming media is that it will make possible applications for location of media packages of interest to end-users, as well as efficient and cost-effective content production environments for those producing streaming media presentations. MPEG-7 would make it entirely feasible to search for video footage on the basis of a query like "the comedy we saw in the '70s with the grouchy hotel owner and the Spanish waiter, where the guest died in the night." One could even whistle the theme tune of a program series and locate all episodes. Similarly, content producers could easily locate archive footage for inclusion in their own programming.

MPEG-4, while leaving a data field for the identification of intellectual property in the descriptor attached to every elementary stream and specifying interfaces to proprietary systems that can manage and protect IP, did not actually codify the intricacies of rights management. For example, the data field in an elemental stream may contain a number of key-value pairs, such as "Composer-Paul Simon." MPEG-4 did not force the key-value pairs to be correct or even present. A criminal might strip the data and still produce a syntactically correct MPEG-4 bit stream. To provide tools to solve this problem, MPEG-21 includes an Intellectual Property Management and Protection (IPMP) element.

MPEG-21 is an initiative aimed at standardizing media transactions or relationships between two media users. Users can be people, like content producers and consumers, for example, or machines, such as media publishing points and content subscriber/aggregators. In short, MPEG-21 is an open framework for the delivery and consumption of media. The vision for MPEG-21 is to define a multimedia framework to enable transparent and augmented use of multimedia resources across a wide range of networks and devices. MPEG-21 goes far beyond mere rights management. It is a media commerce facilitator, as well as a framework for many other media access applications. MPEG-21 standardizes the flow of media information and services all the way along the delivery and value chain, from content creators to end-users. To support this, the content has to be identified, described, managed, and protected. Transport and delivery of content will occur over a variety of networks, between a variety of machines and devices (e.g., servers and players). Events will occur and require reporting. Reporting will include reliable delivery, management of personal data and preferences (respecting privacy), and the management of (financial) transactions. The overarching goal is to provide a framework to ensure that systems delivering multimedia content are interoperable (so that providers can buy components from any vendor they choose) and that transactions are simplified, unambiguous and, ideally, automated.

The seven key elements defined in MPEG-21, shown in Figure 2.22, are:

- **Digital Item Declaration**—A uniform and flexible abstraction for declaring (i.e., announcing the existence of) digital items.
- **Digital Item Identification and Description**—A framework for description and identification of any entity, regardless of its nature, type, or granularity (you should, for example, be able to identify a full movie, a scene, or an individual shot, for example).
- **Content Handling and Usage**—Providing interfaces and protocols that enable the creation, manipulation, search for, access, storage, delivery, and reuse of content across the content distribution and consumption chain.
- **Intellectual Property Management and Protection**—Enabling content to be persistently and reliably managed and protected across a range of networks and devices (digital rights management, in other words).

Figure 2.22
MPEG-21
interactions.

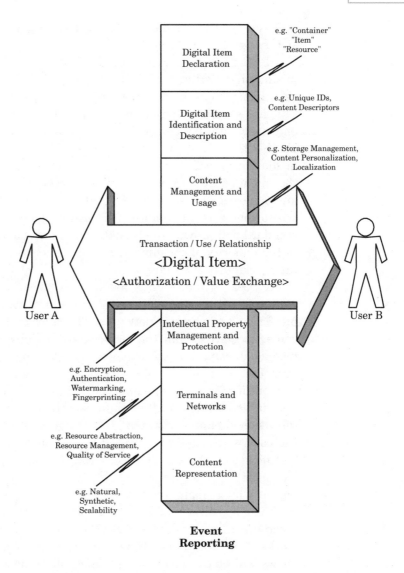

- **Terminals and Networks**—Providing interoperable and transparent access to content across a variety of networks and terminals (e.g., servers and players), including control over quality of service.
- **Content Representation**—How media resources are represented.
- **Event Reporting**—Including the metrics and interfaces to enable users (at all points in the delivery chain) to understand precisely the performance of all reportable events within the frameworks (e.g., did my streaming media ad really get watched?).

As part of the standard, a Rights Data Dictionary and Rights Expression Language will be specified.

MPEG-21 is a work in progress, but is highly relevant to streaming media, since without standardization, digital rights management has been nonexistent or implemented with proprietary and closed schemes. This has severely limited content providers' interest in the medium and end-users' ability to receive and play any media they want. When MPEG-21 is finalized, some time in 2003, applications to streaming media will provide interoperability between media delivery systems, so that media played from a compliant server can be viewed on any compliant playback device, ensuring that the rights of the end-user and content owner are not compromised at any point in the transaction. This standard will provide the breakthrough needed to take streaming media to the majority of consumers. It is a standard highly significant for the future viability of streaming media.

MPEG-21 will also unify media delivery via broadcast television networks, satellite systems, terrestrial digital radio, Internet streaming platforms, and wireless networks. The rules and descriptions will be the same, no matter what the media and the delivery route. For the home media gateway, this technology will enable end-users to see a single pool of media, available for playback on any device with the right playback capabilities for a given media type. Billing and licensing will also be unified and simplified. Rather than establishing agreements with several television channels to watch all the content available to you, as is often the case with satellite-delivered television content today, the end-user will ultimately be able to access and pay for any content he or she can locate, with the transaction to obtain and pay for licenses handled transparently by a single transaction engine in a user's device. Television, as we know it, will be nothing more than a source of media and a delivery mechanism, equivalent to all other media sources and delivery networks, from the point of view of the end-user. Prime-time television shows will be competing for audience share head-to-head with all other streaming media presentations, offered to the end-user as equivalent choices on the same player. Bearing in mind that there are approximately 36 million Web sites in the world today and that as streaming media takes root, many of these will offer streaming content, this is a daunting prospect for even the most popular television show.

Broadcast television operators, of course, would rather not compete on this basis, since they already have a more captive audience. They will, therefore, resist the move toward MPEG-21 compliance, in all likelihood. However, if an MPEG-21–driven media marketplace establishes itself, in isolation from and in parallel with the broadcast networks, then over the very long term the options available to the broadcaster become stark: comply or die.

As the ISMA claims, adoption of a standard such as MPEG-4 can help to accelerate the uptake of streaming media delivered over the Internet, but not everybody is pleased to see the MPEG-4 standard. In the first place, production houses and broadcasters who have recently upgraded to MPEG-2 digital production have a vested interest in not making their equipment obsolete (even though MPEG-2 data and tools are usable in MPEG-4). Also, MPEG-4 provides a new level of complexity to content authors and there are as yet few tools to support MPEG-4 content production. In addition, strong streaming media system vendors have invested much time and money in their own proprietary solutions, which in some respects outperform the standard MPEG-4 system. The sheer weight of numbers of the proprietary players that have already been installed may make it hard for MPEG-4 players to establish dominance, even though downloading new player software does not present all that onerous an obstacle to most users. Time will tell if the pressure to standardize outweighs the vested interests to stay with proprietary or semi-proprietary systems.

MPEG-4 was designed to be used on television broadcast carriers as well as the Internet (or a mix of both). If MPEG-4 content eventually finds its way onto the digital television broadcast networks, the television will have to be a computer with considerable power, since the standard leaves lots of computational work to the client application. Even though DTV delivers great quality of service today and even though DVB-delivered AVOs can be freely mixed with Internet-delivered AVOs, the broadcast nature of digital television limits the choice of programs to just the few carried by the system. IP delivery of all AVOs over a broadband network will provide an almost limitless choice of on-demand programming. Even where multicast IP delivery is chosen, to minimize bandwidth costs, the choice of multicast IP channels is likely to exceed the offerings of even the most well-provisioned digital television system. Therefore, if broadband networks begin to offer quality of service at a cost comparable to that in digital television networks, IP delivery is likely to displace MPEG-4 delivered via MPEG-2 transport streams.

Content Delivery Networks

Content delivery networks (CDNs) are schemes devised by companies to get streaming media content, in particular, to consumers, with a better quality of service than from the raw public Internet. The public Internet is prone to outages, congestion, queues, uncontrolled cascaded router hops, denial of service attacks, flash-floods of people requesting content from a particular siten and other behavior that can prevent a media stream from continuing to stream. Content delivery networks exist to get around these problems. Paying customers use CDN providers to transport their data from origin to the edge of the network, whether that edge is within their enterprise or serving the general public.

There are three significant approaches adopted by content delivery networks:

- Pick the best route through the public Internet
- Go around the public Internet entirely, using a dedicated network
- Use periods of relative inactivity on the Internet opportunistically to move content to servers located physically closer to the consumer (in terms of both geographical and network proximity)

All these approaches make use of *edge caches* (explained later in this chapter) to serve content transparently from a proxy located closer to the end-user than the origin server. The end-user logs on to the origin server, but the stream is actually delivered from a local proxy, with the end-user none the wiser.

Edge caching is done to improve the quality of playback, since with fewer network segments between the end-user and the stream source, there is less likelihood of network congestion, packet loss, and traffic queues. The other benefit is that once the content has been moved to the edge, multiple end-users can access that content in a locale using much less expensive bandwidth than if a connection to the origin server had been opened for each player. It costs less to serve a stream from a server located in your city than from a server located in another country. Local calls are cheaper than long distance ones.

In the first content delivery scheme mentioned, content is delivered across the public Internet, but a map of network congestion is maintained in real time, allowing the content requested to be routed from the origin server to the requesting consumer, avoiding congested network segments and unnecessary router hops. This is, in essence, the approach adopted by Akamai. The scale of the machinery needed to maintain a real-time map of activity on the entire Internet is monumental, to say the least. However, such a map is necessary if streams are going to be delivered by avoiding all network roadblocks as they occur. In addition, traffic routing optimization algorithms reflect a sublime understanding of the mathematics of graph theory and chaotic behavior. Deciding how to reroute a media stream in response to sudden network congestion is not a trivial task and a sound and rigorous theoretical mathematical basis is required if those algorithms are going to be provably optimal. Suboptimal algorithms merely move the congestion around, choose routes that are not the least congested or shortest, or else cause congestion of their own. Content delivery networks that adopt this approach are useful for live streaming, on-demand streaming, or static Web content.

The second approach to content delivery, avoiding the public Internet entirely, is to build a dedicated network, either by laying it in the ground, using satellite links, or buying dedicated capacity from existing bandwidth suppliers. This allows a content delivery company to keep all traffic other than paying customers' traffic off their network. A closed, private highway is built from origin servers to edge servers. When the end-user requests content, whether live or on demand, the content delivery network can guarantee fairly good quality of service, simply because there is little else on the network to get in the way. This only works while there is sufficient bandwidth the cover the peaks in demand. If such a private network were to experience sudden extreme demand, because it was the only network carrying a particular breaking news item, for instance, the network would saturate and quality of service would be comparable to that of the public Internet. The more popular and successful this style of network, the more traffic it will have to deal with and the less likely it will be to maintain quality of service for all users. Providers like Digital Island have adopted this approach.

The third variety of content delivery network can be used with enterprise WANs, dedicated content delivery networks, or the public Internet. It comprises software to detect when network activity is low or bandwidth is cheaper, and then replicates content from origin servers to edge servers. This works well for content that is updated infrequently, but is no use at all for live streaming during peak demand periods, except as a simple way to load balance, through replication to the edge. Cisco provides a software suite that can perform this kind of content delivery.

For content delivery networks to deliver content to end-users, the CDN company's edge servers must be colocated with Internet access points (called POPs or Points Of Presence) provided by Internet Service Providers (ISPs). Typically, a CDN company enters into an agreement with an ISP to install an edge server into the ISP's equipment racks, to provide short and direct connection between dial-up or always-on subscribers and the CDN. The quality of the CDN is highly dependent on the number of such colocation agreements that the CDN can negotiate.

For enterprise applications, the CDN's edge server is colocated behind the enterprise firewall, in the data center or an enterprise proxy server.

Edge Servers

As has already been suggested, networks with edge servers/caches/proxies bring quality of service and playback advantages for consumers of streaming media, as well as bandwidth savings for the content delivery networks (who bill the content owner, ordinarily). There are other advantages of edge caching. Round trip delay is one of the key factors which determine the crispness of the response to interactivity that requires server-side code. The fastest that a signal can make its way around the world on any network, limited by the speed of light, is about 130 milliseconds. Routing and switching delays extend this, of course. What this means is that if a consumer were trying to stream a clip from a server on the other side of the planet (antipodal to the user), it would take a minimum of a fraction of a second or so just to provide the consumer with a noticeable response at all (in fact, it would take considerably longer, since this is the minimum time it would take just to negotiate protocols before buffering could start). This might not be significant for start-up delay, when playing a stream, but it would be a noticeable lag when trying to fast forward or rewind to a specific place in the stream. In contrast, if the stream is served from an edge cache located 40 km from the user, and is already resident in that cache, it takes one thousandth of the time to respond.

The other significant advantage of replicating content to a network of edge servers is that each individual edge server can accommodate about as many simultaneous users as the origin server. So edge networks allow for greater stream-serving capacity, as well as providing the ability to load balance, serving an end-user from the next closest edge server when the closest edge server saturates.

Edge cache devices can be either regular servers (Dells or Compaqs or Suns, for example), with software running on them to offer the caching functionality, or else they can be dedicated, purpose-built, hardware-based Internet appliances. CacheFlow, InfoLibria, and NetworkAppliance are among the companies that build dedicated edge devices.

The tendency for CDNs to concentrate their investments at the edge of the network mirrors what the telecommunications companies (i.e., the phone companies) are doing with their networks. In order to deliver the elusive quality-of-service improvements that data customers require, they are moving to a network architecture that has all the intelligence on the edge, with the core serving only as great big data "fire hoses," delivering bulk data from edge to edge. Whereas once the core of the telecommunications network consisted of large, expensive crosspoint switching equipment, with the advent of VLSI (Very Large Scale Integration) technology, switching and routing devices located on the customers' premises are increasingly providing the bulk of the packet-switching intelligence.

For edge caching to work, a raft of new protocols is required, so that edge proxies can transparently satisfy requests to origin servers and so that cache contents are refreshed appropriately. There is also a need to refresh edge caches from other edge caches, instead of from the origin server, and to pass requests to the nearest, yet least-loaded proxy. The IETF (Internet Engineering Task Force) had a "Web Replication and Caching" working group (known as WREC), which has now shut down, whose activities were taken up by the "Web Intermediaries" (known as WEBI) group. There has been a slew of Internet Drafts, many of which have now expired, as well as an attempt to create a taxonomy of Web caching and replication terms and a listing of known problems with Web caching. The attention of the WEBI group is now focused on developing a generic Resource Update Protocol (RUP), whose use cases (usage scenarios) include intra-CDN applications, inter-CDN applications, content provider to CDN, content provider to arbitrary Web intermediary, content location update, content prefetch hinting, content updating, metadata updating, client-driven invalidation, and server-driven invalidation. This work is informed by earlier drafts such as:

- **OPES (Open Pluggable Web Services)**—Aims to allow the plug-in inclusion of code which runs on an edge server (an "edge-side include"), analogous to server-side includes, code modules that run on the server, such as CGI (Common Gateway Interface) scripts.
- **ICAP (Internet Content Adaptation Protocol)**—A protocol to allow edge servers to perform value-added services, such as language translation, compression standards transcoding, virus checking, content filtering, local ad insertion, wireless protocol (WAP) translation, anonymizing, image enhancement, image magnification (for those with sight problems), and batch downloading of Web content.
- **WCCP (Web Cache Coordination/Communication/Control Protocol, depending who you believe)**—Primarily for redirecting requests away from the origin server and to the edge cache.
- **ICP (Internet Cache Protocol)**—A UDP-based protocol used for locating instances of cached responses in neighbor caches.

- **WPAD (Web Proxy Auto Discovery)**—A mechanism that enables Web browsers and other clients automatically to locate an appropriate proxy cache within their domains (an alternative to "transparent connection hijacking").
- **NECP (Network Element Control Protocol)**—A lightweight protocol for signaling between servers and network elements that forward traffic to them, primarily to perform load balancing.
- **EPSF (Extensible Proxy Services Framework)**—A successor to the NECP and ICAP protocols, which seeks to provide a general-purpose framework for deploying services on edge servers.
- **WCIP (Web Cache Invalidation Protocol)**—Aims to provide methods to guarantee cache consistency and synchronization with the content on the origin server.
- **IRML (Intermediary Rule Markup Language)**—An XML-based language used to describe service-specific execution rules on an edge server.
- **CARP (Cache Array Routing Protocol)**—An alternative to ICP, using hash functions to decide which cache a request is to be forwarded to, in order to maximize hit ratios and minimize the duplication of content among a set of caches.
- **ICCP (Inter Cache Cooperation Protocol)**—An extension to HTTP and ICP which allows purging of cached objects, tracing of HTTP requests through a sequence of proxies, and the removal of URLs from ICP replies.

Clearly edge caching is experiencing a torrid period of innovation, as new problems are discovered and solutions proposed.

Unfortunately, edge caching is not a universal panacea for streaming quality of service, bandwidth usage, load balancing, and streaming scalability. The colocation of edge servers with POPs requires agreements with companies, who may act to limit the access by one CDN and encourage access by another. Placing the edge server in some other company's premises also carries with it a risk of content theft or unauthorized redistribution. Cache refreshing policies are currently loose, so content may have changed on the origin, yet the end-user is served stale content from the edge. Cache refreshing policy is often left to the cache owner, not the content owner. Edge networks are harder to maintain, because all the equipment is no longer conveniently located in a single network operations center. Content owners are very nervous about leaving their precious media assets on servers located in far flung places, which can be accessed by personnel with whom they have no relationship (and hence no way of trusting). Unless content is rights managed and encrypted, content owners run the risk of losing control of their media asset, once it replicates to edge caches. Search engines can mistakenly index content to edge servers rather than origin servers. This leads to the problem of indexing out-of-date content or else of storing links that are likely to break, as the cache is flushed for other uses.

From the point of view of end-users, cached content is suspect, since there is no way to ensure that the content is presented as the content producer intended it. Edge services can modify content transparently, adding anything they like to the Web page or media stream, as it flows through the edge proxy. Malicious access to cached contents can lead to end-users' being served counterfeit content. Cache abuse and "cache poisoning" are of great concern to content providers and content consumers alike. There is also the issue of transcoding quality. If the streamed content is compressed in a format that your player does not recognize, an edge proxy can decompress the content and recompress it into a format your player can play. However, transcoding can be performed with varying levels of quality. If the cache owner chooses to do a poor job of transcoding one content provider's material, but an excellent job with another that he has a business agreement with, the end-user will be served degraded content or excellent content, depending on the whims of the cache owner. A CDN that delivers Web search engine results could, for example, filter out hits that reveal the CDN's trade secrets or modify financial reports served via their CDN to obscure information damaging to the company. With MPEG-4 streams, it would theoretically be possible to remove an inconvenient individual from a scene entirely, in Stalinesque fashion, simply by blocking the elementary stream representing that person as an AVO. Similarly, what a person said, on record, could be transparently modified to imply meanings never expressed by the speaker. Indeed, somebody else's face entirely could be texture-mapped onto a talking face. Edge cache abuse such as this is not only possible, but also likely, unless the industry faces and solves trust issues associated with such tampering.

Quality of Service

Quality of Service (QoS) on IP networks is another important subject. It relates to the ability to get packets through the Internet reliably, with time-sensitive packets (such as those comprising media streams) able to jump queues and preallocate or reserve network resources, to avoid congestion. It also refers to the ability to guarantee time-sensitive packet delivery, with no packet loss.

This is a critical network performance parameter for streaming media, since any delay in packet delivery or any packet loss manifests itself as stuttering playback and rebuffering delays. Both are unacceptable to the end-user (and content producer, for that matter). In other words, the quality of playback experience is highly dependent on the network's quality of service. Today, the Internet is undifferentiated. All packets get treated more or less the same way (except that UDP packets cannot be resent if lost). The Internet's routers make a "best effort" at sending each packet. There is no guarantee that each packet will follow the same path or that packets will arrive in the order they were sent. So, for the public Internet, there aren't any QoS guarantees offered, save the Service Level Agreements (promises not to let the purchaser of content delivery services down, without paying some money) offered by some specialist CDNs. Sending data into the public Internet does not guarantee its timely delivery, or even its delivery at all.

We have already touched on the differences between UDP transmission and TCP transmission. The summary is that UDP is prone to data loss, but is also fast and simple, because packet transmission is not error checked. Firewalls tend to prohibit UDP traffic by default, unless the network administrator allows UDP to pass. So, people behind corporate firewalls can't receive UDP transmission. TCP, on the other hand, provides reliable packet delivery, since lost packets may be resent. TCP is the carrier for all HTTP Web page traffic, so all firewalls admit it. TCP is a best-effort protocol, but it achieves this through a slow-start algorithm that progressively increases the rate of packet sending, until packet loss is experienced. This is bad for streaming, because initial buffering does not make best use of the available bandwidth and the connection is eventually driven into loss, requiring data to be resent. We have also mentioned RTP, which differs from TCP in that it does not offer any form of reliability or a protocol for flow and congestion control. RTP cannot ensure real-time delivery, since it has no influence over the lower layers of the network that control resources in network switches and routers, but it is called a real-time protocol because packets contain a timestamp and it provides control mechanisms for synchronizing different streams with timing properties. The timestamp is used to place the incoming audio and video packets in the correct timing order. The synchronization properties are to ensure that several receivers can get the data at the same time, which is particularly important in video conferencing. Whereas TCP uses the lower-level IP for transport, RTP can use either UDP or TCP. The TCP protocol ensures that the total amount of data sent is received correctly at the other end, whereas the UDP protocol, its cousin, just contains the source port number, the destination port number, the data itself, and a checksum. The IP protocol provides the routing mechanism for both TCP and UDP by including the address of the destination network as well as the destination station.

We also described the differences between HTTP and RTSP delivery when we discussed QuickTime serving. To summarize, RTSP is rejected by many corporate firewalls, whereas HTTP will pass through as a plain file. RTSP is suitable for high-volume, high-availability streaming, such as live events, long events, and large files. HTTP is better suited to smaller data transfers and interactivity. RTSP allows the end-user to play back the media on the server effectively, while he or she watches it. HTTP is more like downloading a piece of media and playing it on the client machine. From the end-user's point of view, RTSP looks like the file is playing from a central location, rather like a broadcast, whereas HTTP feels more like getting a video from a video library and playing it on a home machine. From a quality-of-service point of view, for streaming, RTSP is the better experience. RTSP provides VCR-like control over the media, such as pause, fast forward, reverse, and absolute positioning. With HTTP delivery, the player software must simulate this experience, after the entire stream has downloaded. RTSP control is often used in conjunction with RTP to carry the actual media data with best quality of service, though it can use TCP or UDP.

LC-RTP (Loss Collection) is a protocol for reliable video caching in edge servers. It is a multicast protocol, compliant with RTP, to provide lossless transmission of streaming media content into edge cache servers, while concurrently delivering lossy RTP packets to end-users using multicast. It achieves reliable cache loading by retransmission. The bet that LC-RTP makes is that the RTP segment from edge cache to end-user will not be as lossy as the path between origin server and edge cache. It is not clear if there are going to be commercially available embodiments of LC-RTP.

For quality of service on the Internet to be realized, it is necessary to provide guarantees of preferential treatment of "important" packets of data, compared to all the other Internet traffic. Whether mechanisms are even needed is hotly debated, since some argue that Dense Wave Division Multiplexing (DWDM) over optical fibers will make bandwidth so abundant and cheap that QoS will be automatically delivered. QoS proponents counter that no matter how much bandwidth the networks can provide, new applications will be invented to consume it (the discussion about how much bandwidth we need for streaming, earlier in this chapter, came down conclusively on this side). Even if bandwidth eventually becomes abundant and cheap, it won't happen soon. In the mean time, we need QoS mechanisms. QoS research owes much to the pioneering work carried out in the development of ATM (Asynchronous Transfer Mode) networks for the telecommunications industry.

A quality-of-service guarantee (i.e., Service Level Agreement or SLA) is somewhat like buying a first-class airline ticket. For "passenger," read "data packet." For "airplane seating capacity," read "bandwidth available." If you have a first-class ticket, you'll get to your destination with fewer delays compared to the "economy-class" passengers, provided the plane flies at all. Also, if the flight is oversold, it is the economy-class passenger who gets bumped, not the first-class passenger. People with standby tickets might not get on at all. However, if the flight is lightly booked, even economy-class passengers reach their destination without much delay, whether or not they get free upgrades to first class. On the other hand, if everyone on board has a first-class ticket, they will be just as delayed in getting to their destinations as if everybody onboard had economy class tickets (though many will never fly with that airline again). How can the Internet provide different service classes? Just as air traffic control allocates slots for aircraft and manages the handoff of flights from an air traffic controller in one segment to the next, so too the routers and switches on the Internet can create opportunities for data to be routed in the most efficient way on its path through the Internet.

Having stretched the analogy as far as I dare, let us talk about how quality of service mechanisms will work on the Internet (there aren't many of these mechanisms widely deployed yet). The IETF has proposed many mechanisms to meet the demand for QoS. Chief among these are:

- The integrated services or RSVP model
- The differentiated services model
- Traffic engineering
- Content-based routing

The integrated services model is synonymous with resource reservation, where applications set up paths and reserve resources prior to playing back streaming media. RSVP (Resource Reservation Protocol) is the signaling protocol used to set up the reservations. In differentiated services, packets are marked differently (sold different classes of airline ticket) to create several service classes. MPLS (Multi-Protocol Label Switching) is the forwarding scheme that looks at the packet's service class and directs the data according to its priority. Traffic engineering is the process of arranging how traffic flows through the network. Constraint-based routing aims to find routes that meet bandwidth and delay requirements. These four techniques differ from, relate to, and work with each other to deliver QoS on the Internet.

The integrated services model proposes three levels of service:

- The existing best-effort service, working just as the Internet works today
- Guaranteed service for applications requiring fixed delay
- Predictive (or controlled load) service for applications requiring probable delay

Under this model, there is an inescapable requirement for routers to be able to reserve resources in order to provide QoS for specific user packet streams, or flows. This, in turn, requires flow-specific state in the routers. (Unfortunately, it may be that a majority of the routers currently deployed in the world don't meet this requirement.)

With RSVP, the sender sends a PATH message to the receiver specifying the characteristics of the traffic to be sent. Every intermediate router along the path forwards the PATH message to the next hop determined by the routing protocol. Upon receipt of the PATH message, the receiver responds with a RESV message to request resources for the stream. Every intermediate router along the path can either reject or accept the reservation. If a router rejects the RESV message, the router sends an error message to the receiver and the signaling process terminates. If the request is accepted, however, link bandwidth and buffer space are allocated for the future stream and the related flow state information is installed in the router. When the stream commences, any protocols and transports can be used, since the path is somewhat dedicated. However, RTSP/RTP is a good choice for streaming. In fact, in recent revisions of the RSVP protocol, there are several ways to reserve resources for aggregated streams (important for MPEG-4 streaming, in particular). Using these protocol extensions, Explicit Routes (ERs) are established, with each having a particular QoS requirement.

Integrated service is implemented on servers, routers, edge caches, and receivers by four components:

- The signaling protocol (i.e., RSVP)
- The admission control routine (think of a nightclub bouncer, since these routines will decide whether a request for resources can be granted)

- The classifier (puts packets in the correct routing queues)
- The packet scheduler (schedules the packet for delivery so that it meets its QoS requirements)

The problem with integrated services is that the amount of state information increases linearly with the number of flows, putting a huge storage and processing burden on routers. For the model to work, all routers must implement integrated services, since any one router in the flow can break the chain.

Because only dedicated CDNs can realistically hope to implement integrated services, since the noncompliant routers in the public Internet are not going to disappear anytime soon, differentiated services were introduced. Every IP packet has a TOS (Type of Service) byte defined. Differentiated services define the layout of the TOS byte. In order for a customer to receive differentiated services from an ISP, it must have an SLA. This can be static (negotiated every year, for example) or dynamic (negotiated with a signaling protocol such as RSVP, on demand). At the entry to the ISP's network, packets are classified, policed, and sometimes shaped (padded) according to the SLA. (Note that when a packet leaves one domain and enters another—e.g., another ISP's network—the TOS bytes can be remarked, according to the SLA between the two domains.)

Some of the service levels provided by differentiated services are:

- **Premium service**—For applications requiring low delay and low jitter
- **Assured service**—For applications requiring better reliability than best-effort service
- **Olympic service**—Provides three tiers of service with decreasing quality (gold, silver, and bronze)

The ISP decides what services to provide. Since there are only a limited number of classes, and class of service applies to an entire flow, the amount of state information is proportional to the number of classes, not the number of flows. Differentiated services are therefore more scalable than integrated services. Second, classification, marking, and policing are only needed at the boundary of the network, not at every router hop.

Another weapon in the QoS arsenal is MPLS (Multi Protocol Label Switching). The motivation for MPLS was to use a fixed-length label to decide packet handling by routers, so that they could be handled more rapidly and simply than with the older routing protocols, such as BGP (Border Gateway Protocol). MPLS is also a useful tool for traffic engineering.

MPLS is a forwarding scheme that evolved from Cisco's tag switching. Each MPLS packet has a header consisting of a 20-bit label, a 3-bit class of service field, a 1-bit label stack indicator, and an 8-bit TTL (Time To Live) field. A router that implements MPLS is called a Label Switched Router (LSR). It works by examining only the label in the forwarding packet. The network protocol can be IP or others, hence the name Multi-Protocol Label Switching. An LSR uses a Label Distribution Protocol (LDP) to set up Label Switched Paths (LSPs). An LSP is similar to an ATM Virtual Circuit (VC). There is some debate about whether or not RSVP should be extended to act as an LDP. The proposal is called RSVP-TE (Resource Reservation Protocol with Traffic Engineered tunnels). An alternative that has been proposed is the CR-LDP (Constraint-based Routing Label Distribution Protocol), but this has had less support so far. RSVP-TE provides bandwidth reservation capabilities on top of the MPLS core protocol.

MPLS LSPs can be used as tunnels. When a packet enters the tunnel, its path is completely determined by the label assigned by the ingress LSR. The packet is guaranteed to emerge at the end of the tunnel. MPLS is significant because it provides faster packet classification and forwarding and an efficient tunneling mechanism. Along with routing extensions such as OSPF/TE (Open Shortest Path First protocol for Traffic Engineering), which allow routing around congested paths rather than always choosing the shortest path, and which meet varying SLAs, MPLS LSPs form a complete package for edge and core routers and switches to provide the infrastructure to support integrated services, when coupled with the data classification facilities provided by differentiated services.

Integrated and differentiated services provide graceful degradation of performance when traffic load is heavy. However, as is the case when a flight is undersold, when traffic is light there is little difference between integrated/differentiated service and best-effort service. Why not avoid congestion in the first place? This is the motivation behind traffic engineering. Traffic is the result of insufficient network infrastructure, or, more commonly, uneven traffic distribution on the network. Uneven distribution can be caused by the current dynamic routing protocols, such as RIP (Routing Information Protocol), OSPF, and IS-IS (Intermediate System to Intermediate System), because they always select the shortest paths to forward packets. As a result, the short paths may clog while links along a longer path are idle. The Equal Cost Multi-Path option of OSPF is useful in distributing load to several shortest paths, but if there is only one shortest path, it doesn't help. Traffic engineering is the process of arranging how traffic flows through networks so that congestion caused by uneven network utilization can be avoided. Constraint-based routing is an important tool for making the traffic engineering process automatic. Avoiding congestion complements graceful degradation of the network under load, so traffic engineering complements differentiated services.

Constraint-based routing is used to compute routes that are subject to many constraints. The goals are to select routes that can meet QoS requirements and increase utilization of the network. Constraint-based routing considers not just the topology of the network, but also the requirement of the flow, the availability of resources on links, and other possible policies specified by the network administrators. A constraint-based router may find a longer, but lightly loaded path "better" than the heavily loaded shortest path. As a result, network traffic is distributed more evenly. OSPF-TE and IS-IS/TE distribute bandwidth information as extensions to their link-state advertisements. Constraint-based routing better meets the need of QoS requirements for unimpeded flows of data and improves network utilization, but increases communication and computation overhead and the size of routing tables, and introduces possible routing instability. Also, longer paths may consume more resources. Because constraint-based routing algorithms recompute routing tables more frequently than dynamic routing algorithms do, they can produce instability, shifting traffic fruitlessly, again and again, to try to avoid congestion. Constraint-based routing is similar to the dynamic/adaptive routing used in telephone and ATM networks. It helps differentiated services to be delivered better. RSVP and constraint-based routing are independent but complementary. When MPLS and constraint-based routing are used together, they make each other more useful. With these protocols and technologies, companies are now building QoS features on top of existing IP infrastructures. The challenges now are in finding methods to design, manage, and operate these networks. In other words, the final frontier in QoS build-out is in provisioning.

There are often questions about whether or not multicast provides better end-user stream playback than unicast. Multicast streaming has no real linkage to QoS provision, other than reducing the amount of bandwidth and other network resources required to supply streaming media to a large number of consumers.

Ultimately, delivering QoS will not be so much a technical challenge as a human resources and people-management issue. Maintaining network performance and integrity boils down to finding, maintaining, and motivating dedicated, skilled network administrators and technicians, as much as it requires resource redundancy for instantaneous fail-over. The addition of edge caching to scale the network and the buildout of more and more fiber serves only to improve QoS as well.

With a combination of self-aware edge server clusters, overlaid with network QoS techniques, forward error correction techniques, real-time packet-delivery protocols, multilevel buffer-filling techniques that burst-load both edge caches and player buffers as fast as available bandwidth will allow, and packet-delivery synchronization techniques like the DMIF layer described in MPEG-4, it ought to be possible to provide near-perfect media streaming over IP networks, at least in principle. Unlike the dedicated routing of DVB transport streams, however, IP-based streaming is more resilient to network equipment failure and network segment faults.

It is interesting to consider the current state of IP network provision in phone and telecommunication companies. While they are likely to migrate to IPv6 over time, adopting MPLS to overlay QoS features, they are, nevertheless, saddled with much SONET/SDH synchronous optical networking infrastructure and ATM equipment. ATM and SONET/SDH already deliver QoS features. For example, ATM beats differentiated services over MPLS in data forwarding speed. Traffic engineering is also possible over legacy ATM networks. Why should the telecommunications companies drop ATM and SONET/SDH? The short-term answer is that they won't. However, with more and more of the bandwidth carrying data, as opposed to simple voice calls, there is pressure to convert the essentially switched-circuit nature of the phone system to the packet-switched IP architecture. ATM cell headers are large, so they waste bandwidth, compared to IP. Also, routers at the network boundaries must be used, rather than more cost-effective network switches. Connectionless approaches, like IP, are also more resilient in failure modes, as we have already noted. This property was the reason for inventing the Internet in the first place. Hence, it is almost inevitable that IP networks will eventually displace ATM and SONET/SDH.

The Achilles heel of network QoS provision could potentially turn out to be the SLAs. When two ISPs negotiate to transport each other's customers' data, money changes hands if one ISP winds up transporting more of their competitors' data than the other. For this reason, network operators, by necessity, impose limits on the amount of data transferred over the interconnections between their domains and neighboring domains. Turning off a network interconnection, or degrading it, because the terms of a SLA turn out to be unfavorable to one of the ISPs party to a particular SLA, is tantamount to denial of service (DoS). Lessscrupulous operators, in principle, could even launch deliberate DoS attacks in order to gain competitive advantage over other ISPs (of course, if this were to occur, companies that monitor the state of the Internet by mapping it in real time would be among the first to notice such abusive behavior). From the content providers' point of view, if service levels that they have paid for are not met, it is difficult to pin the blame on any individual ISP, unless the SLA provides for such remedies, the ISP chosen provides end-to-end data delivery from source to end-users, or there are SLAs negotiated and in place with every ISP involved in serving their consumers. It would be an irony indeed if QoS remained an elusive deliverable merely because of the inability of the ISPs' lawyers and commercial negotiators to strike satisfactory SLA deals. The future of streaming media could, in a very real sense, be limited by the contractual terms contained in these inter-ISP SLAs.

Real Video and Real Audio

We deferred treating RealNetworks' streaming media components and system until now because the latest version of their platform incorporates features that make RealNetworks' solution a content delivery network as well as a network of edge servers. We couldn't discuss the system without first understanding the fundamentals of content delivery networks and edge caching. At the time of writing, version 8 was the current incarnation of the system.

In the simplest form, the RealNetworks solution has all the same major components as the Microsoft Windows Media Technologies offering. There is a media encoder, called Real Producer, a media server called Real Server and a software-based player application called Real Player. However, how these individual components work "under the hood" is different. RealMedia uses different file formats, different compression technology, different approaches to serving, different media transport protocols, different rights management techniques and the company's business model and licensing terms are very different from Microsoft's. Developers must embrace completely different software development kits (SDKs), with different software interfaces, which achieve their underlying functions in ways different from Microsoft's. Broadly speaking, the system does what Microsoft's Windows Media Technologies does, but almost completely differently. In defense of RealNetworks, however, it must be said that they staked out the territory first. It was Microsoft that went a different direction from RealNetworks, not vice versa.

Today, none of the systems offered by Microsoft, Apple, or Real-Networks is even vaguely interoperable. This may be changing, however. Both Apple and RealNetworks have announced that they are lining up behind MPEG-4 compression and file format technology. ISO compliance will bring a level of interoperability that has been missing to date. RealNetworks has also paid consistent attention to improving the quality of streaming service delivered with their system, introducing several new approaches with every release.

RealNetworks currently markets its platform under the RealOne trademark. The four main parts of the RealOne platform are:

- The RealOne Player (an update and amalgamation of RealPlayer and RealJukebox)
- The RealOne SDK
- The RealOne Services Infrastructure (which provides just-in-time delivery of plug-ins, account, e-commerce, and member communications services)
- The RealOne Service (which offers to subscribers a wide range of premium content, including downloadable music and streaming of major label artists)

The RealOne platform is built on top of the company's RealSystem iQ universal media delivery platform. RealOne is the consumer's view of the offering. RealSystem iQ is the underlying delivery technology. We will examine this technology presently.

RealOne aims to unify the consumers' media experience, presenting downloadable content, local playback, CD ripping and burning, and the ability to extend to portable devices in a single application and interface. From the content producer's point of view, the system provides simple, powerful authoring tools, better control of the media and playback experience, tighter content security, higher quality delivery, and sustained revenue opportunities. It is anticipated that MPEG-4 support will be available end-to-end, in the fullness of time, as version increments to the software components in the Real System are released.

RealNetworks' encoding application is, as mentioned earlier, called Producer, available in a free "Basic" version and a "Plus" version that you must pay for. These applications convert audio and video into RealNetworks' proprietary compressed streaming files with RealNetworks' .RM extension. Editing programs also have plug-ins available, which can encode into RealMedia format. Real Producer can also live encode. The application allows users to encode at eight different bit rates, with deinterlacing (removing the artifacts of PAL and NTSC television systems), video scaling (to change the image size) and inverse telecine (removing the frames that are inserted when a film shot at 24 fps is converted for playback on television at 30 fps). The current version of Producer is 8.5. Some of the features of Real Producer include two-pass video compression, variable-bit rate compression, and SureStream.

Two-pass video compression increases the quality of the video output by analyzing the video data for the entire file before encoding the clip. The algorithm looks for transitions and complexity, using this information in the second pass to optimize the encoding. Without the second pass, the codec may be set suboptimally at a transition, whereas with it, the encoder can be set optimally at every point in the video stream. Single-pass encoding requires some compromise in codec efficiency, whereas two-pass enables optimal codec utilization. Two-pass encoding, of course, cannot be applied to live sources.

Variable-bit rate encoding is very often used in conjunction with two-pass encoding. This feature enables the codec to vary the bit rate throughout the clip, depending on the type of content being encoded. More bits are spent on high-action scenes, taking away bits from low-action scenes. This motion-sensitive encoding improves the overall perceived quality of the encoded clip.

SureStream is a feature unique to Real System, which allows the Real Server dynamically to adjust the stream's bit rate for each unique consumer, depending on the dynamic network conditions and congestion between the server and the consumer. If a network path becomes congested, RealSystem "downshifts," sending a stream of a lower bit rate, until congestion clears, at which time it automatically "upshifts" again. In order for the server to do this, Real Producer must produce a multi-bit rate file.

As part of the Neuralcast technology in Real System iQ (which will be explained below), Real Producer can actually produce redundant live streams, which can be sent to multiple Real Servers to provide a failover feed, in the event of a network or equipment fault.

In December 2001, RealNetworks announced that it was supporting MPEG-4 in its Real System iQ technology, through server and client plug-ins from Envivio. Unfortunately, at the time of writing Real Producer was not available with MPEG-4 support, though the Real Player and RealOne Player were.

As with Microsoft's Windows Media Technologies, there is a need for a metadata redirection file to launch the Real Player from a Web browser. In Microsoft Windows Media Technologies, this is the .ASX file. In

Real System, the equivalent file is the .RAM file. This file is necessary to launch the Real Player application, but also to provide an RTSP URL (address) for the clips on Real Server. Clips on Real Server are delivered with RTSP protocol, rather than HTTP, so that the URL used to request clips must start with rtsp:// rather than http://. Since browsers cannot make RTSP requests, the linkage from the Web page that launches the media to the actual media stream is the .RAM file, which the content producer writes with a simple text editor. Finally, the .RAM file can pass parameters to Real Player.

Other media creation tools available from RealNetworks are Real Presenter and Real Slideshow. Both are available in free "Basic" versions, as well as paid-for "Plus" versions. There are third-party tools, such as GriNS Editor Pro available, to add synchronized multimedia elements to the stream, using SMIL 2.0 markup language (which will be discussed in a subsequent section of this book). Real Presenter is currently at version 8 and adds audio and video streams to a Microsoft PowerPoint presentation. With Real Slideshow, you can combine digital pictures and images with audio to create dynamic slideshows. Real Slideshow is currently at version 2. GriNS Editor Pro allows the creation and editing of streaming SMIL presentations that combine audio, video, images and text. SMIL presentations are played back with Real Player. Real System also lets content authors embed Macromedia Flash content into synchronized multimedia presentations.

The beating heart of Real System iQ is Real Server, available in three versions: Professional, Plus and Intranet. There is also a free trial version of Real Server designated "Basic," which supports 25 simultaneous users. Real System Proxy, a new addition to Real System, provides edge caching and replication capabilities. Real System iQ employs a new technology, called Neuralcast, which allows Real Servers to work in tandem to provide improved quality of service and system reliability.

Real Server Plus is the entry-level server platform. It is priced to allow enterprises to get started in streaming media, serving both the Internet and their own intranets. It supports up to 60 simultaneous users, but has limited administration options compared to the more costly versions. Also, multicast is not scalable on the Plus version and there is no support for advertising insertion or user authentication.

Real Server Professional is the company's streaming media server that delivers media content to consumers on the public Internet. Each server supports between 100 and 2,000 concurrent users, depending on how much you pay RealNetworks (licensing is on a per-stream basis). The Professional server supports advertising and user authentication extensions, as well as IP-based client connection control. The Professional server also supports stream capacity segmentation to scale the streaming load-bearing capacity across machines. Content can be delivered to end-users over the public Internet, via a corporate Wide Area Network (WAN), using a specialist content delivery network or via satellite. This version of the server supports both subscription-based and pay-per-view media commerce. The Professional version also allows advertising-supported business models for media delivery. Because it is integrated with RealNetworks' Media Commerce Suite, Real Server Professional encrypts streams.

Real Server Intranet is designed for enterprises that wish to replicate and serve content cost-effectively within an organization. Interestingly, a content delivery company could potentially use Real Server Intranet to create a global content delivery network serving individual consumers, colocating edge proxies with local ISPs. It is the company's scalable multicast platform. It also supports all the features of the Professional version. The headline feature of the Intranet version is capacity sharing. This provides a dynamic and intelligent way for a cluster of Real Servers to share the total number of allowed consumer connections specified in a license key to be distributed across physical machines. For capacity sharing to work, one server is configured as a "publisher." The publisher dynamically allocates its licensed stream capacity to connected server "subscribers" that do not use a locally licensed stream capacity, but instead acquire their capacity from the publisher. This solves two problems. It handles unpredictable demand and it guards against vulnerability to a single point of failure.

Real Proxy is the company's edge cache and live stream repeater, which can offload media stream requests from the origin server, fanning out to many more connected users. Real Proxy software can be installed on a network or ISP gateway; it aggregates and handles client requests for media streamed from Real System Server. Real Proxy software has been incorporated in dedicated edge server appliances, too, such as those made by CacheFlow. Real Proxy accepts live streams from other servers and re-serves the data stream to end-users. Acting as a live streaming repeater, it reduces network traffic by eliminating redundant requests to the origin for streaming media. By eliminating the redundant traffic between the server and the proxy it requires less bandwidth or many more consumers to receive the same amount of content. Lower bandwidth usage translates directly into cost savings. Real Proxy also allows the inbound and outbound bandwidth to be capped, ensuring that mission-critical applications are not adversely affected by streaming activities. The Proxy also authenticates every client request at the origin, so control of the content always stays at the origin. IP masking of internal users is also a feature of Real Proxy. Finally, because the content can be served to end-users from a server physically closer to them, transmission problems due to network congestion at router hops is minimized, improving the quality of the playback experience.

Neuralcast is the software that coordinates the activities of Real Servers and Proxies to replicate content automatically, to load balance, to provide failover intelligence in the event of network or equipment faults and to ensure high quality of service. Before Neuralcast, media delivery was based on one-way communication—server to player, origin to edge. During heavy use periods, this method was less than satisfactory as bottlenecks occurred and wait times increased. With Neuralcast, the network of servers and proxies is configured as a self-aware "honeycomb," in which all servers talk to each other, making instantaneous decisions about capacity sharing, optimization, and redundancy. This effectively turns any node in the network into both an origin and an edge. Content can be injected at any node and served to the rest of the network of servers, and then on to end-users, wherever they are. In this topology a much larger audience can be served, while removing bottlenecks and waits, thus improving the end-users' quality-of-streaming experience.

Neuralcast provides:

- Capacity failover in the case of network or equipment faults
- Capacity allocation to respond dynamically to demand
- Guaranteed live broadcast delivery from server to server through redundant network paths
- Less costly administration through on-the-fly remote configuration and monitoring
- Zero points of failure for live broadcast transmission

Neuralcast Communications Protocol enables all servers to act as one, by providing information about capacity allocation and capacity failover. Whereas systems that use TCP for stream delivery cannot resume and reconnect a stream once a connection is broken, Neuralcast's use of UDP for both unicast and multicast allows for connectionless transmission, with no back channel. This makes the Neuralcast system ideal for satellite delivery. Hence, Neuralcast can support both unidirectional (connectionless) and bidirectional (with state held on the server) transmission.

Forward error correction, as we have discussed previously, is a coding technique that allows data to be lost in transmission, yet permits the original stream to be reconstructed perfectly. With Real System iQ, the forward error correction scheme is designed to tolerate up to 10% packet loss, without any degradation of the stream. When coupled with robust packet resending and SureStream technology, Real System iQ is able to promise almost 100% flawless playback. As with Microsoft Windows Media Technologies, Real System iQ delivers CD-quality audio at just 64 kilobits per second and VHS-quality video at mainstream broadband rates. When connectionless transmission is in use, where there is no back channel to the server, packet loss is expected and UDP has no method of resending those lost packets. What the player software relies on in those cases is forward error correction, so that packet loss is irrelevant.

Real Server uses two connections, known as channels, to communicate with clients: one for communication with the client and the other for the actual streaming data. The communication channel is known as the control channel, since this is the line over which Real Server requests and receives passwords and the client Real Player sends instructions such as play, pause, and stop. The audio and video media are actually streamed over a separate data channel. Real Server uses two sets of protocols in transmitting its data. For the control connection, it uses bidirectional TCP. The TCP protocol guarantees packet delivery, which is important for control over streaming and for error checking. Though it has built-in congestion control, it responds too slowly to changing network conditions, so is a poor choice for media delivery. For the data connection, Real Server uses UDP by default, unless this is blocked by a firewall, in which case Real Server will use TCP instead. UDP packets are sent in one direction only. Because UDP involves no error checking, it can deliver packets faster than TCP. Real Server uses two main application-level protocols to communicate with clients: RTSP and the legacy PNA (Progressive Networks Audio). These protocols work with the two-way TCP connection to send commands from the client, such as start and pause, and from the Real Server to clients to convey information such as clip titles. A third protocol, HTTP, is used in sending other types of data.

RTSP is designed specifically for serving multimedia presentations. Only RTSP can deliver SureStream files, which use multiple-bit rate encoding to compensate for network congestion. PNA is a proprietary protocol supported solely for backward compatibility reasons. HTTP is used for .RAM metafiles and for HTML pages served by Real Server. It may also be used to deliver clips to clients located behind firewalls. RTSP differs from HTTP in that it has methods not available in HTTP, RTSP servers need to maintain state in almost all cases, both RTSP servers and clients can issue requests, out-of-band RTSP data can be carried by a different protocol (RTP for example), and the RTSP Request URI (Universal Resource Identifier) contains the absolute URI.

We have already stated that media travels over UDP in preference to TCP, unless UDP traffic is blocked by a firewall, but it is worth mentioning that Real Server uses one of two packet formats for sending media data to an RTSP client. These are standard RTP and RealNetworks' Real Data Transport (RDT). When data must be sent using TCP rather than UDP, data is interleaved with the RTSP control stream. When data is being delivered via UDP, this can either be multicast UDP or unicast UDP.

When the RTSP client selects RTP delivery over UDP, it sets up three network channels with the RTSP server (Figure 2.23). A fully bidirectional TCP connection is used for control and negotiation. A unidirectional channel is used for media delivery using RTP packet format. A third full-duplex UDP channel called RTCP (Real Time Control Protocol) is used to provide synchronization to the client and packet loss information to the server.

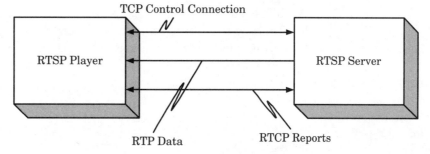

Figure 2.23
RTSP communications with UDP.

When data is delivered using RDT, the RTSP client again sets up three network channels with the RTSP server (Figure 2.24). The only difference between this configuration and the one previously described is that the third channel is a unidirectional UDP path from the client to the server to request that the server re-send lost UDP media data packets.

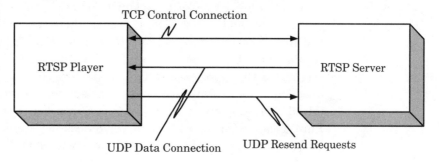

Figure 2.24
RTSP communications with RDT.

In the case where UDP transport of media data is impossible, media data may be formed into packets using RTP or RDT and carried using TCP. In this scenario, a single full-duplex TCP connection is used for both control and media delivery from the RTSP server to the client (Figure 2.25). The data stream and the RTSP control stream are merely interleaved.

Figure 2.25
RTSP communications with TCP.

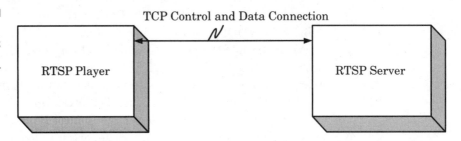

TCP Control and Data Connection

RTSP Player   RTSP Server

An RTSP server typically holds three states about a client (Figure 2.26). There is the initial state, where there is no client connected; the ready state, where the client has established a connection but is not playing the media; and finally the self-evident playing state.

Figure 2.26
RTSP state machine.

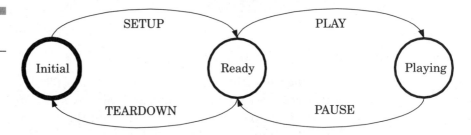

SETUP   PLAY

Initial   Ready   Playing

TEARDOWN   PAUSE

Real Servers can also serve QuickTime content, as well as supporting .MOV, .AVI, and .WAV files. However, there are some limitations with the Real Server's support of QuickTime. Real Server 8 can stream Apple's QuickTime 4 clips encoded with any codec and hinted for streaming to Apple's QuickTime 4 player. Hinted QuickTime 4 clips encoded with a standards-based codec can stream to RealPlayer 8 (H.261, H.263, and .MP3, but not Sorenson, Cinepak, Qualcomm PureVoice, or Qdesign). RealServer 8 does not support earlier versions of QuickTime (e.g., QuickTime 3). Real Server ignores any QuickTime tracks other than video or audio. Scripting commands, for example, are discarded. Live broadcasts of QuickTime can be supported using Sorenson's Broadcaster product, but archiving of live broadcasts in QuickTime is not possible. It was not clear whether or not QuickTime 5 was supported, though this is, at the time of writing, the current version.

There are no less than five flavors of multicast support in all Real Servers. These are:

- IP multicast
- RTP multicast
- UDP multicast
- TCP/IP multicast
- HTTP multicast

At present, due to the lack of standardized multicast support on the public Internet, multicast is useful only for intracasting applications, or in situations where the entire network supports multicasting. Real Servers run on Windows NT, Linux, Sun's Solaris, HP/UX, IBM/AIX, and Compaq Tru64. All Real Servers support live and on-demand broadcasting, SMIL, SureStream, bandwidth management and Java.

RealNetworks has a service called RBN (Real Broadcast Network), which is a streaming media hosting outsource for content producers who do not wish to host their own streaming media, whether a live one-time event, or around-the-clock live or on-demand programming. As part of the company's RealOne platform, the RealOne Services include RealOne Music (a digital music subscription service) and the RealOne news-to-sports to soufflés multimedia programming portal. One presumes that the RealOne offerings are hosted on the company's Real Broadcast Network.

At present, RealNetworks supports two players: Real Player 8 and the newly released RealOne Player. RealNetworks claims to have distributed 200 million copies of Real Player. RealOne player is a three-pane experience, including a media playback window, a related information window, and a media browser. This allows users to find content, play streaming and downloaded content, organize their audio and video in a personalized library, and take music away from the desktop either on CD-R as a compatible audio CD or MP3 collection, or else download music from the desktop to a portable MP3 player. RealOne Player integrates with the RealOne subscription service, which offers music, sports, entertainment, and news programming. Most of the content is repurposed broadcast television programs. The RealOne Music subscription service also allows consumers to safely and legally download and stream music from major record labels and artists (though not all of them). RealOne Player presentations are authored in HTML, using standard Web page development tools. What RealOne Player is, in reality, is a Web browser crossed with a media center. A Web site can live entirely in the RealOne Player environment. If a site wanted to offer streaming media in the past, the options were to embed a media player in the page and to put contextual and advertising information around the video window. This didn't allow management of the windows, video resizing, or user control of the playback. The content was disjointed from the rest of the site. With RealOne Player, these elements are integrated.

As a possible prototype model for how an MPEG-4 player with MPEG-7 searching capabilities and MPEG-21 rights management might appear, look no further than RealOne Player. One can only speculate about the degree to which the MPEG standards will be embraced by RealNetworks, since they already have many simpler, though more limited, proprietary techniques working to create a user experience not completely unlike that obtainable with the very best MPEG-4 player implementation imaginable (3D spatializations and synthetic objects excepted).

Unlike the previous Real Player or the equivalent embedded ActiveX media player that was often embedded into Web pages, RealOne Player offers video controls, audio equalization controls, and the ability to add media clips to a favorites list or playlist. With the old players and embedded controls, a user couldn't continue to browse in other pages while media was playing. With RealOne Player this is possible. RealOne player also solves the problem of nonstandard or missing media controls, often a difficulty with players embedded into Web pages. RealOne Player supports SMIL 2.0, so that sophisticated synchronized multimedia presentations can be presented. RealOne Player can also display Macromedia Flash content. Most importantly, decryption software and rights management software are now included in RealOne Player and can be added on as a plug-in to Real Player 8.

For rights management, RealNetworks offers a Media Commerce Suite. It consists of four software applications:

- RealSystem Packager
- RealSystem License Server
- Media Commerce Upgrade for RealPlayer
- RealSystem RealServer secure file format plug-in

Packager is a utility that lets content providers securely package media files prior to distribution. License Server is an HTTP server that accepts requests for and generates licenses that permit access to secured media. Media Commerce Upgrade for the Player is a trusted client that recognizes secured RealMedia files (with a .RMS extension) and enforces overall system integrity of the client-side media engine, ensuring that content can only be played back in a trusted, tamper-resistant environment. Real Server plug-in enables Real Server to stream secured media packages seamlessly.

These components interact with existing content delivery mechanisms, a retail Web server, and a back-end database. Secure media files can be transported on virtually any delivery mechanism, including FTP downloads, peer-to-peer networks, multicasting or physical media like CDs or DVDs. A content database stores the secured content key and globally unique identifier (GID) for each content file. It makes the data from Real Packager available to the retail Web server during content licensing. The retail Web server serves the front-end Web site through which consumers request licenses to secured content. The retail Web server sends these requests to the License server and returns the licenses thus generated back to consumers. Only when the license is received can the trusted Player play the file. With the Media Commerce suite, rights are stored separately from content and content is encrypted using strong encryption techniques. Packager allows metadata to be stored with the encrypted package. Only the retail Web server can interact with the License server, and the keys are always delivered as encrypted data, never as clear text. Content rights are stored in RealNetworks' XMCL (Extensible Media Commerce Language) and the rights allow for very flexible licensing options. The License server has the capacity to act as a revocation agent, so that rights can be revoked if the user violates license terms. It can also revoke a compromised component on the client or revoke all content by the content provider.

RealNetworks provides four SDKs for independent software vendors. These are: RealSystem SDK (formerly known as the RMA SDK or Real Media Audio SDK), RealOne Visualization SDK, RealOne Digital Distribution SDK, and RealOne Metadata Package Toolkit. The company also provides an SDK for Real Producer, as well as embeddable ActiveX controls for both the encoder (Producer) and player.

RealSystem SDK is the architecture upon which RealOne Player and Real System Server are built. Developers who need to build client or server applications compatible with Real System, or who have previously used RealPlayer or RealAudio SDKs, now use this SDK. Every interface in the system is demonstrated via header files. RealOne Visualization SDK allows developers to reskin RealOne Player and to create visualizations for it. RealOne Digital Distribution SDK provides third-party device manufacturers an API to transfer music data from RealOne Player to portable devices, like MP3 players and removable storage peripherals. RealOne Metadata Package Toolkit allows content providers to create .RMP metafiles, which can improve the consumers' music experience by incorporating other relevant content with music downloaded or streamed.

Streaming Media Servers

Since we have discussed many of the more popular streaming media servers already in this chapter, this discussion might seem superfluous. The point with streaming media servers is that they are optimized for streaming, meaning that they make better use of network bandwidth, hard drive bandwidth, and server resources than do regular Web servers, when streaming media. They tend to have features that Web servers don't have, such as support for RTP/RTSP protocol based on UDP transport. They also tend to have inbuilt schemes for dealing with Internet congestion. Increasingly, they act in clusters rather than as standalone servers, allowing more streaming media users to be supported simultaneously, with less exposure to the failure of any one server or network link.

A streaming media-serving farm capable of serving several million players is a physically large entity. Such operation centers can occupy a vast amount of floor space and draw tremendous amounts of power. In addition, there are very few locations today that can provide sufficient connectivity to Internet backbones to serve all the streams necessary. It is for this reason that globally distributed networks of servers are now becoming a popular solution, since these servers can usually be colocated in ISPs' network operations centers, or even in large enterprises' IT departments.

For streaming media serving, server density is a significant problem. There needs to be maximum use made of rack space. Servers need to have a small footprint and draw little power, while requiring little cooling. Some streaming media server farms have begin to experiment with one-rack-unit servers, while others are employing servers built for the telecommunications industry based on a compact PCI form factor. The latter can squeeze an incredible number of servers into a small amount of rack space. Since these compact PCI servers are based on laptop components, they are inherently low-powered and require less cooling than their more traditionally based one-rack-unit brethren. However, processor speeds tend to lag behind those of the traditional servers. Media serving therefore poses significant operational difficulties, since operators must always balance space and power usage against number of streams served.

In the streaming media serving world, operators flinch at Real-Networks' stream-based licensing model. However, operators like the range of servers supported. While the Microsoft Windows Media Server is free, the Windows Server software with which it is bundled costs significantly more than the popular Linux operating system and reliability and availability are reputed (or believed) to be poorer than on Linux-based Web servers running the open-source Apache server. Apple's QuickTime Streaming Server is rare at present, since few servers run Mac OS X. However, the open-source Darwin may find wide application, particularly since it runs on Linux and Sun's Solaris, which are popular choices for ISPs. Of course, many streaming media producers will not serve their own media, but will inject their publications into the networks of content delivery companies, like Akamai, RBN, Yahoo! Broadcast, and Digital Island. These companies take care of hosting the media, operating all the networks, servers and edge caches that their particular systems require.

Multicasting

We have, by necessity, mentioned multicasting throughout this chapter, without explaining what it is or how it works. The reason for delaying the discussion until now is that multicasting really doesn't exist on the public Internet. Multicasting requires all the routers in the multicast network to support multicasting protocols, and multicast routing is far more complex than regular unicast or point-to-point routing. Reliable multicasting requires more complexity still, so that the only applications of multicasting that exist in reality occur within contained enterprise LANs or WANs. It will be years before the public Internet supports routine multicasting, if it ever does. Many specialist CDNs also do not support multicast. Multicasting raises issues of control over digital rights, just as edge networks and caching do, since with multicast, routers make digital copies of content in order to forward them to other subnets. Hence rights management is just as crucial in multicasting as it is in edge caching.

Many streaming applications are one-to-many, where many receivers simultaneously receive one stream (like a scheduled broadcast), or many-to-many, where multiple streaming sources send to multiple receivers (multiperson video conferencing, for example). These applications could be realized using a point-to-point unicast for each source to receiver path, but the amount of bandwidth and network resources required to support, say, 1000 users would be very expensive indeed. For a start, in a one-to-many application, the server alone would need to support 1000 simultaneous connections. Multicasting provides an alternative method of delivering these data packets to everyone who wants them (i.e., to every player), with an apparent server load of just one.

Multicast enables sources to send a single copy of a message to multiple recipients who explicitly want to receive the information by using the intervening routers to replicate the content, where required, in order to serve all the requesting clients. It does not simple-mindedly send a copy of the message to every node in the network. Rather, it sends the multicast data only to those nodes supporting subnets containing requesting receivers, since many nodes may not want the message.

Multicast is a receiver-based idea. Receivers join what are called *multicast session groups*. Data is delivered to all members of that group by the network. The sender doesn't need to know the addresses of all of the receivers, since multicasting is a connectionless distribution method. Only one copy of a multicast stream will pass over any given link in the network and copies of the stream are made only where paths diverge at a routing node. For example, if there are a thousand people in Australia wanting to receive a particular multicast stream from the US, only one stream will be routed through the trans-Pacific link, instead of 1000.

The main advantage of multicast is its scalability. There are people seriously considering multicast as an emergency service, so that streams of significant breaking news (live updates on the World Trade Center atrocity, for example) could be disseminated, without individual streaming news sites and servers going down under the weight of a sudden, incredible load. Multicast service would enable emergency broadcasts and communications. The cost savings to mass-audience streaming media providers, in terms of bandwidth charges, are also significant.

One of the earliest implementations of multicasting was the MBone (Multicast Backbone) a project; an outgrowth of the first two IETF "audiocast" experiments, in which live audio and video were multicast from the IETF meeting site to destinations around the world, via volunteer (mostly academic) networks who cooperated in the project. The MBone still exists. It is the IETF's semi-permanent test bed for IP multicast developments. Today, there are organizations like the International Webcasting Association (www.Webcasters.org.uk) and the IP Multicasting Initiative (IPMI at www.ipmulticast.com), which are working to establish international multicasting infrastructures for streaming content delivery.

IP multicast is an extension to the standard IP network transport protocol dating back to 1989. It is defined as the transmission of an IP datagram to a "host group," a set of zero or more hosts identified by a single IP destination address. A multicast datagram is delivered to all members of its destination host group with the same "best-effort" reliability as regular unicast IP datagrams. Membership in a host group is dynamic. Hosts may join and leave at any time. The reason that all members of the group are called hosts is that multicast is designed to encompass the many-to-many scenario, in which each receiver is also a server (or host). In actual fact, the majority of the hosts in the group may be mere receivers, as would be the case in the one-to-many scenario. There are no restrictions on the number of hosts in a group, or on the location of the hosts. Hosts may also be members of more than one group at a time. At the application level, a single multicast host group IP address may have multiple streams on different port numbers, on different sockets, in multiple applications ("sockets" and "ports" refer to the notional connections to the Internet offered by the network adaptor protocol stack—software responsible for all communications on the network). You can deliver multiple streams at once in a single host group. Also, on any given host, multiple applications may share a single group address (for example, if there are two different players on a PC simultaneously playing a stream from a single multicast source).

To support native IP multicast, all nodes, including the sending, receiving, and intermediate routing nodes must be multicast-enabled. This means they must have support for IP Multicast transmission and reception in their TCP/IP protocol stacks, they must support the Internet Group Management Protocol (IGMP) to communicate requests to join a multicast group or groups and receive multicast traffic, and they must have network interface hardware that filters data packets from the range of IP multicast addresses. In addition, servers and receivers must have IP Multicast applications (e.g., a streaming server must support multicasting and the streaming media player must be able to receive multicast streams). Firewalls also need to be configured to permit multicast traffic. Many new routers have support for IP multicast, but older routers may require memory upgrades before they can support multicasting.

IP multicast uses Class D IP addresses to specify host groups. Any address in the reserved range 224.0.0.0 to 239.255.255.255 is a multicast group. Some of these addresses are permanent groups (for example 224.0.0.2 addresses all routers on a LAN). Other nondedicated addresses in the reserved multicast groups address range can be used for temporary groups. To send an IP multicast datagram on a group, the sender specifies the appropriate temporary group address and sends the datagram using the same Send IP operation used with unicast datagrams.

Compared to sending a multicast datagram, receiving one is much more complex, particularly over a WAN. To receive the datagrams, a host application (e.g., a streaming media player) requests membership in a multicast host group (e.g., today's one o'clock live press conference with the President). This membership request is communicated to the LAN router and, if needed, to all other intermediate routers between the sender and receiver. The receiver's network interface hardware now starts to filter for the data packets tagged with the right addresses, corresponding to the host group's IP address. As the network interface hardware detects packets of interest, it passes them to the TCP/IP protocol stack, which makes them available to the user's application, such as a streaming media viewer.

Each multicast packet uses the Time To Live (TTL) field of the IP header to limit propagation of individual packets. Every time a router touches a packet, it decreases the TTL value. Hence, the TTL value counts router hops. Any packet that has a TTL value of zero is dropped, without an error being signaled to the sender. If the router encounters a packet with a TTL of 1, it knows to confine multicast transmission to the local area network only. For values greater than 1, the packet is forwarded to other multicast routers, which will multicast the data on their subnets (provided they are reachable within the TTL), if there are any members of that host group attached. Several standards for TTL are specified for the MBone, for example. On the MBone, 1 confines multicasting to the local area, 15 to the site, 63 to the region, and 127 to the entire world.

Most routers block multicast traffic by default. IP tunneling is an interim mechanism used to connect islands of multicast routers separated by vanilla unicast routers. In IP tunneling, multicast messages are wrapped in point-to-point unicast datagrams and sent. This was how the MBone achieved multicast operation globally.

For multicast-enabled routers, multicast packets from remote sources must be relayed, only being forwarded to the local network if there is a recipient for the multicast host group on the LAN. IGMP is used by multicast routers to learn of the existence of host group members on their directly attached subnets. It does so by sending IGMP queries and having IP hosts report their host group memberships. In other words, the router asks each media player, for example, "Which multicast groups are you listening to?" Each one responds with the list of "channels" being received. IGMP is carried by IP transport and there are only two kinds of packets: Host Membership Query and Host Membership Report.

To determine if any receivers on a local subnet belong to a multicast group, one multicast router per subnet periodically sends a hardware-generated multicast IGMP Host Membership Query to all IP end nodes on its LAN, asking them to report back on the host group memberships of their processes. This query is sent to the reserved "All Hosts Group" whose address is 224.0.0.1. A TTL of 1 is used to prevent propagation of the request to hosts outside the confines of the LAN. Every host that sends a report back also sends it on the all hosts group, so all group members see it. Thus, only one member need report membership for the router to continue sending multicast traffic to receivers.

When a media player, for example, wants to tune in to a particular multicast channel, it asks to join a host group. The hardware driver on the network interface card creates a multicast address mask for its packet-sniffing filter and an IGMP Host Membership Report, which is sent immediately. When the last host member of a group leaves the group, the router only finds out about it because it gets no reply to its Host Membership Queries. Thus, routers poll for host memberships and hosts that are members send their reports at random time intervals, if they have not seen a report by another member of the group. The random interval prevents multiple hosts from reporting all at once, which would cause network collisions.

Routers use IGMP updates to communicate host memberships to their neighboring routers. Thus group membership information is propagated throughout the entire Internet (eventually).

How does a multicast packet find a route from source to destination(s)? With unicast, the IP protocol contains the single IP address of the recipient, which includes information about its physical network location, including the subnet number and host number on that subnet. Routers periodically send information about the connected devices they can see on their particular subnet to other routers, so that data can be forwarded correctly according to the entries in each router's routing tables. Things are not so simple with multicast. A multicast address specifies a particular transmission session, rather than a specific physical destination. A naïve approach to routing multicast packets would be to send a separate copy of the data to each receiving subnet (as notified to the router by IGMP messages). However, this would be grossly inefficient, since many of the data streams would follow the same path throughout much of the network. Instead, the multicast router must know how to translate multicast addresses into host addresses. In multicast routing, a single router is selected from all its neighboring routers, by IGMP, to be the designated router for each physical network.

Designated routers construct a spanning tree that connects all members of an IP multicast group (Figure 2.27). A spanning tree has just enough connectivity so that there is only one path between every pair of routers and it is loop free. If each router knows which of its lines belong to the spanning tree, it can copy an incoming multicast datagram onto all of its outgoing branches, generating only the minimum number of copies. Messages are only replicated when the spanning tree branches, thus minimizing the number of copies of the messages that are transmitted through the network. This spanning tree must be dynamically updated at the designated router, since members are joining and leaving groups all the time. Branches that no longer have recipients must be pruned. The spanning algorithm used and how multicast routers interact depends on the objectives of the routing protocol.

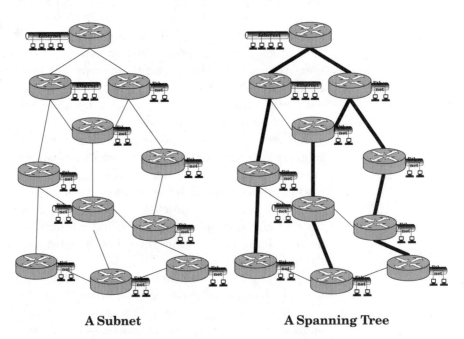

Figure 2.27
Subnet spanning tree in multicast routing.

A Subnet   **A Spanning Tree**

If almost all the hosts belong to the same group, then the group members are said to be densely distributed throughout the network. When this is the case, and bandwidth is plentiful, the best approach to routing is "dense mode," where the network is periodically flooded with multicast traffic to set up and maintain the spanning tree. Dense-mode routing protocols include Distance Vector Multicast Routing Protocol (DVMRP), Multicast Open Shortest Path First (MOSPF), and Protocol Independent Multicast—Dense Mode (PIM-DM).

When multicast group members are sparsely distributed throughout the network and bandwidth is not widely available, then "sparse mode" routing is the best choice. Sparse mode does not mean that there are fewer group members, just that they are more widely dispersed. Rather than flooding the network, which in this case would be unnecessarily wasteful of bandwidth, sparse-mode routing protocols rely on more selective techniques to set up and maintain spanning trees. Sparse-mode routing protocols include Core Based Trees (CBT) and Protocol Independent Multicast—Sparse Mode (PIM-SM).

Because of the complexity of establishing spanning trees in many-to-many applications and because even router software can contain bugs, multicast routing is prone to the danger of feedback loops (where packets recirculate on the network ad infinitum) and cascade failures (where the packets don't reach all the intended recipients). Such a bug once took out all the IP routers in New Zealand. To avoid these problems, there have been proposals for simplified routing for particular applications and scenarios. One is the Single Source Multicast proposal, which addresses the one-to-many application scenario.

IP multicast is primarily a routing protocol, rather than a transport protocol. Therefore, all the protocols relevant to streaming can be used with IP multicast, including UDP, RTP, RTCP, RTSP, and the QoS protocol RSVP. Note that the inclusion of QoS considerations adds complexity to the already complex task of establishing optimum multicast routings, since paths chosen in the spanning tree must also meet jitter, delay, and packet-loss specifications. Currently there is a great deal of research being done on reliable multicasting. There are no IETF standards for reliable multicast, but working groups are examining various commercial solutions, evaluating the technologies for possible standardization.

We have discussed how a member joins a multicast group, but how does a user or application learn about forthcoming IP multicast sessions? There are out-of-band methods for announcing sessions, like e-mail and Web sites, but there is also a need for mechanisms to:

- Announce sessions
- Determine temporary multicast addresses and ports for those sessions
- Issue invitations (for example, to conferences)
- Negotiate parameters such as memberships, rights data, media encoding, and encryption keys

- Add and delete members during the session
- Provide other control functions

The IETF has a working group charged with designing protocols for the management and coordination of multiple sessions and their multiple users, in multiple media (for example, audio and video). The working group is called the Multiparty Multimedia Session Control Working Group (known as mmusic). It has defined a number of draft protocols, such as Session Description Protocol (SDP), Session Directory Announcement Protocol (SDAP), Session Announcement Protocol (SAP), Simple Conference Control Protocol (SCCP), and Session Initiation Protocol (SIP). None is yet standardized.

Clearly, multicasting is an important technique for many streaming media applications, but it remains a work in progress, which has not yet been integrated into many production routers.

Audio and Video Cleaning

Professional audio and video producers have developed techniques, over decades, for improving the sound and appearance of audio and video respectively. Sometimes called "sweetening," it can be seen as a method of codec steering, in a streaming media context. If the producer emphasizes some aspects of a program, while de-emphasizing others, through judicious use of color correctors, graphic equalizers, and other signal processors, the compression codec, whose job it is to throw away information you won't mind losing, is presented with signals that have already been graded to eliminate unwanted signal artifacts. In this case, the codec can use its bandwidth budget to encode aspects of the sound and pictures that the producer wants to be noticed.

A full catalog of the possible audio and video processing techniques that may yield better encoded and compressed streaming media is beyond the scope of this book, but these techniques are in widespread and daily use at the best streaming media encoding labs. (Some of the more popular techniques are introduced in Appendix C.) Grass Valley Group's Aqua streaming media encoder includes many signal processing options, which enable skilled operators to present the encoding software with the best possible signals. Popular signal processing techniques include deinterlacing, where the television scan is converted into successive frames of video, which register exactly on one another, rather than moving by the video line typical of interlaced transmissions. Progressively scanned images have less motion jitter, so encoders can do a better job of predictive coding from frame to frame. Another technique to remove motion jitter is inverse telecine. When a film is made, it is typically shot at 24 frames per second. To show that film on television, which has a rate of 25 or 30 frames per second, depending on where you live, the machine transferring the film to video inserts duplicate frames, from time to time, in a process known as "pull-down." (In fact, for 25-fps television, the 24-fps film is often just played back at the faster 25 fps rate, so that everything moves more quickly and the voices all sound higher. The movie's running time is also shortened.) The result is motion jitter, since something that was moving slowly across the screen now takes momentary pauses, as a result of the duplicated frames. Before encoding to streaming media formats, these duplicate frames must be removed. When motion jitter is minimized, the encoding software can spend more of its bits encoding dynamic features of the picture, rather than poorly rendering each frame. The fewer pixels that change from frame to frame, the better the compression algorithm can do its job of reducing the data rate, without changing the look of the images.

Skilled audio and video producers can use audio equalization and color correction techniques to enhance the material, so that the colors are more vibrant than they could ever be in real life and the sounds are crisper and more clearly defined than they were when recorded. Hollywood has used these techniques to enhance the audience's experience of feature films for decades. With streaming media, ordinary content producers can use desktop tools to sweeten their own productions with the same facility as big Hollywood studios do. With skill, independent content producers can create streaming media works that rival the very best movies in terms of look and sound.

Synchronized Multimedia

There are some simple but effective solutions for delivering a synchronized multimedia experience to consumers today. While not as powerful as MPEG-4's BIFS, they are nevertheless illustrative of the way forward in streaming media interactivity.

Synchronized multimedia allows applications such as closed captions tied to the playback of video, additional information about a song and artist to be displayed while the music plays, stock tickers to be burnt into the playing video, etc. In essence, the appearance of graphical, text, hyperlink, or related audio and video material is timed against another multimedia element.

When synchronized multimedia was first proposed for streaming media, there were competing formats for describing the synchronized elements. Microsoft promoted HTML+Time, which was related to Dynamic HTML, and RealNetworks promoted SMIL (Synchronized Multimedia Integration Language). Eventually the two approaches coalesced into the IETF-sponsored SMIL 2.0.

In digital television production, the titles and graphics are overlaid on the video during the production phase. They are actually "burnt into" the video program material. If you think the stock ticker at the bottom of the screen on CNBC is annoying, you either have to change channels and watch something else, or put up with it. If you happen to speak Chinese, the English-language titles cannot be translated to something in your native tongue. In fact, when international subtitles are added to the program, they often compete for screen real estate with the burnt-in titles. Also, with television, if you wish to zoom in to the sidebar text that accompanies the video so that you can read it, you can't. It's static.

With synchronized multimedia, however, if the player software permits, you may turn elements on and off, or focus on the video or text or any other HTML panel that is synchronized to the other multimedia. Hyperlinks can be embedded as hotspots on the video, so that clicking on part of the picture launches a Web page alongside the video, for example. SMIL can even be used to launch other audio tracks or even a second video window that plays in tandem with the first one. The possibilities are almost endless. What limits the usefulness of SMIL in practice is the lack of bandwidth to the player, the lack of support for resynchronization of packets delivered from a variety of distributed servers, and the primitive support for SMIL tricks in the player software. Authoring tools are available, but are not yet widely used. Besides dedicated SMIL editing tools like GriNS, which we talked about when we discussed RealNetworks' tools, there are tools like Microsoft PowerPoint Producer for PowerPoint 2002 and Real Slideshow, which actually use SMIL to create the synchronized multimedia presentations.

Other things that can be used to create synchronized multimedia presentations, which are to a greater or lesser degree alternative to SMIL, are Apple's QuickTime and Microsoft's Synchronized Accessible Media Interchange (SAMI). SAMI was specifically created to simplify the creation of synchronized captions for multimedia content.

As ever, the digital television broadcast people invented their own technologies for adding synchronized multimedia elements at the set-top box. With these schemes, multimedia elements are hidden in the broadcast video data and rendered by the set-top box. This, of course, imposes limits on how often content can be updated, how frequently content must be redelivered to accommodate people who have just tuned in, how much pre-roll time is needed to deliver the content to the set-top box before it is rendered to the screen, how complex the synchronized multimedia can be before the rendering engine runs out of processing power or memory, how individualized the content may be and so on. Communications and interactive requests from the viewer back to the source require use of a phone line, so response times to user interactivity are painfully slow. There are a number of open and proprietary standards for delivering an interactive synchronized multimedia experience to the set-top box using broadcast networks. Popular ones are OpenTV, PowerTV, RespondTV, ATVEF (Advanced Television Enhancement Forum), MHP (Multimedia Home Platform), and MHEG-5 (a standard produced by the Multimedia and Hypermedia Experts Group). Some of these standards for synchronized multimedia also allow delivery via IP as well as broadcast networks.

Internet-delivered synchronized multimedia has the potential to be better synchronized, richer, and more responsive to user interactivity, since broadband networks can ultimately provide much more bandwidth in which to deliver multimedia elements (so that they arrive sooner); elements are downloadable on demand, not on a timed carousel basis; there is usually more computing power and memory available on a PC to render the material for the viewer; and the back channel is implicit in the IP connection.

If I were pressed to choose a favorite technology, I think the flexibility and controlled quality of service of MPEG-4's BIFS with DMIF would attract me most.

Peer-to-Peer Replication

If every streaming media receiver could also act as a streaming media cache and proxy, then content could be served from user one to user two, without resort to a stream delivered from either the origin server or an edge cache colocated at a POP. In other words, every consumer could become a streaming server as well. The edge of the network would, in essence, move to the very fringe of the network: the end-users' machines.

Peer-to-peer (P2P) replication by downloading was made popular by applications like Napster. Unfortunately, although P2P was tremendously convenient for end-users, since they could find any number of sources for media content they wished to download, it was a disaster for content owners, since they lost control over their assets, with consumers freely and illegally copying their precious content without paying royalties. Eventually, content owners used the law to curtail the illegal peer-to-peer copying of digital media assets.

However, if strong encryption and rights management solutions are successfully developed, there is little to prevent content from being streamed once from the origin server, and then subsequently streamed from one consumer to another. There are a number of companies actively developing peer-to-peer replication solutions today, including vTrails, Groove, Kontiki, and Allcast. The potential exists for all the quality-of-service advantages and bandwidth and storage cost reduction advantages that have driven the growth of edge caching to extend all the way to the individual consumer's machine, provided that broadband networks become the norm.

Rights Management

We have touched on rights management in our discussions of Microsoft's Windows Media Technologies and RealNetworks' media commerce solutions. These are proprietary answers to the rights management problem. Apple's QuickTime does not at present provide a comprehensive rights management solution. The problem with all the proprietary systems is that they don't readily interoperate and they don't protect the rights of everybody involved in the media transaction, instead providing sticks with which content producers may beat consumers, should they choose to do so.

As we will discuss in a future chapter, there are efforts to standardize digital rights management, and the leading one seems to be the MPEG-21 initiative. The proprietary digital rights management systems proposed work by encrypting or otherwise locking streaming media data, so that it cannot be played or copied until you obtain a digital key. This means that the player has to be tamper resistant. It also means that the transaction required to play a piece of streaming media now involves more parties than the client and the server. In addition, there is more computational overhead on the player software, since it must decrypt on the fly, as well as render the content for playback.

Widevine Technologies Inc. is taking an interesting approach to digital rights management. This involves encrypting UDP and TCP/IP packets after they leave the streaming media server and decrypting the packets between the network stack and the player software application on the client side. This scheme has the advantage of working with any stream that uses UDP or TCP transport (i.e., all of them). Other vendors, such as SecureMedia and PassEdge, have similar products. The company calls this "on-the-fly" encryption, in contrast to Microsoft's approach of encrypting the media files on the media server, which Widevine calls "pre-encryption."

Pre-encryption, such as Windows Media DRM, secures media files sitting on a server from unauthorized theft. Someone who hacks into the server and copies the media file can't play it without a key. However, pre-encryption is not suitable for live broadcasts, since the media can't be encrypted until the entire file is complete. Pre-encryption also implies that each file can be decrypted with a single key, meaning that copies of the file broadcast to users are identical. If the key becomes known, anyone who has the file can decrypt it. Keys are not usually released illegally by brute force hacking, since strong encryption techniques make this sort of cracking uneconomical. Rather, keys usually find their way into the public domain when an employee or trusted party leaks one or more keys. This could obviously be disastrous if all a particular content owner's media assets were encrypted using a single key.

Pre-encryption is fine as far as it goes, but a better solution is to add on-the-fly encryption as well. In this scenario, the client player negotiates a secure channel (using public-key exchange) at the beginning of each streaming transmission. The file's actual key can be transmitted over this secure channel. This way, the key is stored only in the computer's working memory, making it much harder to determine (although a dedicated hacker could find it). However, because each stream uses a new key, even if one stream's key is obtained, the key is useless to others who might want to view the stream, or even for a later instance of the same stream. Thus, with two levels of encryption, using both pre-encryption and on-the-fly encryption, the media file resident on the media server would be protected against illegal copying, and stream-unique keys would individually protect the streams.

Unfortunately, the security of Microsoft's Windows Media DRM is highly dependent on the assumption that the application on the end-user's machine that claims to be Windows Media Player is, in fact, the trusted application and not an impostor designed specifically to defeat the DRM security measures. Since the Windows Media Player includes the code to manage the encrypted licenses, it is subject to reverse-engineering attacks. The downloadable nature of the media files, combined with the static encryption needed to make the client license model viable, makes it simple to mount a large-scale brute-force attack. So, Windows Media DRM is not only a flawed streaming security system, it is also a flawed pre-encryption security system.

It is instructive, at this point, to consider the ways in which a stream can be stolen. There are three points during the content streaming process at which media can become vulnerable. The first is at the producer's facility, the second is as the data travels across the public Internet, and the third is on the premises, wiring, and computer equipment of the consumer of the streaming media content.

At the production facility, it is a sad but true fact that employees may steal the content and later distribute it over the Internet. Employees at the streaming media serving facility also have access to the servers and can make illegal copies of media assets. Digital watermarking can be used at various stages of production to allow any leaks to be traced back to their source. However, physical security and use of only highly trusted and closely monitored personnel, with legally enforceable security agreements, are commonly the best protections. All of these same issues are currently faced in DVD duplication plants and solved in similar ways.

Any organization which carries data between a media server and the consumer of the content, could, in theory, intercept the data as it streams through its facility on the way to the client. In practice, this is extremely difficult, simply because of the sheer volume of data that constantly passes through even a modest sized ISP. However, strong encryption of the data payload is the obvious solution to the issue.

Most viewers watch streaming media content on computers designed to make information easy to share, not to prevent sharing. Some viewers will be highly motivated to remove any measures broadcasters put in place to protect their content. This third "territory of vulnerability" is where a multitude of techniques can be used to steal streaming media content. It is important to remember that, while the likelihood of any particular stream's being pirated by a randomly selected user may be very small, the fact that the media is digital means that if even one viewer is able to gain control of the content, he or she can make and distribute perfect digital copies to anyone in the world, over the Internet. The goal of streaming media security systems must be to make the cost and effort required to steal content greater than the value of the content. In fact, some content's value declines rapidly with time (news reports, for example), so another goal is to make the time required to breach the security system sufficiently long that the content, when finally cracked, is effectively valueless.

Let us now turn our attention to ways in which an individual with a computer can steal streaming media. There are eight popular techniques:

- Client hacking
- Client spoofing
- IP stack hacking
- Packet sniffing
- Proxy/firewall/router compromising
- Screen scraping (with speaker sucking)
- Analog copying
- Man-in-the-middle cryptographic attacks

The simplest way to steal streaming media, called *client hacking*, is to defeat the security measures that prevent the streaming media player from saving the stream to disk. All the hacker needs to do is find the locations in the player's software code that test whether or not the player is allowed to save the file and insert patches to skip the test. Software pirates have long used similar techniques to disable serial-number validation code in copy-protected software. Windows Media Player and QuickTime Player do not have built-in code to save streams to disk, so adding the capability is not a simple matter of patching a few bytes of code. These players must be attacked in other ways. The only feasible solution to the problem of hacking the client is client-integrity checking, where the software is repeatedly checked to ensure that the checking routines are unmodified. This checking code can be inserted hundreds of times in nonfunctional routines to amuse the more dedicated hacker. Of course, the code that checks the integrity of the checks could also be hacked, so this too must be included several times in the software. In fact, infinite levels of nesting, where the checker checks the checking checker, and so on, could be employed.

The Digital Millennium Copyright Act makes it illegal to circumvent security technology intended to protect digital media. Hence, another approach is to take legal action against individuals who remove protection features from client software. Of course, identifying these individuals is often difficult or impossible and the cost of the legal case may be prohibitive.

Another simple way to steal streaming media is to write software that pretends to be a standard streaming media player, but has, as its main or secondary purpose, the saving of streaming media to disk, for later playback and redistribution. Spoofing clients are already available for RealMedia and Windows Media. The software looks, to the streaming media server, like a regular player, but the spoofing client is designed to write the streamed data to disk. Streambox VCR was a product that allowed any RealMedia stream to be saved and was the subject of a lawsuit by RealNetworks. The suit was settled out of court and the feature was removed from the software, but older versions of the software still circulate. Another utility called ASFRecorder allows Windows Media streams to be saved to disk. To date, Microsoft has taken no action to attempt to remove ASFRecorder from circulation. There is currently little demand for QuickTime spoofing clients, since QuickTime security is almost nonexistent.

Encryption provides a measure of protection against client spoofing, since it adds a significant engineering challenge to authors of spoofing clients to implement the decryption software. The software must now include key exchange software and encryption routines compatible with the originals, which must be written or else extracted from the target player and used in the new software. This represents more work than most casual hackers are willing to expend, although it isn't completely impossible. On-the-fly encryption systems need to take steps to ensure that the client requesting the data is a known, legitimate client. This can be done by inspecting various characteristics of the calling application and the runtime environment of the computer, or even by deliberately forcing errors and expecting signature errors as a response. Since the test that the on-the-fly decrypter uses to validate the client could eventually be reverse engineered and defeated by the spoofing client, the on-the-fly encryption system should allow tests to be updated, to keep one step ahead of the hackers.

The term *IP stack hacking* refers to attacks on the part of the operating system that deals with the TCP/IP and other Internet protocols. In most operating systems it is possible for an application to insert a "callback" into the IP stack, allowing a bit of code, external to the stack, to be called each time a packet is sent or received. This allows an external program to copy data from streams as they pass through the IP stack, saving the data to disk as it is collected. Encryption is the first line of defense against extracting streaming data from the IP stack. It is theoretically possible to defeat on-the-fly decrypters simply by extracting the data after the IP data in the stack has been decrypted, but most on-the-fly decrypters remove themselves from the calling chain and reinsert themselves at the end as the stream begins. They can also be designed to detect any irregularities in the IP stack and shut down the stream if the IP stack is compromised.

Another technique for stealing streaming media is called *packet sniffing*. Here the IP stack is hacked outside the client machine. The stream is monitored as it moves across the LAN to the client and the packets are snatched as they pass by a second machine. Packet-sniffing programs are widely available for UNIX, Windows, and the Mac OS. Once the packets are captured to a file, you can massage the data to recover the streaming media payload. The best defense against this is strong encryption that renders the captured stream useless without the appropriate key.

Firewalls, proxies, and routers can be compromised in order to steal streaming media. For example, many proxy servers already contain code to cache Internet documents locally. Such proxies might be capable of being configured to save TCP/IP streaming video and audio to disk, since they are already saving other types of files for the local cache. If the proxy (or firewall or router) is software based, it might be possible to obtain the source code (assuming it is "open source," as many programs are today) and design the software so that it saves files to the bridge machine's hard disk. Once stored there, of course, the file can be copied anywhere. It is important to note that this need not take place in an office network environment where the use of a proxy is required. A hacker might set up a compromised proxy or network address translation router at home, specifically to capture streaming media as it passes through on its way to a second box. One way to prevent this theft technique is not to allow streaming at all via TCP/IP, but this leaves streams unable to penetrate firewalls. Encryption is a far better deterrent, since a well-designed encryption system allows streams to pass through firewalls, routers, and proxies unhindered while providing security against capturing a usable version of the stream.

Screen scraping is the vernacular term used to describe the grabbing of uncompressed video as it is written to the screen, or immediately after it has been written. This can be done either by intercepting the actual API calls that write the uncompressed bitmap to the screen, or by running a background task that periodically (i.e., every frame) interrupts the computer to copy the particular segment of display memory, which the player is using to write the uncompressed video data to disk. Because most streaming media players write directly to display memory, for speed reasons, it can be difficult to get good-quality screen scrapes. However, we can expect screen-scraping techniques to be refined as other security loopholes get closed. A number of commercial utilities are already available to capture a screen session to a video file. There are only a few ways to defeat screen scrapers (and their audio brothers, *speaker suckers*). One way is for player software to deny any other processes any CPU time while running. In most modern operating systems, however, application-level software cannot control CPU utilization. Another method is to design hardware so that the CPU does not have access to display memory. DVD playback cards work this way. The DVD stream is decoded and written to a display memory using another processor entirely. The host CPU never sees the data and the decoded video is "overlaid" on the computer's normal video display. However, for streaming, this is a totally unrealistic solution, since virtually all general-purpose computer display cards are designed for reading as well as writing. Limited protection can be achieved if the player watches for known screen-scraping utilities, halting streaming if any are detected. Naturally the player would have to contain protection against the detection software's being defeated and allow for easy updates as new screen-scraping utilities are released.

As a last resort, the determined stream thief can simply connect a good-quality tape deck to a device that converts the computer's display output to a standard video format (called a scan converter) and do the same with the audio, then simply make a recording. If the quality of the stream is high to begin with, the quality lost in making this kind of copy is negligible. DVD players guard against this by including Macrovision copy-protection circuitry on their video outputs. With this technology, the picture can be displayed by a television set, but narrow pulses are inserted into the video signal that drive the usual VCR's automatic gain control circuitry bananas. Using signal processing to remove the pulses can defeat Macrovision copy protection. Another method is to disable the automatic gain control of consumer video machines by tinkering with the electronics, or else by using a professional video deck. There is no chance of this form of copy protection's appearing on general purpose PCs. Consumer resistance would be too great, since users would have no reason to buy the new video cards, and there is a huge installed base of unprotected video cards. In the final analysis, analog copies cannot be prevented, since in the extreme, the determined pirate could still get acceptable results from filming the video screen with a good-quality camera and placing microphones in front of the speakers.

The final method of stealing streaming media in the home is the "man-in-the-middle" technique. Any streaming media system that uses cryptography can be attacked this way. In fact, any secure sockets layer (SSL) Internet connection can be attacked this way. In short, an eavesdropper inserts himself or herself into the channel between the client and the server, negotiates keys separately for talking to the client and the server, and pretends to be the client when talking to the server and to be the server when talking to the client. Since the eavesdropper was "in the middle" during the exchange of keys, he has both sets of keys and can easily decrypt any messages sent by either side—then re-encrypt them for transmission to the intended recipient, so that neither side is the wiser. In a traditional key-exchange scenario, the eavesdropper is a third party unknown to either of the two intended parties to the communication. However, with streaming media, the eavesdropper may be the same person as the consumer of the stream. For example, there might be a compromised proxy attack, described above, with the addition of the "man-in-the-middle" technique. Public-key encryption relies on a certain level of trust between the communicating parties. However, trust may be entirely absent if the consumer is actively trying to steal the stream being provided by the server. The only protection against this attack (and it isn't much) is the level of technical sophistication required to mount it.

The three most popular streaming media systems today are RealMedia, Windows Media, and QuickTime. All have serious security flaws. Most are vulnerable to nearly all the attacks described above. For example, RealMedia allows the author of a stream to specify that the RealPlayer Plus software should not allow the stream to be saved to disk. The more popular RealPlayer Basic does not have the capability to write to disk at all. Essentially, then, RealMedia security boils down to a single bit in the stream header. Little wonder that the company has embarked on initiatives to create viable rights management to enable digital media commerce with its latest offerings.

Windows Media offers the most robust rights management functionality of any current streaming media architecture. Files can be protected by strong encryption. Each user who wants to use a file receives a key that works only on his or her computer. However, we have already noted the vulnerabilities of pre-encryption systems and their inability to protect live streams.

QuickTime security, at the time of writing, is almost entirely absent. With Fast Start, the pseudo-streaming file delivered by HTTP can have embedded header information that instructs the QuickTime player not to allow the user to save the file to disk for later viewing, but this constraint is easily circumvented. To play the Fast Start file at all, the entire file must eventually be downloaded into the Web browser's cache folder. It takes seconds to look for a large, recent file in the cache folder with a .MOV extension and copy it to a more permanent location. Whereas there are no known spoofing clients available for QuickTime, this is only because of low demand. As more and more valuable content appears in QuickTime format, there will be greater incentive for these spoofing clients to be written, and QuickTime provides no technical barriers whatsoever to the creation of such software.

Of course, much of the streaming content available on the Internet will have only limited value. It is entirely possible that media properties of limited value will be offered under the honor system, analogous to the shareware movement in computer software. This invitation to pay what the consumer thinks the media is worth has proven quite successful in the software application arena and could succeed for certain types of streaming media content. This system can make content producers very wealthy, yet requires no rights management solution at all.

The unacknowledged truth about digital media is that the determined thief can steal it, no matter what measures are taken to prevent theft. This is as true for DVD movies, shrink-wrapped software applications, and digital television as it is for Internet streaming media. What the industry must do is to make theft difficult, costly, and slow, through pre- and on-the-fly encryption, using tamper-resistant player hardware. This level of technical protection must be achieved without alienating honest consumers. However, even with these measures in place, imperfect, but acceptable, copies of the digital media cannot be prevented. As with the war on software virus writers, resort to legal enforcement is needed, as is public education. Stealing streaming media from people who spend much time and money producing it is akin to killing the goose that laid the golden egg. The most sought-after media is the most entertaining and best-made media: precisely the sort of digital content that will disappear entirely, in the future, if it is stolen widely and often.

Other Things That Go "Stream" In the Night

Other ways of streaming digital media are already in widespread use. However, these tend to be related to digital television and therefore conform to the "we play, you watch" broadcasting model. Digital Video Broadcast streams MPEG-2 encoded digital video from point to multipoint. DVD players stream MPEG-2 encoded video from the surface of the disk to the player's video output. PSIP (Program and System Information Protocol) streams digital metadata about the accompanying video and audio data, which allows set-top boxes to construct interactive electronic program guides to aid channel surfing. Extraordinarily, prior to PSIP, television channels were just broadcast, with no metadata description other than what could be embedded in the program stream itself. ATVEF (Advanced Television Enhancement Forum) adds interactive features to digital television, providing a synchronized multimedia experience that emulates on a television set what is possible with Web content. There are many other proprietary schemes for adding interactivity and conditional access to television signals, but all these systems run into the sometimes-severe constraints and limitations of the asymmetric nature of broadcast content.

While acknowledging the outstanding technical achievements of those involved with these forms of streaming media, this book discusses streaming digital media that is somewhat free from the constraints of closed-access systems such as broadcast television. Rather, we have focused on a discussion of technologies that rival television technology and in many cases exceed its capabilities and freedoms. There are many fine sources of information on digital video technology. An in-depth discussion here could add little to the literature. The reader is referred to those other works when considering digital media streaming in a broadcast television context.

Why Was Streaming Media Invented?

The short answer to why streaming media was invented is that it was recognized that people learn to observe visually and audibly long before they can talk, read, or write. People are good at assimilating information when it is presented as video and audio. Technologies to do this existed, but the Internet, a great system for distributing digital media, was invented. People wanted to find a way to get audio and video across this giant, resilient, fast-growing, and ubiquitous network. If a company found a way to get audio and video content across the Internet, riches should follow, either by controlling advertising to this large group of viewers, by controlling access to content, by selling the machinery to make video and audio available on the Internet, or by stealing viewers from other audio and video technologies.

It was realized early on that consumers wanted an experience like television, which already was a satisfactory, if limited, way of getting audio and video. That meant the pictures and sound had to be as good as television and flipping from program to program had to be nearly instantaneous. This meant that users could not be expected to download a video or audio asset before playing it. They had to be able to view as soon as it was selected. In other words, the media had to stream. Unfortunately, the bandwidth available on the public Internet is a fraction of what is required to render raw video and audio in acceptable quality (i.e., like television). This limitation led designers of streaming media technology to concentrate on data reduction and compression techniques and to focus early streaming media offerings on corporate applications, where the streaming could be contained within a single LAN. If this could be achieved, there would be enough bandwidth to render decent pictures and sound.

Real Networks (formerly Progressive Networks, named after the term for streaming, "progressive download") was one of the first companies to offer a publicly available streaming media solution. It initially concentrated on delivering streaming audio, but eventually added video that was so heavily compressed, it was possible to use it on the low-bandwidth public Internet. Compression techniques were under development at the same time, under the auspices of the JPEG (Joint Picture Experts Group) and the MPEG (Motion Picture Experts Group), since getting digital audio and video across other networks than the Internet presented bandwidth and cost problems that professional media industries needed to solve just as urgently. In some sense the streaming media industry initially consisted of a number of disparate industries trying to get audio and video through digital networks more cost effectively, and developments therefore leaked from one industry to another.

In the early days, companies like Xing and Streamworks presented alternative methods of streaming media across the Internet, but they were eventually absorbed or defeated by vendors who aggressively marketed and promoted their free streaming media players to the Internet's growing population of users. The streaming media companies that still compete for market share on the Internet tend to be those that established early dominance in terms of the number of players installed. Various industry claims put this figure in the hundreds of millions (RealNetworks, for example, claims over 200 million of their players).

In this chapter, we discuss some of the early innovators in streaming media, not necessarily in chronological order. Our aim is to answer the question of why streaming media was invented.

Corporate Communications

One of the applications initially targeted by streaming media vendors was corporate communications. There are many situations in corporate life today where a video is played in order to communicate with or to train staff. It was envisaged that real-time video communications could be added to this mix and the entire thing delivered as streaming video and audio, across a corporate IP network. There is no doubt that corporate communications greatly benefit from being able to routinely include audio and video content. However this was not so widely recognized or even cost effective when the first companies began addressing their streaming media products to corporate clients.

Before there was the idea of streaming media, some companies had already experimented with delivering digital media files in entirety, before playing them. Microsoft's first version of a streaming media system, called NetShow, was intended to show lowish-bandwidth multimedia presentations to individual desktops on a corporate LAN, either on a broadcast schedule or on demand. Microsoft's own multimedia had existed for some time before this, but it did not have powerful compression, so a greater proportion of the corporations IT resources was used to play a video clip than most IT managers were happy with. Microsoft had Video for Windows and later AVI (Audio Video Interleaved) files to deliver the audio and video, in a single file, for playback on Microsoft Windows desktops. At the time that these technologies were introduced, few corporate PCs had the memory, processor power, graphics, and sound cards needed to render the video and audio for playback. Corporations resisted the higher specification required. Video and audio was not considered compelling enough to add more network bandwidth, more powerful servers, and more expensive hardware to every desktop. Indeed, it wasn't until the advent of graphics cards that connected to the PC via the AGP (Accelerated Graphics Port) bus that decent video rendering to a desktop computer screen was even possible. Similarly the Creative Labs SoundBlaster led the way in PC audio. Before that, there was little standardization in PC audio and little audible quality.

One of the early video compression schemes made use of a compression scheme developed to compress still images for display on computer networks. JPEG compression allowed detailed photographs to be delivered in a few kilobytes, well within the capabilities of even the most rudimentary network and desktop machine. For video, all you had to do was compress each frame of the video as if it were a still. Thus, motion JPEG or M-JPEG was born. In machines with enough power, you could decompress the video, frame by frame and manage to display that to a desktop computer screen at a regular frame rate. This was not simple, since each frame compressed to a different size. Thus, to play video reliably, it was necessary to be able to cope with several worst-case file sizes in succession. Elaborate decompression buffering schemes were developed to solve the problem. For M-JPEG, most desktop PCs had insufficient processing power to decompress images at an acceptable rate to create a video presentation. A company called C-Cube and IBM produced silicon coprocessors that could do the decompression on the graphics card, prior to display. These parts were initially targeted at the broadcast television industry, as they were already interested in compressing digital video. However, some desktop machines were fitted with silicon M-JPEG decompressors. To say that they were extremely rare in the average corporate office environment would be a gross understatement. Corporate users struggled with jerky, tiny video for years.

WANs Are Cheaper than Airlines

One of the killer applications that has motivated the development of streaming media technologies, like video compression, is video conferencing. Virtual meetings ought to be cheaper than dragging people to one place at one time, to discuss matters. WANs are cheaper, in the long run, than airline tickets and hotel rooms. Virtual meeting software, like Microsoft's NetMeeting, for example, can allow users connected to the right server to see a number of their colleagues on their screen, in real time, and carry on live conversations, share and update documents, and organize follow-up meetings with a shared electronic diary. The problem with these solutions is that the connection process is sometimes fraught with difficulty, and everybody needs access to the same resources—a camera and microphone, plus the server that hosts the virtual meeting. In practice, NetMeetings can be hard to get working.

Another vendor of virtual meeting technology is WebEx. Cybermeetings hosted by WebEx provide a cost-effective and timely way to maintain a personal touch in a business relationship while enabling customers to tap into the expertise they need. Not only can a customer be served without the expense of an in-person visit, but electronic meeting technology improves the availability of highly skilled people.

The eventual dream is ubiquitous teleworking, where people can live wherever they wish, yet participate in the workings of a virtual corporation. The head office is then everywhere.

Distance Learning and Interactive Learning

Interactive learning is a proven technology. Producers of interactive tutorials have used CD-ROM and tools like Macromedia Director for years to produce training packages that are compelling, engaging, interesting, and informative. What streaming media makes possible is the ability to deliver those training materials over an IP network, instead of on physical disks. The beauty of this method of distribution is that training materials can be updated more readily, making it possible for online training to be more current and relevant than materials that must be frozen in time and pushed through a CD-production and distribution process.

There has been strong demand from companies supplying interactive training materials for streaming media technology innovations. Various educational organizations, some already conducting distance-learning programs, have begun to use streaming to augment their online offerings. The Open University in the UK is an example of a learning institution that could make good use of streaming media by repurposing its extensive catalog of video learning materials developed for broadcast.

Many schools are now offering undergraduates the chance to sign up as students from anywhere in the world and view lectures and tutorials online. Streaming technology makes it possible to host live lectures and get questions and feedback from the students watching. There are difficulties with the economics of providing broadband streams to distant students. Online learning organizations need to supply enough server bandwidth and transmission bandwidth to support their far-flung student body. However, there are already learning organizations making money with streaming distance learning applications. As the cost of bandwidth falls, the economics will improve.

IPTV

Cisco's IP/TV delivers live, high-quality video content to desktops, classrooms, and meeting rooms over today's enterprise networks. It enables organizations to provide high-impact instruction, communications, seminars, and more, directly to the desktops of employees, partners, and students.

Originally developed by startup company Precept and bought by Cisco in April 1998, the system was targeted at streaming applications within an enterprise. This strategy allowed the problems of the lack of available bandwidth on the public Internet to be sidestepped. The basics of the system include a streaming media server (sold as a turnkey system, incorporating the necessary hardware and software) and a media player application that is loaded onto every desktop machine. The server allows video to be captured and encoded into MPEG-1 video streams. There is another piece of server-side software that manages the media available and routes it to various multicast IP channels for desktop consumption.

Cisco, being primarily a hardware vendor, did not unbundle the server software from the hardware, so corporations that did not wish to invest in a Cisco server passed on IP/TV. The player installation on every desktop presented some obstacles for adoption as well. Since it was based on MPEG-1 and an early generation of product, there were initial stability and smoothness-of-playback issues. However, the product is still sold and supported and has been through several revisions. It is now a part of Cisco's Content Networking Architecture, along with acquisitions from companies like Sightpath.

Some IT managers are still concerned with the amount of network load that a system like IP/TV places on their corporate networks. There are also concerns about whether or not IP/TV can bring a network down, if it has problems. Multicast traffic is something that many IT managers will only have tried for the first time with an installation of IP/TV, so they have concerns about the learning curve required, multicast compatibility with their routers, and being able to support the technology while maintaining the network's integrity for existing users. Since they are people who generally prefer safety to new network services, IT managers do not like to install anything that they perceive risks denial of service to a class of network users who don't care about video on the desktop, or that could topple the entire network. These suspicions and fears are enough to keep IT departments from embracing systems like IP/TV, whether or not their fears are founded. All enterprise applications of streaming media face these obstacles.

Microsoft Video for Windows

One of Microsoft's first efforts at streaming media was Video for Windows. It didn't stream, as such, but it was a piece of software that allowed video files to be played as moving video on the desktop. Shipped in 1991, Video for Windows was part of the company's multimedia services first offered on the Windows 3.1 platform. IBM and Microsoft were looking for ways to enrich the desktop user experience, and the incorporation of multimedia elements in their PC operating system products seemed a way to do that.

At the time, Video for Windows required a 386SX with a VGA card, plus a sound card and large capacity hard disk to work. In reality, a 386DX was required to play video at anything approaching 25 frames per second. The system had rudimentary video compression plug-ins and introduced the Audio Video Interleave (AVI) file format, still in use today. The machine required to handle Video for Windows was not cheap, in its day. Multimedia PCs were also not particularly common in offices. Nevertheless, it did blaze the trail for desktop video. Video for Windows worked well enough to stimulate other developments.

Microsoft NetShow

NetShow was first shipped with Microsoft's Internet Information Services (IIS) version 3.0. This dates it to about 1997, for it appeared on Service Pack 3 of Windows NT 4.0. It introduced the Advanced Streaming Format (ASF) and was the first server that Microsoft made that could stream files to the desktop. Microsoft entered the streaming scene well after RealNetworks had gained a majority of the multimedia on-demand market share.

By the time NetShow version 3.0 shipped, in 1998, this free server not only threatened the competition, who were pricing their servers on the basis of the number of streams they could serve, but it could also play content in the competition's formats, though not their most up-to-date formats. Microsoft also fought aggressively on the basis of audio and video quality, matching that of their competitors. Microsoft's product was codec-agnostic, supporting a plug-in architecture so that their own and other companies' codecs for audio and video compression could be used for encoding and playback.

Microsoft's advantage was that they could bundle their streaming media server with their NT operating system and their player with Internet Explorer, their Web browser. This became a bone of contention and was part of some celebrated legal activity against the company. Competitors claimed that it undermined their business, particularly since, at one point, the Windows Media player, when installed, associated itself as the default player for the competitors' file extensions. This meant that the player software would play streams in the competitor's format, instead of playing them with the competitor's player.

To encode video for NetShow (and NetShow Theatre, which was targeted at higher-bandwidth video delivery on enterprise LANs), there were two manual steps involved. First the video had to be encoded (meaning compressed) and then the ASF file created, using an ASF Indexer.

Having, in some senses, pioneered desktop multimedia on the Windows platform, Microsoft found itself in a race to be a significant player in streaming media to the desktop. NetShow was its response. NetShow later became the basis for Windows Media Technologies.

Real Audio

One of the first companies to stream media to desktops over a network was Progressive Networks, which became RealNetworks. As an aggressive, fast-moving startup company, it stole a march on other more-established companies already working with digital multimedia, releasing RealAudio Server 2.0 in 1996 (I was unable to find information about version 1.0), running on the Windows NT 3.5.1 operating system and a host of others, some of which no longer exist. The initial product streamed audio at an astonishingly low bandwidth. A stream of audio could be downloaded in real time using what was then considered to be a fast modem. Users could hear audio on their desktop machines when connected with either a 14.4 or 2.8 kilobit-per-second modem. Audio could be broadcast live to the Internet, as on a radio station, or else users could get audio clips on demand.

Later in 1996, the company followed up with the release of RealAudio Server 3.0, which simplified support for the live broadcasting of audio. It also had features to recover lost data packets by requesting that they be resent. The server added another eight codec choices, presenting CD-quality audio on ISDN and LAN connections.

In 1997, the company released RealServer 4.0, which was its first product capable of delivering streaming video. It also supported unicast and multicast modes, using UDP (User Datagram Protocol), TCP (Transmission Control Protocol), or HTTP (Hyper Text Transport Protocol) protocols. UDP, upon which the earlier products were based, does not provide a service to divide a message into packets (datagrams) and reassemble them at the other end, like TCP does. It is used when applications want to save processing time, because they have very small data units to exchange. HTTP differs from both TCP and UDP in that if the file being transported contains links to other files, transfers of those linked files are automatically initiated as well.

RealServer 5.0 followed in late 1997, adding facilities to stream Macromedia's Flash content (called RealFlash), ad-insertion capabilities, and authentication features to identify the recipient of a media stream. In 1998, the company produced RealSystem G2, describing an entire system for streaming media production and distribution. The RealServer G2 product introduced RTSP (Real Time Streaming Protocol) and SMIL (Synchronized Multimedia Integration Language), better licensing support for people deploying streaming media servers in server farms, and SureStream technology, which allowed media streams to thin when the network connection was poor. RealServer 7.0 followed in 1999, with enhanced support for SMIL (you could View Source on SMIL files) and enhanced SureStream support. The company's current server product is RealSystem iQ Server 8.0, which supports streaming of MP3 files and Apple's QuickTime 4 files. It also supports redundant encoders, to ensure that live streaming is uninterrupted, and distributed licensing. The company's current system has fail-safe redundancy, in the case of equipment or network failure, can dynamically allocate capacity to respond to demand, uses error correcting codes to guarantee 100% reliable distribution, uses redundant network paths to guarantee broadcast delivery from server to server, and claims to have zero points of failure for live broadcast. The company's NeuralCast Communications Protocol allows RealServers to exchange information and make decisions, so that capacity can be shared and servers can take the load from failed servers and network connections.

The company continues to create innovative solutions to allow devices to receive streaming media. RealNetworks is active in defining how media streams on third-generation mobile phone networks. Its player technology is ported to various mobile device operating systems. The company continues to improve image and audio quality and embrace standards.

Liquid Audio

Funded as a startup by Hummer Winblad Venture Partners, among others, in early 1996, its mission was to solve the sound quality, copyright, and copy protection issues that needed to be addressed so that streaming audio would appeal to the music industry. While RealAudio produced something akin to radio-quality audio, Liquid Audio teamed up with Dolby Laboratories to create a better-sounding streaming audio codec. The company became a content delivery network as well as a vendor of infrastructure for secure audio playback.

Liquid Audio has adopted other people's standards and technology as well as created plug-ins for other streaming media players. In March

2001, the company announced it was producing a plug-in for RealPlayer, though it had its own Liquid Audio player available for free download for years. Liquid Audio was one of the early adopters of RealNetworks' RTSP protocol in 1996 and announced that it had adopted Microsoft's NetShow server and ASF file format in late 1997. In November 2001, the company announced its latest effort to incorporate Dolby Laboratories' Advanced Audio Coding (AAC) technology. AAC is said to produce better audio quality with less bandwidth than the currently ubiquitous MP3 standard.

Liquid Audio continues to innovate and actively promote the idea of streaming the best available music content, in good quality, securely. With its technology, it believes it can best support e-commerce with music.

How It Panned Out

As mentioned earlier in this chapter, in corporate LAN environments, fears about excessive network loading, the wide variety of platforms that must be supported, and the belief of some managers that watching TV at the desk is "not real work" have meant that one of the earliest targets for the development of streaming media has so far been a resistant market. Few corporations are keen to upgrade their IT infrastructure to support streaming, even when a return on investment can be predicted. The investment required to stream within an enterprise is not just an investment in hardware. IT departments must be trained to maintain and support the streaming infrastructure. The corporation also needs to learn how to make content professionally, for it to be engaging, interesting, informative, and useful. This is a new skill for many corporations and one that has not traditionally been associated with the corporate world, having been one of the creative arts more typically allied to the entertainment industry.

Streaming on the public Internet, directly to consumers, still hasn't fulfilled its promise, since broadband connections are still relatively sparse (although there are currently 21 million US homes that have DSL connections). Players also need to be able to "flip channels" of streaming content, just as on television (Microsoft's latest "Corona" Windows Media Technologies offering claims to be able to do just that).

Hence, a large number of the target applications that drove the initial development of streaming media have not been fully satisfied with the technology so far available. Developments are just now reaching sufficient maturity to address those applications in wholly satisfactory ways. In the next few years, we can expect to see a sudden and dramatic uptake of streaming media technology, as initial concerns and obstacles to adoption are addressed and solved by streaming media products. The early adopters have given their feedback and helped vendors improve their technology to the point where the products are beginning to be robust and ready for prime time. We are now at the very beginning of mass-market adoption.

Why Is Streaming Media Better?

The views expressed in this section are just that. It is a largely impossible to prove the proposition that streaming media is "better" than another medium. However, I wish to set forth the case for asserting that streaming media, indeed, is new and an improvement over existing communications media. It is my opinion, but there is compelling evidence to support my assertions.

The main reason why streaming media is an important and improved technology for communication has everything to do with the information glut in today's society. There is so much information, people have a hard time working out what to pay attention to. Whatever they want to know, they need to navigate to the most concise, most easily assimilated information available, wherever they happen to be and whenever they need to access it. Navigation and assimilation take time and every human on earth only has a maximum of a lifetime's worth of time to give. People need to access the information significant to them without giving up too much of their life. Streaming media technology fits the bill, because of its richness, searchability, wide audience reach, and scope, and because video is easier to assimilate than other forms of information.

Unlike other media, streaming media is inherently interactive and, because of its ability to use variable bandwidths in transmission, the message can be delivered cost effectively. There is little bandwidth wasted. Few people remember the liberating effect of the first transistor radios. Suddenly, news and information were available wherever you were. The fact that streaming media will be received on a new breed of receivers, some of which haven't yet been conceived or designed, makes streaming media just as exciting and groundbreaking as those early transistor radios. With streaming media, you can receive anything, any time, anywhere.

Much of this chapter contrasts streaming media with broadcast media, including radio, television, videotape, and DVD. We'll consider both analog and digital versions of these in making our comparisons.

Better Than Text

The reason Hollywood makes films, not slide shows, is that the medium is immersive, emotionally engaging, and a great way to impart messages powerfully and quickly. Even when a speaker uses a slide show to accompany a lecture, it's the talking and the body language of the presenter that conveys the most information, as anyone who has ever read a PowerPoint deck in the absence of the lecture transcript will know.

We learn to see and hear long before we read or write or even talk. As a species, we are adapted to audiovisual queues and nuance. We absorb audiovisual information quickly. This evolved as a survival skill, so that our ancestors could outrun or outwit a predator.

A medium that allows us to make interactive use of our natural affinity for audiovisual information gathering is going to help people absorb information as quickly as they are able. In other words, streaming media presents information in a way that is optimized for human assimilation, consumption, and learning.

It is often argued that the written word is more effective than video, or at least more engaging, since it allows readers to gather the information at their own pace, filling in the details with their own imaginations and pondering key points at their leisure. As mentioned in the introduction, books have some features that are difficult to replicate with streaming media, such as rapid random access and low-power portability. However, the narrative power of synchronized multimedia and other features of streaming media are starting to make books lose some of their superiority as a medium for delivering serious research and information.

When even serious technical authors and researchers become familiar with streaming media, so that rather than writing a book, they choose to create a streaming media presentation, the information will be much easier and faster for an audience to understand and learn. For example, the bulk of the research for this book was carried out online. If this book could include the hyperlinks, rather than mere citations, the reader could directly verify what I have been saying and, more importantly, consult my sources to see what I have left out. Meanwhile my voice could give you the narrative, illustrating key concepts with animated diagrams, while you continue to flip through my cited sources. Researchers have noted that the rate of reading text from a page is about 55 baud, as has been previously noted. If you compare that to the scenario I have just described, the amount of information that a reader could absorb every second from a well-crafted streaming media presentation on the same subject is potentially far higher. With streaming media, information can be assimilated from a number of senses and sources at the same time. This is its advantage.

On-Demand Viewing

Television tells you what you can watch and when you can watch it. Broadcast schedules exist so that the broadcaster can gather the most audience share with the media assets available. For this reason, you often find that when one channel is carrying something of great interest to you, the other channels are carrying similar programming in exactly the same time slot. The audience for each is therefore diluted, in the name of attracting audience share. Ironically, if the broadcasters separated their offerings, so that a program that appealed to a particular demographic segment was never on at the same time as one appealing to the same demographic group, chances are that both broadcasters would achieve better audience figures in both time slots.

Who wants freedom of choice restricted when it doesn't have to be? Because streaming media can offer any program ever made at any time of the day or night, the notion of a broadcast schedule is somewhat obsolete. Indeed, there may still be programs that the content owners wish to schedule for playback at particular times, even delivered as streaming media, but this in no way precludes the viewer from opting out of the scheduled programming and seeking on-demand content.

At the time of writing, this nirvana of on-demand viewing hasn't materialized. This is mainly due to fears over media piracy, lack of business models that let content owners get fairly recompensed for offering their media for consumption, the lack of a truly ubiquitous broadband network, and the fact that the only available players tend to be PC based. None of these obstacles is insurmountable and widespread on-demand delivery is undoubtedly likely in the long term.

A Universe of Choice

Anyone who has ever traveled internationally on business will know the frustration of not being able to watch programs from back home on the hotel television set, or worse still, of getting interested in a series of programs while away, but not being able to access the remainder of these back home. Students of foreign languages will also attest to how difficult it is to find programming in the language they are studying, so crucially necessary to achieving immersion and thus accelerated language learning, unless they happen to be in the country whose language they are learning.

Streaming media makes distance irrelevant. For a viewer of streaming media with a receiver that can accept media from any compliant media server on the network, wherever it happens to be located, the possibility of accessing foreign programming is a realistic one.

Distance is no barrier to access and neither is time. Streaming media allows content owners to make deep archives of media available for public viewing, cost effectively. Indeed, with some media archives, streaming makes it possible for media owners to make money on even the most ancient assets, since streaming media technology makes these assets directly accessible, with little human intervention by the media owner. Today, much of that historical media is totally inaccessible to viewers, since it is held on shelves in darkened vaults, with its physical media too delicate to leave on a player. Digital copies of this archived material need only be made once and then left on media servers for access at any future date. The choice for historical archive content owners is simple. Make little or no money by hiding those assets in vaults, or make some money making those assets available for streaming media audiences who may be interested.

Niche programming is also more readily made available to an audience using streaming media technology than it is with broadcast media. Even if only a few dozen people are interested in a particular streaming media presentation, it may still be cost effective to make it available on demand. The same cannot be said if the decision is whether or not to occupy a valuable and scarce broadcast channel transmitting that same video.

Global and General

Streaming media assets can be accessed from anywhere in the world, as previously noted. They represent a truly global system for delivering media, comparable in speed to the internal networks used by today's global news-gathering organizations. Breaking news can be streamed to a global audience at a pace at least as fast, if not faster, than the broadcast networks can achieve. For the first time, we have a globally universal video standard. There is no need to convert from PAL to NTSC, as there is in broadcast television. A receiver of streaming media can roam globally.

Ironically, one disadvantageous situation for streaming media, compared to broadcast media, is addressing a mass audience. In broadcast, once you have transmitted the signal, it doesn't matter how many people watch it. Each additional viewer costs no more to serve, yet returns more to the broadcaster in advertising revenue, since advertising rates are related to the number of people watching. With streaming media, each new viewer demands a stream's worth of bandwidth. Thus the millionth viewer costs as much as the tenth. The millionth viewer may actually cost more than the tenth, since the millionth viewer may saturate the server and require a load-balanced alternative source.

Solutions to the mass-audience serving problem include multicasting, which has not yet been widely deployed, as well as schemes to fan out media-serving capacity from a central server to several edge caches, or at the extreme, among peers viewing the media. These approaches have merit, but are not yet widely available. It remains to be seen whether or not programs scheduled to play to mass audiences will be the dominant form of programming, once streaming media on demand reaches critical mass. However, there will always need to be the capacity for such events (such as when some significant news event of global significance is in progress). The streaming media industry will need to address the problem of cost-effectively serving mass audiences. It may be that traditional digital broadcast carriers become the best way of delivering a media stream to a mass audience.

The nice thing about being able to access media from all over the world is that the national biases in reporting particular events of global significance can be unveiled. There can be little doubt that the tragic events of September 11, 2001 in the US were reported with differing viewpoints around the globe. Allowing an audience to make its own direct comparisons between global news reports of the same event, for example, can serve to enlighten and inform. Only streaming media can do this universally today, even though some satellite transmissions often allow viewers from some territories to view programming from others. Streaming media has the added advantage of allowing the same pictures to be accompanied by voiceover and closed captioning in several languages at once. In other words, the same stream can contain multiple language soundtracks and on-screen captions. Broadcast finds this trick difficult.

Wide Reach

With streaming media, any person anywhere in the world can freely watch programs that no local broadcaster could ever find an audience large enough to justify scheduling. In fact, program makers can aggregate an audience large enough to be of interest to sponsors, merely by drawing the audience from a global catchment area. Individuals comprising that audience may be few and far between, but nevertheless represent significant numbers, when aggregated across geographies. It is almost impossible to do the same thing with the existing broadcast television infrastructure.

Interactivity

Unlike broadcast media, streaming content can be designed to empower viewers in deciding how to use their time. With a television program, the narrative dictates the pace and the scheduled and linear nature of the medium dictates the order in which information will be discussed. With appropriately authored streaming media, the viewer can choose to "cut to the chase," as if fast forwarding through a program, or else search for information in greater depth, via hyperlinked sidebar video, for example. The viewer takes charge of the narrative and the order, pace, and depth of the information delivered.

Interactivity also aids learning and information retention. The very reason schools make learning active is so that knowledge sticks. The viewer of streaming media equipped with synchronized interactivity features can ask questions, search for answers, interact with other viewers, take self-assessment quizzes and so on, while watching the streaming media presentation. Streaming media encourages multitasking while viewing.

Enriched User Experiences

When consumers understand that streaming media presentations can be crafted to offer more to do and see, with multiple paths to explore, more people will be attracted to the medium. The user experience is as rich as a computer game, in many senses. Compared to watching television passively, the act of viewing a streaming media program is much more engaging and active. This is especially true for content that incorporates integrated interpersonal interaction features, such as live chat.

Content producers are only just beginning to envisage the possibilities afforded by the medium. Mimicking television production is not the best use of the medium, since it ignores some of the more enticing and enriching features that would appeal to audiences. A creative streaming media producer can incorporate all manner of surround-sound tricks and effects, three-dimensional interactivity, and virtual reality into programs. These elements can be streamed and groups of viewers can interact in real time via these quasi-computer gaming elements, by connecting with each other over the Internet.

When used to best effect, streaming media programs can offer a much more exciting and involving user experience than either computer games or television.

Targeted Advertising

Unlike broadcast television, streaming media permits you to receive only advertising that is relevant and interesting to you. It might be necessary to pay for the privilege, but streaming media technology makes such advertising possible. From the point of view of an individual streaming media player, it makes no difference if the ad is served from one server or another. If the consumer gives preference information to the content owner, advertising can be selected to fill the gaps so that it isn't annoying or irrelevant. Consumers can have their own customized menus of advertisements.

This ought to be a very attractive proposition to advertisers. What could be more ideal than playing ads to an audience that actually wants to listen to them? With today's broadcast television, advertisers often do themselves more harm than good. Not only is much of their money wasted in delivering advertisements to people who have no interest in them, but the annoyance they cause and the imposition on the unwilling viewer's time leads consumers to think worse of the sponsor than they otherwise might, if left alone. In other words, the effect of much advertising today is to create dissatisfaction with the sponsor.

Just because consumers opt for a particular company's ads doesn't mean they don't want to be surprised. Permission to deliver ads is granted only because it makes access to content cheaper. Clever advertisers will realize that having obtained permission to market to a group of consumers, they need to entertain and delight them, if they are going to make a sale. They also need to be sensitive to the fact that the permission the consumer has granted is conditional and can be withdrawn, without explanation or notice, at any time. Again, smart advertisers will make opting out and in very easy. There may be times, for example, when the consumer is willing to pay for freedom from any advertising whatsoever.

The future of advertising on streaming media technology is actually very bright, because it offers options that are satisfactory to consumer and advertiser alike. It is increasingly difficult to understand how advertisers who insist on imposing their messages on the public and wantonly wasting their time will be able to continue to do so, once streaming media achieves critical audience mass. Yet intrusive advertising is all that broadcast television will ever be able to offer.

Immediately Measurable Response

Another thing that is impossible with broadcast television, but eminently possible with streaming media advertising delivered via a carrier with a back channel, is that the click stream for every viewer can be analyzed. The advertiser can tell if you have agreed to take the advertising and then skipped over it. So, not only does the consumer give permission to the advertiser to sell him something, the advertiser can immediately know when a sale has been made. The return on advertising investment can be calculated in real time.

With data mining and database correlation, the advertiser could make inferences about the person buying, in response to any streaming media ad, but in all likelihood this would be seen as an abuse of the permission granted. It is more likely that the advertiser would request details of the purchaser in exchange for gifts or discounts, so that the relationship is maintained and the consumer does repeat business with the company. This would enable advertisers to know their customer bases and preferences in fine-grained detail and with high confidence. This advantage could be decisive in a highly competitive environment.

Enhanced E-commerce

Another application enabled with streaming media is allowing viewers of a streaming media program to commence a dialog seamlessly with actual people on an e-commerce Web site onto which they may have strayed from the main program, perhaps in response to a link embedded in the entertainment stream, advertising some object on screen in the video program. With audio and video streaming, an actual dialog with a real sales assistant could commence as soon as a consumer enters the e-commerce site. With streaming media, not only are consumers directed to enter the e-commerce site by an ad embedded in the entertainment stream, they are now telepresent in a virtual store, talking with real salespeople. The salesperson could know who they are, if they are repeat customers, from data already provided on previous visits. Not only that, customers are pre-qualified, since they would not have even seen the ad, had they not fallen into the category of people who wanted to see the ad. It is difficult to be a virtual timewaster.

Having entered the store, the consumer may watch more entertaining streaming media content that makes the sales pitch and demonstrates the product. This kind of rich interaction is likely to have a very much higher rate of conversion of inquiries to sales than any other e-commerce technique linked to advertising.

When e-commerce was first posited, the prevailing view was that its purpose was to eliminate human contact and save money through not having to have huge numbers of people on staff. This turned out to be wrong.

An e-commerce presence is only useful for getting a company's store-front into every town on the planet for much less than the cost of bricks-and-mortar shops. Having attracted customers to an e-commerce online store, companies still ought to serve their customers the way they would serve customers walking in from the street to a real store. That means using natural modes of communication, which, as we have established means audiovisual streaming media. After all, how many stores on Main Street would expect you to transact business by filling out forms for goods you can't even see, without the opportunity to ask questions? Yet this is precisely the experience offered by most e-commerce sites today.

With streaming media, the line between entertainment and selling blurs and the experience of buying something online has the potential to be a form of recreation and entertainment itself, using natural modes of human communication, not rigid Stalinesque forms and procedures. Even interactive variants of broadcast television cannot hope to offer this seamless segue from entertainment to shopping and still make the experience an enjoyable one for the consumer.

Mobile and Portable

Streaming media receivers are relatively easy to make as portable devices. Silicon solutions for streaming encoding and decoding are well within technical feasibility and some of these devices already exist. The same applies to streaming media cameras. They can be made light-weight, small, and battery powered. With the advent of broadband wireless networks, this makes it feasible to have mobile streaming receivers and mobile streaming sources. In fact, third-generation cellular networks are explicitly designed to allow mobile devices to move while receiving IP traffic. For this reason, streaming to and from the car is achievable.

Contrast this to the receiver requirements for digital television. The dish required to receive satellite television is between 60 and 90 cm (24 and 36 inches) in size and it must be kept in careful alignment with the satellite. A terrestrial digital television antenna is also large, when compared with a third-generation cell phone antenna. Also, the circuitry required for receiving and decoding a satellite signal or a terrestrial DVB (Digital Video Broadcast) signal is significantly larger, in terms of transistors and circuit-board real estate, than a streaming media receiver. Lastly, it is difficult to maintain signal reception from either of these digital television sources when the receiver is in motion. For these reasons, mobile digital television receivers will probably never materialize.

Distribution

The availability and ownership of television channels is regulated because airwaves are a scarce resource. So much of the available spectrum is required to broadcast television that it is not possible for anybody who wants a channel to stake a portion of the radio frequencies and begin transmitting. Chaos would ensue. Broadcast television is very wasteful of radio frequency spectrum. Because bandwidth for television transmission is scarce, the broadcast industry has geared itself always to try to maximize the number of people viewing the content delivered through that scarce channel. It is an inherently industrial-age, mass-market medium.

However, the truth is that bandwidth isn't scarce. Every time another optical fiber is laid, the amount of available bandwidth increases. There is no practical limit to how much fiber can be installed. The limit on bandwidth provision is really a limit on human capital, because installing and maintaining that bandwidth needs people, and on the cost of the equipment that converts electrical signals to optical ones and back. The number of streaming media channels that can be served is only further limited by the number and power of the streaming media servers that can be deployed worldwide.

The important point is that, compared to broadcast television, the number of distinct program streams that can be delivered at the same time is far greater. The potential exists to provide a virtually limitless choice of streaming media programs. In fact, streaming media can be cached at the edge of the Internet, enabling the content to be delivered reliably to consumers, with little delay. This will ensure a high quality of experience, as well as a vast range of choice.

With broadcast television, while the territory that the distribution channels covers is vast, it is not completely global. Few television broadcast networks reach a global audience. With streaming media, the reach is global, to the extent that wireline transmissions have "global" reach. However, IP streams can also be delivered over satellite to the unwired world. This is as true for broadcast television as it is for streaming media. The difference is in the efficiency of the use of the bandwidth. The compression techniques used in streaming media tend to use less bandwidth for a given image quality than those used for digital television.

It would be naïve to think that there will be no gatekeepers controlling access to the streaming media distribution infrastructure. Infrastructure, whether broadcast or streaming, will always cost money and need to make a return. Someone, at some level, will always own parts of the streaming media distribution infrastructure. However, the broadband Internet will be harder to monopolize than the airwaves, since there are fewer barriers to entry and no regulated scarcity of bandwidth. The potential for a wider choice of streaming media distribution providers exists.

Ownership of the major streaming media distribution backbones, gateways permitting access to the home, and of the streaming media servers may be concentrated in a handful of powerful companies. That is always a risk. However, the potential for greater diversity in ownership of these key resources is far greater than it is for broadcast television. There is no monopoly enshrined in licenses to use the airwaves as there is in broadcast television. With streaming media, there are fewer national legislative boundaries that limit the reach of the distribution, as they do with broadcast television today.

Freedom of entry into streaming media distribution is less restricted than it is for broadcast television and incumbent companies do not have such strong advantages. Whether this will benefit the industry overall or hinder it is debatable. What it really does is make possible far greater choice, with the consumer able to use competition to get cheaper access to media. Competition is usually thought to be beneficial to any market.

Content Production Costs

Content can potentially be produced more cheaply for streaming media than for television. The reason television production has typically demanded such high technical quality is because of losses in the analog television production and distribution chain. If you start with pristine pictures and sound in the television studio, by the time the program has reached the viewer's television set, the picture has had so much noise and distortion added by the analog distribution process, that it is barely acceptable. High-quality sources are necessary to preserve acceptable quality at the receiver.

When television production converted to digital, the same high quality levels had to be maintained, for backward compatibility reasons. Even digitally produced programs get transmitted to a vast audience via analog transmission systems. This is significant because the cost of digital production equipment capable of such high quality levels is enormous. The data rates required to produce such high image quality push some technologies to the very limit and hence cost more.

The benefit to consumers has been better picture quality at the receiver, because with digital production, there are fewer losses introduced within the television production plant. However, even for digital television transmission, the high quality that can be obtained at the production stage is not maintained. All digital television distribution systems (cable, satellite, and terrestrial) use compressed images. What the viewer watches is a necessarily lesser-quality image than what was captured by the cameras in the studio. As an example, many production systems are capable of producing images at 270 megabits per second. Even DV (Digital Video) format cameras produce images at 25 to 50 megabits per second. Yet, digital television channels often allocate no more than 4 megabits per second for transmission. A large percentage of the data is thrown away before it gets to the consumer.

What this means is that, were it not for the need to inject high-quality video into the extremely lossy legacy analog broadcast network, digital television could start with pictures of lower quality and still have no discernable effect on the quality of the pictures received by digital television viewers. Of course, it is true that if you start with better pictures and sound, the compression process tends to add fewer artifacts, but how much quality is worth paying for? In other words, if all television production and distribution were digital, production could take place with cheaper, lower-quality equipment and the digital television viewer would not lose.

Television set manufacturers have insisted for years that the public demands higher picture quality. HDTV (High Definition TeleVision) developments and initiatives had their beginnings in this assertion. Yet the evidence is just not there to support the claim. The viewing public has accepted highly compressed digital television images and even VHS-quality pictures for years without complaint. The claim that all viewers want and need high-definition pictures for everything they watch is somewhat dubious, as is the fanciful notion that high-definition pictures will significantly change viewing habits, leading to people watching more hours of television, on average. The consumer demand case for very high picture quality, in a totally digital television production and distribution system, has not been proven.

Streaming media production and distribution is digital, from end to end. There is no analog transmission system involved and picture quality requirements are no higher than for digital television. Production of streaming media can use cheaper equipment right away. This ought to be an advantage.

Most video production equipment falls into one of two categories: professional broadcast production equipment and consumer-grade video equipment. Unfortunately, what is required for streaming media production is a hybrid of the two. Consumer-grade equipment has too much foolproof automation, such as auto focus and auto iris features, which limits its usefulness to serious video producers, whereas the professional equipment available is overspecified for streaming media production. The streaming media producer is therefore not well served by video production equipment makers.

That said, the cost of the technology for digital video production continues to fall. Non-linear editing equipment costs approximately a fifth of what it cost just five years ago, with a well-specified system costing on the order of $10,000. Digital video cameras that work in low light conditions, capable of producing very good pictures, are available for $1000–$2000. Hard disk storage has plummeted in price. In 1995, a state-of-the-art 9-GB drive cost several thousand dollars. Today, 150-GB drives are available in smaller form factors and cost only a few hundred dollars. Desktop software packages that can create Hollywood-quality video effects and animation are available for between a few hundred and a few thousand dollars. Producers can use these cheaper-than-broadcast-quality machines to produce streaming media programs of a quality acceptable to viewers.

Democracy and Media Control

The sheer range of choice of programming and diversity of editorial opinion that can be accommodated by streaming media means that the medium is potentially more democratic. Streaming media technology allows anybody with the right software and hardware to be a desktop broadcaster. At the very least, with streaming media, individuals with strong opinions or dissenting voices can post their view of stories on the Web, while the news is breaking, in full-motion video, if they wish. Indeed, the production values for the rebuttal can be equivalent to those of the news. Streaming media programs could easily encourage these dissenting voices or else link to sites that carry differing editorial commentary. Diversity of opinion can be a very healthy force in a democracy, where freedom of speech and the right of free expression are held to be values worth upholding. Under more repressive regimes, dissenting works may still be posted without directly imperiling the dissident, by use of anonymizers, which disguise the identity of the person publicly posting the content.

Some dissidents quite correctly insist that in a world where most people in the Third World have not even made a phone call, the cost of the technology needed to produce streaming media effectively bars access to the poor and oppressed. They assert that it is, therefore, undemocratic, since only the world's elite may voice their views via streaming media. However, my argument is that this technology is still cheaper than broadcast equipment and comparable to printing equipment, so while not extending democracy universally to all men, it nevertheless extends it somewhat further than has been possible so far. Thus, streaming media can potentially strengthen democracy.

Setting the Agenda

When channels of information dissemination are limited, as they are with broadcast television, the range of opinions expressed on any given subject tends not to vary greatly. In fact, because news programs in a locality monitor each other's output, in a quest to ensure no loss of audience share, news reporting becomes remarkably uniform. There is a tendency for the first network to break the news to set the agenda for all competitive networks, almost by default. Unlike print media, broadcast networks carry almost the same stories and take very similar editorial stances on issues of the day. Indeed, a media owner could exert control over the editorial agenda for all the broadcast channels in a locale with little risk of discovery and perhaps even unwittingly, just by being the first to carry a story.

Diversity of opinion and fragmentation of the audience into smaller interest groups tends to dilute this effect. Although content owners that are large corporations will still be able to exert more influence over opinion than a single journalist working from a desktop, their power to control events or sway the minds of the general public is more limited when other viewpoints are so readily available.

The other important feature of streaming media technology is the speed with which desktop video producers can get a story to the world. In many cases, they can outrun the broadcast networks in breaking a news story. Already, many news stories have broken on the Internet, as text, before they appeared on television news reports. Before the Internet, the world heard about the events in Tiananmen Square in 1989 through reports faxed to the West from China. Imagine the power and impact of being able to publish to the entire world video footage of an atrocity in mere seconds. With a laptop computer, a DV camera, some software, and a phone line of some description, this is precisely what streaming media makes possible. During the Afghanistan war, reporters were filing stories to broadcast networks with streaming satellite videophones. In the future, they could post directly to the broadband Internet, permitting viewers to see the report immediately, without having to wait until the next scheduled news bulletin aired. Although the speed at which something can be published can mean the danger of reporting events before they are adequately understood, the technology of streaming media allows the news agenda to be set, yet also encourages diversity of opinion. Broadcast television is, in contrast, both slower to react and more limited in its variety.

Encryption

In common with conditional access digital television systems, streaming media allows content to be encrypted. This is useful for content producers who want to protect their media from unauthorized access and allows recipients to verify from whom the media came. Streaming media technology potentially allows content distributors to trace who accessed content and when, even when copies are made and redistributed. There are many civil liberties issues with this particular practice, but broadcast encrypted content could also face similar issues.

The important difference between broadcast television encryption and streaming media encryption is that cost-effective tools for encrypting content can be made available to desktop streaming media producers. With the conditional access technologies used by digital television companies today, the cost of the encryption tools is high, and the broadcasters have tried to keep tight control over access to their set-top boxes. In other words, they have used conditional access to effectively preclude other companies from delivering digital media via their set-top boxes. They have used it not only as a tool to prevent copyright abuse and theft of media assets, but also as a gatekeeping tool on the set-top boxes they supply, thus barring access to viewers by other digital media distributors. With streaming media, on the other hand, there has already been some effort toward making the conditional access and encryption techniques an open standard.

The Joys of Unregulated Media

For better or worse, broadcast television is regulated in many countries to protect viewers from content deemed inappropriate or harmful. Many believe that a free society should allow freedom of expression, so long as other laws are not violated. Streaming media, on the other hand, is largely unregulated. People can and do produce content of questionable moral value. Indeed, an interesting case has arisen where photorealistic child pornography has been created and published. The images are generated entirely by computer. No children were involved in its making. However, some hold that such explicit content, whether real or not, has the potential to deprave and corrupt individuals viewing these images. It will be interesting to see how lawmakers respond to streaming media content that is globally deployed and may be entirely synthetic. The question of who has jurisdiction over such cases is also yet to be satisfactorily resolved.

Freedom of expression is, in general, a concept worthy of protection. On a global Internet, this should mean that people are responsible for making their own choices about what they want to view, without censorship. Schemes to categorize and rate media content in a standardized way, software that allows parental control over access and standards for expressing viewer privacy preferences have already been established. These put the onus on the individual to protect family members from material considered offensive. However, government agencies also monitor users who view child pornography, so that the state already works in cooperation with transnational authorities to detect and arrest these individuals. Media may not be regulated, but offenders certainly are.

Streaming media has the ability to cross national borders, allowing citizens (at some personal risk) to view prohibited material from beyond their boundaries. Political regimes that seek to restrict the flow of information to their citizens from sources other than those officially sanctioned may have a harder time doing so with streaming media. When streaming media is used to commit acts of treason, espionage, unlawful surveillance, passing of classified information, or acts of cyberterrorism, who should act? Who has jurisdiction? What laws apply? Some believe that global laws and authorities should address such questions and that the legal framework needs to be established with some urgency. Others hold that existing legal systems should coordinate their laws on the matter. What is clear is that the question remains unresolved and that while it does, streaming media presents dangers that broadcast media no longer does. On the other hand, it promises to liberate in ways that broadcast television cannot.

Play It Again, Sam

Personal Video Recorders (PVRs) have been developed to time-shift digital television programs and to shuttle through programs at high speed, skipping the ads, if that is desired. One of the best-known machines is the TiVo, which records programs digitally to a hard disk. Today, most of the streaming media players disallow recording of streaming content to a hard drive, but this is a commercial choice, not a technical impossibility.

If media can be streamed on demand more cheaply than the cost of storing it, people may abandon the very idea of keeping recordings of programs. Today, it is still cheaper to store a program than to stream it again, which seems absurd. It is a matter of time before streaming media players allow the recording of streamed content to a hard drive, especially when digital rights management schemes emerge to effectively prevent replay and copying not authorized by the copyright owner. By that time, however, the price of bandwidth may have dropped sufficiently to make this moot.

Searching and Filtering

Digital television has introduced electronic program guides, to allow viewers to trawl through electronic program schedules, rather than paper-based listings, in order to find things they want to watch. With the proliferation of channels, however, finding what you want to watch with a program guide becomes an unwieldy, time-consuming, and frustrating exercise. Streaming media provides an alternative solution.

With the vast number of streaming programs that may one day be available, a program guide would be a hopelessly inefficient way of navigating to content of interest. The Internet faced the same problem and solved it by the use of directory guides, such as Yahoo!, and with search engines like Google. With these tools, users can narrow their search criteria and quickly locate material of interest. With streaming media, search engines have a harder time classifying media. There are some specialist engines being developed that can recognize actual pictures and sounds, but the user needs to be able to specify the search in terms of pictures and sounds as well. This presents some practical difficulties.

A more promising approach is to tag the streaming media assets with metadata. Metadata is descriptive text data about the streaming media asset. With a metadata-enabled media search engine, particular streaming media assets could be located, just as if the user were searching for text on Web pages. Today, there are very few search engines that can index streaming media metadata effectively. For streaming media users of the future, this situation will need to be remedied.

In theory, search engines could be developed to read the electronic program guide that accompanies digital television programs, index it, and then allow users to locate media by searching on that data. However, I am not aware of the development of any such search engines to date. Also, because of the asymmetrical nature of most digital television broadcast systems, it isn't clear how users would effectively interact with a search engine from their set-top box, nor how the results would be delivered quickly to individual users. Being IP based, streaming media has the advantage of having an easy and seamless link to search engine technology.

Copyrights Rule

Copyright is a thorny issue that causes heated debate. Some hold that information needs to be free, so thatpeople who create original works should not expect to control access to those works. Others claim that were it not for copyright protection, there would be much less quality work produced, since producing it requires an investment of time and money. Unless there is a way to protect returns from access to that work, there is little commercial incentive to create new and original pieces.

Streaming media is of concern to copyright holders because a piece of streaming media can be replicated without authorization and redistributed globally in seconds. In fact, with digital television, if everybody in the world had a digital recorder (like a DVD recorder, for example), the owners of digital television programs and feature films would already have a major problem with copyright enforcement. As with audio CDs, pirate copies could take significant revenues away from legitimate distribution channels.

Historically, copyright owners have sought to prohibit or else tax home recording. They have done this by influencing equipment manufacturers to include features in their products that make copying impossible or by setting up region coding, so that media cannot easily cross borders. Paradoxically, despite industry opposition to every new technology that allows content to be copied at home, copyright owners have, in fact, made more money. For example, Hollywood initially vehemently opposed the makers of VHS machines, claiming that home taping would wipe out their industry. Yet, revenues from videocassette sales now comprise a significant proportion of the earnings of feature films.

In my opinion, the truth of the matter is that the customer is king. Customers just want access to media of interest. Whether they make a physical copy or not, their aim is just to access the content. Few people with the means to pay for content begrudge making the payment. Even the most self-interested individuals realize that undermining the earnings of their favorite content producers by stealing their works, results in less output from those producers. On the other hand, there is a ground swell of opinion that counters this, saying that once content producers have made a fair return on their work, grasping for further payment, to fund what often are flagrantly extravagant lifestyles, is pure greed.

Content owners face a difficult choice. On the one hand, they need to expose their work to a large audience in order to build a following. In this case, things like Napster actually help, because they expose an artist's work to a wider public. On the other hand, they need to make enough from selling access to their work to enable them to continue working as artists. Here, Napster merely takes money, to which they feel entitled, from their pockets.

Digital Rights Management (DRM) systems have been designed to augment streaming media. With a DRM system, consumers are free to copy and redistribute digital media at will, but the content owner regulates the conditions under which the media can be played. This gives the content owner the ability to allow free playing and copying of content for perhaps the first ten days after release and then charge a nominal fee per play thereafter. Copies may then only be allowed, provided the recipient of the copy is identified and pays a license fee to play the media. When pirate copies were made and sold, it would be possible to identify the thief, since each copy of the media that gets distributed could be stamped with a unique ID.

The system is not perfect and there are ways of getting around the protection. However, it does provide a tool for the management of copyright to at least some degree. What is far from clear is whether or not consumers will tolerate the inconvenience of needing to obtain a license every time they wish to play a piece of media, or if they will allow their privacy to be compromised in the name of monitoring copyright abuse.

Digital television potentially faces all the same problems, as does any digital medium (computer games, for example). The difference is that the problems are not as easy to manifest with the digital television system and digital television's rights management and conditional access solutions are not as flexible as streaming media's digital rights management solutions for the user or copyright owner.

Fingerprinting and Watermarking

Another technology to protect copyright applies equally well to digital television and streaming media. Both fingerprinting* and watermarking allow individual frames of video to be uniquely identified with a small digital signature and cataloged. With fingerprinting, the signature is created by analyzing the video data. With watermarking, a code is hidden within the digital data comprising the frame of video, to identify it uniquely. Watermarks have the disadvantage of being degraded by some video processes.

Using fingerprints or watermarks, content owners can identify when their material has been used or copied without authorization. However, with streaming media, there is an additional use. If the media servers are crawled by a Web spider that fingerprints media encounters, fingerprints and watermarks allow content to be located using a specialized search engine, given just a few frames of the media. Thus, clips that are

---

*The newly-standardized MPEG-7 provides a rich set of tools for the creation of fingerprints and related machine-generated media meta descriptions, which search engines can readily use as indices. In fact, MPEG-7 can equally well catalog digital television as it can streaming media, but commercial solutions have not yet been widely deployed. However, crawling is still an impossibility on many legacy digital video servers.

offered for on-demand viewing can be located by their fingerprint data. With digital television servers, this crawling process cannot take place, given the design of the majority of today's video servers.

Archiving

Storing broadcast-quality video is expensive because of the sheer amount of data involved, which leads to physically large storage solutions, and because the tapes deteriorate with time and must be recopied. Analog video deteriorates visibly, whereas digital video experiences bit errors and eventually becomes impossible to play back at all. Streaming media, on the other hand, is much less expensive to store, since the compression used creates smaller files for a given playing length, compared to the originally digitized video file. Also, regular archival solutions used in the IT industry can be used to archive streaming media files. There is no need to buy exotic and expensive machines, designed expressly for video archival. Archival can be to DVD-RAM, making it both economical and no longer space intensive.

An additional feature of streaming media is that some machines in existence can take an archive of digital source video and convert these files to streaming formats automatically, as a batch process. As codec technology improves, allowing streams to be created which produce better image quality for less bandwidth, these machines can automatically recompress the source video to any new compression format devised. The lighter-weight nature of the compressed streams makes economical near-line storage for near video on demand (NVOD) applications.

Finally, since streaming can take place without any mass storage device at all, consumers do not need to store videos and DVDs in their homes, as mentioned earlier. The archival feature of streaming media that is not so easily replicated with digital television is that if consumers want to watch a program they have seen before, but at a time that suits them, they do not need to have a copy of that program stored in their houses. They only need to stream the program again.

Using Metadata

We touched on metadata earlier in this chapter when we talked about searching for content, but didn't elaborate on what kinds of metadata can be stored and where it can be stored. Descriptions of the content can actually be embedded in the file container that carries the streaming media payload, under several of the existing open and proprietary streaming media file format standards. It is possible to include information about the content owner, who directed the piece, who the main actors are, etc. As we noted earlier, this allows text-based search keys to be used. If the metadata is indexed in a search database, people can enter a search target such as "all films by Alfred Hitchcock starring Tippi Hedren" for example. Such searches are hard to do with digital television content, unless the content has been logged by hand.

Of course, the act of supplying the metadata in the first place falls to the content producer or copyright owner and there is scope for indolence and confused terminology. One look at the metadata that accompanies many MP3 files that could once be freely obtained via Napster demonstrates that much of the metadata is missing, incorrectly spelled, in the wrong place, or badly formed. There are often multiple copies of the same content with different metadata. Standards bodies like the Society of Motion Picture and Television Engineers (SMPTE) have sought to regularize the data tags and their interpretations by compiling metadata dictionaries (the standard is designated SMPTE 335M). MPEG-7 is another initiative to create some semblance of metadata consistency across streaming media content.

Metadata can be embedded in streaming media semiautomatically. It is possible for cameras to insert the time and date and the camera settings, plus GPS (Global Positioning System) coordinates. These cameras have not achieved widespread adoption yet, but the ability to record data about the scene being captured automatically is compelling.

Simulcast Synchronized Multimedia

Digital television has become more interactive, using proprietary standards and more open ones, such as ATVEF (Advanced TeleVision Enhancement Forum), as a platform for interactive content. These technologies add interactive features to digital television broadcasts, synchronizing media elements to moving pictures. Contestants can play along with game shows at home. Educational programs can be supported with a wealth of interesting facts and activities to engage viewers.

Unfortunately, the interactivity for any given digital television program is not written once and then deployed to all set-top boxes. Depending on which interactivity platform particular set-top boxes support, the producers of the show must write a specific interactivity package to accompany their program for each platform they wish to address. End-user experience is often not entirely satisfactory anyway, with slow load times and limited freedom of exploration. To compound this, some interactivity platforms deployed on set-top boxes also lack software robustness, causing them to crash and freeze inexplicably.

In contrast to this, streaming media can already link to a vast array of standard synchronized Web elements; yet allow exploration to the entire Web, starting from when the program is viewed. The walled gardens of interactivity that the broadcasters offer violate Metcalfe's law, which says that the value of a network is in proportion to the number of nodes connected. Technologies like SMIL (Synchronized Multimedia Integration Language), on the other hand, allow streaming media producers to create regular Web-style graphical elements and coordinate them with other audiovisual elements, without resort to specialized software programming.

Indeed, it is technically possible, though not widely done, to create synchronized multimedia presentations that synchronize from device to device. In other words, you can theoretically create content that shows video on one device, displays Web pages on a second, and plays audio from another, yet maintains perfect synchronization of all elements.

Interactivity on digital television systems typically makes use of left-over bandwidth to transmit the interactivity data. For example, in the ATVEF standard, the interactivity data is typically broadcast during the vertical interval of the video signal. This severely limits the amount of interactivity that can be used in creating the content. Streaming media places no such restriction on the interactivity data accompanying a video stream. As long as the carrier bandwidth is sufficient, the amount of interactivity data has no hard and fast limit. Streaming media programs can therefore be more interactive than digital television ones.

Standards Conversions Obsolete

Most people are shocked and amazed when they first learn that videotape bought in the US cannot be played in England, without a specialized video player (or vice versa). Equally alarming is the discovery that a collection of DVDs amassed while living in Australia cannot be played on a European player, without clandestine software hacks. The reason for this incompatibility is regional standards. Whereas the US adopted the NTSC (National Television Standards Committee) television system, most of Europe adopted PAL (Phase Alternating Line), except for countries that adopted SECAM (Sequential Couleur Avec Mémoire). With DVD players, most of those bought in the US support only NTSC, while most European players support both NTSC and PAL. However, even in the case where both regions support PAL, disks bought in one region are not playable by unmodified players in the other, due to the region coding embedded in the disk's data.

With streaming media, it is currently the case that a stream can be played anywhere on the globe, regardless of where it is sourced, provided that the right player is installed. The potential for a truly global standard for audiovisual distribution is tantalizingly close. However, the problem with streaming media is that not all players can play all streams, since companies that create streaming media compression schemes often jealously guard their software as proprietary trade secrets. It is for this reason that coalitions like ISMA (Internet Streaming Media Alliance) are pushing for a single, player-neutral, global standard, based around MPEG-4.

With streaming media content, there are no difficult standards conversion issues, as there are when you need to play some NTSC television program on a PAL receiver. Today, you need to play the tape into a standards converter, which changes the number of lines, the number of pixels per line, and the number of frames per second, to accommodate the other television system's requirements. In essence, the content must be resampled, or sample lattice converted. These issues do not vanish entirely for streaming media, since to be totally correct and introduce no motion artifacts, streaming video ought to be sample lattice converted according to the characteristics of the computer monitor that will display the video. However, sample lattice conversion is much simpler with streaming media, since there is no interlacing of the fields as there is with television signals.

Information Density

The beauty of streaming media, in a time-pressed world, is that lots of information can be delivered in a very short amount of time, using a variety of rich media. For example, with streaming media, it is possible to deliver material edited together with far more cuts and jumps than would be acceptable on a mass medium like broadcast television. Also, it is possible to have several players open at once, all streaming different content. Textual sidebars can be viewed while the video plays. Voiceover sidebars could talk to you in your left ear, while the soundtrack of the video continues in your right ear. People who engage in streaming video chat commonly view multiple conversation partners at the same time. In fact, services like PalTalk default to three video windows simultaneously.

Many people are very comfortable with such multitasking. In many US homes today, people have the television on, while they surf the Net. Children quite comfortably do their homework with music playing, or play computer games while holding a conversation with friends. Attention is captured in fleeting moments. People are not glued to one program or the other, but spend their time noticing bits of each, at different times.

To some, delivery of information with this intensity sounds like hell. To others, it is just the thing to get through lots of perhaps semisuperfluous information very quickly. What is important is that the medium grants control over the density of the information delivered. Television, with its measured and fixed delivery, does not enable the viewer to upload information any more rapidly than the program producer dictates.

Tracing Sources

One of the empowering features of streaming media relates to factual reporting of news events. Today, when you see a news report, you only see what the editor wants you to see. The majority of footage shot never gets broadcast. However, with streaming media, not only can the news editor deliver a concise summary of the news event, he or she can also leave links to the rushes shot. People can choose to view every frame of video that the reporter has shot. This is a nice feature, because it prevents reporters from distorting the story, through selective use of imagery. The story can neither be sensationalized and conflated, nor minimized and whitewashed.

Indeed, it may add credibility to a news organization to include links to all source material, including documents and historical footage. The public may come to trust only those news organizations that allow first-person access to underlying material, without the editorial bias and inevitable distortions that news production brings. Broadcast television will struggle to replicate this feature. It may do so by using companion media, like Web sites tied to the news stories being broadcast, but ultimately streaming media allows seamless access on a single receiving device.

Trust Networks

Beyond letting you check and interpret sources for yourself, streaming media could allow you to establish trust networks, where you collect the names of reporters whose opinions you are willing to believe and respect. With such a trust network, you could verify that a streaming media report originated from a trusted source and had not been tampered with, through the use of digital signatures. As with a favorites list on a Web browser, you could keep track of favorite pundits and commentators. Indeed, with such trust networks, you might be able to chat live with the commentator. You could also meet other people who had the trusted commentator on their lists and through this contact, find other trustworthy sources of information.

To build trust, sources that wish to be seen as trustworthy could disclose their affiliations, biases, business interests, background, influences and those who fund their work. With such disclosures, it would be very much harder to spin news stories or to disguise a vested interest as impartial objective comment. Sources that declined to reveal these salient facts would immediately be seen as less-trustworthy sources. Often, what people require is the ability to tell when they are being manipulated and when they are being genuinely enlightened. Because of streaming media's easy relationship with hypertext, such trust networks are easy to establish and maintain. Again, this could be replicated by the broadcast television news industry, but only by resorting to companion Web presences.

Another aspect of streaming media that can encourage trust in visual reporting is the use of metadata to embed time, date, and location information into footage shot on location. If viewers are able to scrutinize more thoroughly the metadata attached to the streaming media, they can verify that reporters really were where they say there were (and also identify when stock archive footage from a library is being intercut with current footage and presented as news). By correlating video footage of the same event, shot from different locations and viewpoints, but with time and date in common, viewers could reconstruct the event and gain insight into what happened, from several angles. Future media players could support automatic correlation of these images and play them back in lock step. In fact, it may become common practice for camera crews to shoot each other as further proof that they were really there and really catching the action that matters.

Discontinuities in the metadata time stream would instantly reveal when an edit had been made. It would also be possible to ensure that an edited news item presented material in the correct chronological order, as it happened. Sometimes, the manipulation of time through careful editing can give a misleading impression of the events that occurred. Not only that, but material included in an edited package could contain links back to the source material. If the source were another edit, it would be possible to represent the genealogy of the material, all the way back to raw camera footage.

In person-to-person streaming, it is often important to know and trust the person to whom you are talking. One of the problems of text-based chat today is that it is easy for people to misrepresent themselves, using the anonymity of text to hide the truth. With video chat, it is harder to deceive, but not impossible. If streaming video were digitally authenticated with a signature, you might not know all the details about a person, but at least a trusted third-party authority would give you confidence in your interlocator's identity and that their personal details had been verified by somebody. In effect, individuals could be authenticated in the same way as streaming media journalists and reporters are. It would be very much harder for a person to use multiple online identities without this being apparent to the other party. If people wanted to use multiple aliases, to maintain their anonymity or to avoid online pests, at least people talking to them would be aware that this was being done. The important point to make is that this technology is relatively easy to retrofit to streaming media.

Also, trust breeds trust, so people you know and trust could recommend people they trust to join your trust network. In some senses, such groupings of trusted individuals could be forces for societal change. If all the people who participate in a particular trust network share similar interests and views, their mutual trust can be a very powerful thing, if directed positively. It is difficult to imagine how broadcast television, as a medium, could give rise to such groupings. It is equally exciting that streaming media can.

Viewer Reviewers

With appropriately designed media, streaming media allows peer reviews to be attached to a particular streaming media program. If content is found to be either poor or excellent, viewers can post their comments and give the stream a rating. In several non-streaming Web sites today, you can see this kind of thing in action. Amazon.com allows readers of books to post their thoughts on the page that displays the book's details. Many customers put great store in these peer reviews and are happy to volunteer their own thoughts for the benefit of other customers.

Similarly, on Stories.com, amateur authors are not only encouraged to post their work for public viewing, but also to review and constructively criticize literary works posted by other authors. Readers can give an item a star rating, indicating to other readers what might and might not be worthwhile reading. There is no technological impediment to doing something similar with streaming media content. Viewers could be encouraged to tag on ratings and reviews to streaming media content made available for public viewing. Indeed, search engines could make use of this review data as metadata, useful for indexing the media and making it easier for viewers to navigate to items of interest.

Of course, malicious reviews can be posted, as well as honest opinions. It is possible to link the reviews posted to biographical and track record data about the reviewer. In this way, later readers of the reviews could form an opinion of the trustworthiness of the views expressed in the review.

Adding viewer reviews to traditional radio and television programs can be done, but not in a way that permanently associates the reviews with the media, as is possible with streaming media. Indeed, such a democratic rating system makes it much harder for traditional media industries to perpetuate fictions like top ten lists, which have been so thoroughly discredited by revelations of manipulation, over the decades, as to render them virtually useless to all thinking consumers. Because navigating to media of interest imposes search costs, viewer reviews serve to minimize the time and effort spent searching for something to watch. They are potentially a very important feature of streaming media.

Not Dictation

As mentioned earlier, users take control of the pace and direction of the narrative in streaming media, if they choose to do so. By comparison, television and radio demand that you sit, watch, and listen, getting only the information they want to give, at the pace they set. They dictate to you and you are meant to passively note the dictated information verbatim. Because television producers must develop programming to appeal to the largest audience they can, to make best use of the scarce distribution channels, they tend to pitch the narrative's intelligence level at the lowest common denominator. This is an unnatural mode of human communications. In face-to-face communications, speakers adjust their delivery and pace by gauging the intelligence and interest level of the audience at every point.

Streaming media allows better accommodation of the normal mode of interhuman communication, where delivery is tailored to the intelligence, assumed knowledge, and interests of the audience on an individual and continuous basis. The technology allows for individualization of the presentation, guided by the viewer. Jumping ahead and going off on sidebar tangents are easily accommodated. Content producers who understand this fundamental difference between broadcast programming and streaming media programming will make streaming programs that speak to their audience in a more intimate, collegial, and ultimately more satisfactory way. Indeed the entire grammar of the documentary narrative is radically changed with streaming media technology.

Because the quest for mass audiences no longer exists (or at least, need not be important to every streaming media producer), the need to sensationalize or spin the story, using suspense to hold the viewers' interest, is removed. The program maker can make various assumptions about the audience and tailor the story's structure appropriately. All audiovisual presentation is storytelling, even factual programming. If you are addressing an audience of experts with a streaming media program, there is little need to use delaying tactics, as some television documentaries do today, to keep the audience watching for the requisite 50 minutes. Indeed, several versions, edited from common assets, can be released, appealing to novices and experts alike. Streaming media technology allows these distinct story structures to coexist in one media stream, with the viewer making a selection at playback time from available options. Content producers can use streaming media cleverly, to tell the story more like the way a real storyteller would tell a real audience, with appeal individualized to each member of the audience.

The Return of Community

Communities of interest in the subject matter of a streaming media presentation can form; members can regularly interact with each other and share collective wisdom. Associating a chat room with a streaming media clip is not difficult. With a community of interest that is limited by geography, there is often an insufficient critical mass of members to sustain the community. With a global medium, such as streaming media, however, communities of interest can form spontaneously and can be long lasting, since there are enough people to keep up the group's momentum. These communities never sleep, as members enter and leave the discussion from all around the world, following the sun. Members may be physically distant, but very close indeed in mind.

Everyone Is Beautiful—Avatars

One of the more exotic applications of streaming media is in live video chat, where two people talk to each other using a fictional video representation of themselves. MPEG-4 allows for transmission of an avatar, instead of a live face. In other words, you can send a "cartoon character," which may be photorealistic, in place of your own image. For strangers wishing to preserve their anonymity or for roleplaying, this technology can be fun. You can pretend to be as picture-perfect as you like. Implementations are not, as yet, widely available, but the technology of streaming media compression and transmission provides for this application. Of course, the potential for fraud also exists whenever a technology allows people to present themselves as something they are not, but provided the streaming media player makes it explicitly clear when an avatar is being used and when it isn't, such problems should be avoidable.

Content is King

The final word on why streaming media is a superior communications medium is that it enables better content to be created, which in turn creates better, more engaging and enchanting viewer experiences. The better the content and the experience, the more effective it will be at telling its particular story and the more likely it is to be noticed. Streaming media technology also removes the authoritarian power of the gatekeepers to media channels that exist in traditional broadcast media. The public gets what the public wants, not some broadcast network's opinion of what that might or should be.

Who Is Driving Streaming Media's Innovation?

Much of the innovation in streaming media technology is in the hands of small startup companies. A few larger companies have invested heavily in developing the technology, but many of the better-known "old media" companies are not among them. Many of those companies are, instead, supporting the Multimedia Home Platform (MHP), which seeks to converge broadcast and the Internet in a common application programming interface (API) for set-top boxes and other computer-based devices. This technology allows IP content to be embedded in a DVB digital video broadcast stream, but significantly sticks to MPEG-2 compression for the video and does not usually allow video to be streamed as an IP data stream. In other words, it is a play to protect the existing digital television broadcast infrastructure, by adding multimedia and interactive applications. As outlined previously, MHP is a transitional technology that can only compete while the fastest, cheapest, and most reliable bandwidth source to the home is a DVB stream and while embedding a fully-blown PC in a display device or creating a PC-based home digital media gateway costs more than building and writing software for non-PC set-top boxes. When this is not the case, the "walled garden" approach to the provision of multimedia and interactive content becomes unattractive to consumers.

In content provision, many more of the companies normally associated with television are active among the ranks. In terms of distribution and bandwidth provision, the telecommunications companies are represented, but are mostly focused on data transmission, not specifically on streaming.

In streaming media technology, innovations tend to be proprietary at first, with standards bodies following behind to open up the technology to more competitors and to standardize what, in effect, are commercial prototypes. Standards that lead the technology generally founder.

This section will focus on innovators expressly interested in streaming, rather than enablers who also focus on other applications of their broadband delivery technology.

Microsoft

As a company, Microsoft has been committed to streaming media for at least the last half decade and digital media for about a decade. The company has continued to invest in platform technologies that make it possible for application writers to use the Microsoft Windows platform to create digital media offerings.

The company's Digital Media Division is responsible for Windows Media Technologies. Because Microsoft has a core technology called DirectX, which exploits the PC architecture and Windows operating system to make high-performance digital media applications relatively easy to create, both the Windows Media Technologies and the Xbox gaming platform are built on top of that application programming interface (API). It isn't clear whether or not a third party could recreate a Windows Media Technologies suite on top of the public DirectX API, but the core DirectX technologies are used. Suffice it to say that the software development kits available for Windows Media Technologies, DirectX, and Xbox allow independent software vendors (ISVs) to create powerful and innovative digital media applications, using freely available tools and platforms (provided they are from Microsoft, of course).

Microsoft claims to have an MPEG-4-compliant codec included in Windows Media Technologies, but also claims that its codecs have a "special sauce," or proprietary advantage, that allows them to produce better images than competitors' products can. Only a Windows Media player can play Microsoft's MPEG-4 compliant files, at the time of writing. Hence, either MPEG-4 compliance is so loosely defined as to allow incompatibility between competitors' products, or else compliance has many levels, making it possible to comply with the file format or container format, but not necessarily with the codec, for example.

Microsoft has sought to cross-fertilize its other product offerings, using digital media technologies to advantage in a number of its products. Hence, the Windows 2000 Server platform, for example, is shipped with Windows Media Technologies, allowing it to be deployed in media serving applications. PowerPoint 2002 can be used to write synchronized multimedia presentations. Digital media, including streaming audiovisual content, is important to the company in a number of ways. It helps differentiate Microsoft products and provides reasons to buy into the entire Windows philosophy.

The company has begun to port its Windows Media Technologies to mobile devices, via mobile operating systems such as Windows CE. Microsoft is also encouraging companies that make mobile devices to embed Windows Media Codecs into their products. Already, there are Windows Media Technology-compliant portable jukeboxes, video cameras, and soon there will be DVD players. Microsoft has also developed a suite of Digital Rights Management tools and software development kits, designed to protect copyrights and hence kick-start online digital media commerce.

Unlike other companies serving the streaming media technology market, Microsoft has not sought to become a content server or bandwidth provider, but does have a directory site that directs viewers to content in Windows Media format (www.windowsmedia.com).

Given the company's past record in leveraging key core technologies across all of its product lines and offerings, we can predict Microsoft will incorporate its digital media technologies into its .NET executable Internet platform. A betting man might christen the offering "Media.NET" or

"DirectX.NET." The prospect of making digital media services available as software components that can be executed remotely across the Web, in conjunction with all the other Web services that Microsoft is already planning and deploying, is quite intriguing.

Real Networks and Intel Architecture Labs

RealNetworks was one of the pioneers of streaming media across the public Internet and continues to innovate, creating products of increasing maturity and capability. RealNetworks was started by ex-Microsoft visionary Rob Glaser, so the company has always enjoyed a sometimes close, but often adversarial relationship with its strongest competitor. Microsoft once invested in RealNetworks, but also ran into disputes over the behavior of player software that hijacked the RealAudio and RealVideo file types on installation.

Some of the early work on video compression technology, which appeared in Real Networks' products, had its genesis in work done at Intel Architecture Laboratories. IAL was established to help create reasons for people to buy Intel processors of increasing power and speed. The kernel of the compression technology has changed over the years. At one point, the use of wavelets instead of discrete cosine transforms (DCT) was seen as a way forward. According to some opinion in the industry, wavelet compression has a great deal more development potential in it than DCT-based compression. Evidence for this is the adoption of wavelet techniques in the JPEG2000 specification for still image compression.

The company has historically striven to improve the quality of the video and audio that the systems' codecs can produce. It has successfully managed to improve codec quality, in a head-to-head battle with Microsoft. According to informed opinion, there is very little to separate the two companies' offerings.

Recently, RealNetworks introduced its RealOne system, consisting of a new player, a content hosting service and a media serving and distribution platform. It has also joined forces with third generation mobile cellular network companies to port the Real Player to mobile devices and to influence the design of third-generation networks so that they support streaming media well.

The company also remains committed to distribution and media serving solutions that improve quality of service. Real System iQ includes technologies to provide flawless transmission of streaming media over the public Internet, to load balance media servers, to allow streams to bypass faulty connections, to redistribute the load automatically when a server fails, to protect the rights of content owners using XMCL (eXtensible Media Commerce Language) technology and to provide media player software that integrates extremely well with established Web content and content-authoring tools, via JavaScript.

RealNetworks has always provided software development kits (SDKs) to encourage third-party independent software vendors to incorporate RealNetworks' streaming media technology into their own products. The company's attitude is that the more users of their technology that exist, the stronger the company's position in the market. So far, this strategy has served the company well.

At the time of writing, RealNetworks had recently announced its commitment to adopting the MPEG-4 standard in an effort to create a single, global, streaming media format. Although the company's player supported the format, through a plug-in developed in conjunction with Envivio, when the announcement was made, encoding software and media server products did not. It will be a measure of RealNetworks commitment to see how long it will take before the MPEG-4 format is supported across all the company's products.

Apple

Apple was one of the first companies to deliver a digital media container technology that enabled desktop applications of digital media. Called QuickTime, this highly versatile digital media container was codec-agnostic, allowing media compressed in any way to be carried in a QuickTime wrapper and decoded by a QuickTime player with appropriate codec plug-in. QuickTime was initially only available on the Apple Macintosh, but was later ported to run on archrival Microsoft's Windows operating system. QuickTime was a late adopter of streaming technology. Apple initially required users to download the entire media package before playing it. Later, Apple introduced progressive download, in which the download started a long time before playback and there was no guarantee that the download would not be caught by the playback process, leaving the player with nothing to play. Apple now has a streaming solution, developed in conjunction with third-party codec developer Sorenson.

More recently, Apple has joined the Internet Streaming Media Alliance (ISMA), along with Sun Microsystems, Cisco Systems, IBM, Kasenna, and Philips, to promote a standards-based, bandwidth-scalable, player-neutral streaming media format. The goal is to allow a choice of products and solutions from different vendors, assured of interoperability. The preferred technology is MPEG-4, accommodating up to 1.5 megabits per second bit rates, with RTP (Real-time Transport Protocol), RTSP (Real Time Streaming Protocol), and SDP (Session Descriptor Protocol). In fact, ISMA states that it is implementing ISMA 1.0 based on existing MPEG and IETF (Internet Engineering Task Force) standards. It is interesting to note the similarities between QuickTime's media container format and that of MPEG-4.

Apple is positioning its latest desktop computers as multimedia creation tools. The company wants to see desktop video production become as commonplace as desktop publishing already is. To this end, Apple ships Final Cut Pro, a high-specification nonlinear editing package based on QuickTime technology. The company has also provided the streaming media world with its open source Darwin streaming media server. Apple has not, to date, announced a digital rights management solution.

Apple also has some content partnerships, with film trailers from top Hollywood studios and short and live programs from several top television networks, available for viewing (but not copying) on the QuickTime Web site. The company appears firmly committed to the development of applications for digital media, particularly streaming media.

Sorenson

Originally a vendor of what was arguably the best codec available for the QuickTime format, this small company is highly focused on video quality, developing codecs that compete favorably with the best from Microsoft and RealNetworks. In a codec shootout, published by streamingmedia.com, dated November 2001, the reviewer concluded that video quality among these competitors is so close it probably shouldn't be taken into account when choosing a technology. Although Apple is known to have a codec of its own in the works, based on MPEG-4, for my money, what Sorenson has achieved is mightily impressive. In line with many other industry players, Sorenson, has an ISO-compliant MPEG-4 codec in development.

The Moving Picture Experts Group

The developments that take place under the auspices of this body are rapid and could fill a book of their own. The group has an impressive track record, with MPEG-1 and MPEG-2 now firmly established standards (although many pieces of MPEG-2-compliant equipment from different vendors fail to interoperate). MPEG has always managed to create standards that are future-proof, addressing issues that often have not been even addressed, let alone solved, by equipment vendors.

The group's latest offerings, MPEG-4, MPEG-7, and MPEG-21, will provide developers with several years of fun yet, since they allow advanced scene description and capture. Some of the technologies required to fully implement the standard are still laboratory curiosities. If a criticism can be leveled, it is that MPEG's standards are too complex and all encompassing, leaving much scope for design incompatibilities (whether or not intentional) when products are developed (I speak here as a developer). This is one of the reasons why industry alliances like ISMA come into existence: to ensure interoperability in a wider, systems sense, while focusing on standards compliance. That said, the standards tend to be popular and well supported. Only time will tell if this remains true for the streaming media applications relevant to MPEG's standards.

Other Vendors

It would be impossible to list all the small companies that are driving innovation in streaming media. That is not to denigrate their contributions. Small startup companies pioneer many of the key breakthroughs. Companies that are developing MPEG-4 products include Envivio, Dicas, DivX Networks, On2, 3IVX, Fraunhoffer IIS, DiamondBack, Vision and a host of others. A good resource is the MPEG-4 Industry Forum (whose Web site is at www.m4if.org). This body maintains lists of industry players aiming to contribute significantly to streaming media technology.

There are arguably just as many edge and content delivery networks, but not all of them are focused on streaming. Most people familiar with streaming media will have heard of Akamai and its traffic management solutions, but there are a host of others, including Digital Island. Some firms, like Kasenna, are developing video-on-demand delivery platforms. Other innovators, like Digital Fountain and vTrails, are developing novel peer-to-peer networks and multicasting solutions. Third-generation mobile network equipment suppliers, such as Ericsson, are also becoming more interested in streaming media applications.

Streaming media appliance makers in evidence include Sharp, Compaq, Nokia, Creative Labs, and others too numerous to mention. Some, like Oratrix and Sonic Foundry, are focused on tools for the creation of advanced streaming media content. All these companies create pieces to the puzzle, enabling streaming media to exist as a system.

Research

Further into the future, streaming media will greatly benefit from developments in pattern and facial recognition, making possible easy separation of people from scenery. Once this separation between scenes and scene objects is made, very much more efficient compression is possible. Video textures can be stretched over wire frame models generated in the player and extracted by the image recognition software.

Photogrammetry, in which two or more images are analyzed to extract a three-dimensional model, wrapped with the video captured, is a technology used in remote sensing to build three-dimensional terrains from satellite photographs. This technology is increasingly finding its way into video compression and production applications. Once an object can be understood and described as a photogrammetric model, future movements of that object need only simple vectors, not detailed bit maps.

Better compression schemes are in development. The ITU (International Telecommunications Union) has a standard imaginatively titled H.26L in development. This is a wavelet-based video coding scheme using overlapped block motion compensation and image warping prediction. The scheme is designed to be free of blocking artifacts for very high compression ratios.

Improvements in still image compression often help the streaming media industry as well. At present, a vector-based compression standard called JPEG2000 is in development, which, like proprietary rival VFZoom (Vector Format for Zooming), should allow users to zoom in on a compressed image and still get a picture of acceptable quality. Other emerging graphics file formats that may have an impact on streaming media include AT&T Labs DjVu, which produces files five to ten times smaller than JPEG, with seamless zooming and panning; BitJazz lossless photo-quality compression; LuraTech's LuraWave wavelet-based compression, MNG, the multiple image extension to the established PNG Portable Network Graphics format; and SVG Scalable Vector Graphics format, which rivals Macromedia's proprietary Flash and Shockwave formats.

What's Wrong with Streaming Media?

If streaming media is such a wonderful thing, why isn't it everywhere already? Why hasn't it become the dominant medium for audiovisual communications? The answer is that the streaming media system, from end-to-end, is not ready for prime time yet. The technology is well developed in parts, but the overall system needed to make the technology into a communications medium is incomplete. The end-user experience isn't good enough.

As was shown earlier in this book, bandwidth is still priced too high for it to compete effectively with other audiovisual delivery methods. Consumers, in the main, cannot afford to stream television-quality pictures to their homes via an IP network. Even if they could, many consumers do not yet have a way to connect to a broadband network. Those who do must be satisfied with the PC as the receiver of media, since portable and television-like streaming media receivers are not widely available in local electrical retail outlets. If consumers are fortunate enough to be able to afford the bandwidth, connect to a broadband network, and accept the limitations of the PC as a streaming receiver, there is precious little to watch. What is available often does not play flawlessly.

No single industry has been able to drive widespread adoption of streaming media alone. Without industry coordination, broadband vendors fail to attract customers, since there is little to watch and only PCs to watch it on. Receiver manufacturers, including those that make software players that run on PCs, cannot make money on streaming media receivers when there is so little to watch and when broadband connectivity is patchy and expensive. Content owners cannot make money with streamed content because the delivery network is not all there and the PC is not the ideal showcase for much of their content; they also fear widespread theft of their digital media assets, without strong rights management solutions in place. Consumers cannot see a strong reason to adopt streaming media, when bandwidth is difficult to obtain or expensive, there isn't much compelling content to choose from and the only player that realistically works is the PC, which does not produce TV-quality pictures.

Yet streaming media in one disguise *does* sell. The DVD is the fastest-growing consumer electronics product in history, projected to exceed 60 million units by 2004. Thus, in 2004, there will be more DVD players in US homes than there are VCRs. If ever an endorsement of compressed, streamed, digital audiovisual media were required, this is it. A DVD player and a streaming media player differ only in that a DVD gets its media stream from a disk, whereas a streaming media player receives its stream from a broadband network connection. Presentationally, the DVD platform is more limited and less standardized, in terms of authoring interactivity and synchronizable media elements, than the "Web stuff" that can be created to run with streaming media, as part of the same stream. Indeed, Microsoft has developed Windows Media for DVD players. Leading chipmakers for DVD players will support the technology in systems shipped in 2002. With Windows Media's four to one compression ratio advantage over MPEG2, the current compression technology used in DVDs, studios could put all the *Godfather* movies or an entire musician's discography on a single CD. The chipmakers signed up to support Windows Media for DVD include the big five, who supplied 90% of the DVD processors shipped last year.

Today there are over 200 million people communicating with each other using instant messaging services. Streaming media makes it possible for instant messaging to include audio and video. There can be little doubt that this is a strong growth opportunity for the streaming media industry.

Digital media has already achieved widespread consumer acceptance. Digital cameras and portable audio and video devices have all experienced strong growth. Consumers love to play around with digital media. Sales of CD burners, which many use to create personalized music selections or to store pictures they like, have outpaced sales of DVDs and digital cameras.

The success of DVD and instant messaging and the growth in digital media consumer devices proves that streaming media, delivered through IP carriers, can succeed, provided the streaming media industry, as a whole, gets the system-level offering right.

Audience Critical Mass

Even though 50 percent of Web users have seen streaming media on their PCs, the audience for sustained viewing needs to be built. The average number of hours per week that consumers spend watching streaming media programming is minuscule, compared with television viewing hours. This necessarily means that devices other than the PC for receiving media are required, since the PC is not a platform that the family can gather around in the evening and view from the comfort of soft chairs. Most PCs are not situated in the living room like this anyway.

Content producers make content with mass audience assumptions in mind, right down to the pace of the narrative, the structure of the story, the editing, and so on. To create audience critical mass, producers will have to create for niche interests, with individualization and interactivity in mind. Because each additional viewer costs as much to support as the first, since they all use the same amount of bandwidth, there are no economies of scale with achieving a larger audience for a scheduled program, as there is with digital television. With television, once you have paid for the transmission bandwidth, the more viewers that watch, the less each effectively costs to serve with that bandwidth, since no additional bandwidth is required for each additional viewer. With streaming media, unless it is multicast, the economies are only in spreading the cost of production over a larger audience. The cost of transmission scales linearly with the number of viewers.

The streaming media industry needs enough people watching streaming content enough of the time to make it viable. Today, individual consumers only rarely access streaming content.

Profitable Business Models

Unfortunately, the streaming media industry is littered with casualties. Companies with very good ideas have gone out of business. Big names like Pseudo, Hollywood.com, Digital Entertainment Network, and DemandVideo are no longer with us. Why?

The economics of bandwidth provision are often based on phone call pricing, when voice traffic is fast becoming an irrelevance, compared to data. Advertising revenue from streaming media presentations is very small. The content available still mimics television and fails to exploit the exciting features of the streaming medium. The end-user experience is not acceptable, in that it doesn't feel as responsive and seamless as television. All these factors serve to keep audience numbers low. Low audience numbers mean low revenues. With the end-to-end system in its current state, it is very difficult to make money. The ills of the economy have only served to exacerbate the situation.

Until it becomes possible for streaming media service, content, and technology providers to make a profit, streaming media will remain a curiosity, interesting only to technophiles. The economics of serving media will need to change. Most importantly, for streaming media to be profitable, companies need to find ways to add value to the end-user's experience and charge appropriately.

Ubiquitous Broadband Networks

At the time of writing, consumers in many parts of the world are still waiting for connection to broadband networks. Those who already have them don't have entirely satisfactory experiences, or much content available to them that would make good use of the speed. The quality of the broadband service is generally not guaranteed, so it is rare that any consumer has sufficient bandwidth available to watch a video stream so that it looks indistinguishable from television. Even with a cable modem or on DSL, there are buffering problems. Dedicated content delivery networks fail to deliver the goods because they cannot ensure that the edge is close enough to the end-user to entirely avoid the network quality problems they were set up to solve. The last-mile problem still exists.

Digital television uses MPEG-2 compression and between 3 and 12 megabits per second of bandwidth per stream, depending on the quality the network wishes to achieve. With MPEG-4 compression, comparable picture quality can be achieved with around 750 kilobits per second to 2 megabits per second of bandwidth, give or take a bit. Yet today's DSL connections are around 600 kilobits per second, dropping to less than that when many users are connected at once. Eighty percent of users are connected with lower-speed links, with the majority connected only with a 56-kilobit-per-second modem.

Connection to the home is one thing, but redistribution of streaming media around the home, from a home media gateway, also faces bandwidth problems. Right now, home networking is in disarray, since no single home-networking scheme proposed meets the consumer electronics industry's wish list: low cost, long range, little interference, high bandwidth, low power, and interoperability with legacy products. Two standards, 802.11 (WiFi) and 802.15 (dubbed WiMedia) are vying for supremacy. The 802.11 standard has already achieved widespread support as a wireless LAN, but for streaming applications, there are quality-of-service issues. The 802.15 Wireless Personal Area Network specifications, on the other hand, encompass Bluetooth and offer up to 55-megabits-per-second bandwidth over a range of 30 meters. 802.16 (also known as WirelessMAN or Wireless Metropolitan Area Networks) Broadband Wireless Access specifications that support the development of fixed broadband wireless access systems are still under development. Whatever network standard becomes the standard in the home, it will have to meet basic quality-of-service and security requirements, if it is to be useful for streaming media.

Another place where broadband networks are needed to enable adoption of streaming media, is in the car. Until quite recently, there were two leading and opposing standards for in-car broadband networking. One was AMI-C (Automotive Multimedia Interface Collaboration) and the other was MOST (Media Oriented Systems Transfer). Now, the AMI-C body has agreed to work together with MOST to promote their standard for 25-megabit-per-second fiber optical networking in the car. This will pave the way for in-car streaming applications, such as seatback movies, individualized music, satellite-assisted navigation, and more.

The lack of access to very fast broadband networks is undoubtedly a hindrance to widespread streaming media adoption. However, the situation is changing rapidly. This once insurmountable roadblock may be removed in the very near future.

Standards and Lack of Adherence to Them

The streaming media industry has been in a format war almost since streaming media technology was invented. Because individual technology vendors wanted to compete on the basis of their own particular compression systems and streaming file formats, content providers wanting to achieve maximum audience reach were faced with the prospect of having to offer their content in all the standards, at all the bit rates. This situation only served to dissuade content providers from entering the market and confused end-users, who were not particularly interested in whose player the media needed to be played on.

Rays of hope have emerged and tend to converge around a single standard: MPEG-4. The ISMA has worked to promote this idea. Many people benefit when there is a single standard. The World Wide Web is such a success because there is only one format for Web addresses (the URL or Universal Resource Locator) and one for Web page transfer (HTTP or Hyper Text Transfer Protocol). While player vendors try to attract audience share on the basis of their proprietary tweaks, the industry as a whole will suffer, since the overall audience will remain small. Competing for a larger share of a small audience makes far less sense than enabling a very large audience.

Quality of Video Service/ Quality of Experience

We have already tangentially touched on the quality-of-experience issue earlier in this chapter. With most PC-based streaming media players currently available, it is not uncommon to experience momentary playback stuttering due to conflict for resources within the PC itself, poor continuity between back-to-back clips, pauses in playback due to network traffic bottlenecks, and failure of player buffering to recover after loss of data packets. The transport controls often fail to work well, with the response quite unlike what is expected for even the cheapest VCR. Streaming media players haven't produced good results to date (though newer generation players like RealNetworks RealOne and Microsoft's Corona may have solved these problems).

At the encoding stage, poor deinterlacing of television images, motion jitter due to poor inverse telecine operations, and incorrect gamma correction are common faults in streaming video. Standards conversion and changing of frame rates and picture resolutions serve to degrade the streaming video presentation even further. There is often insufficient care taken to normalize audio levels from clip to clip as well, resulting in startling jumps in audio levels on playback.

Using a streaming media player can be quite a painful experience, compared to watching TV. These faults come under the category of quality of experience. The painfulness doesn't apply to the technical quality of the video and audio played back, though this is important. Rather, it refers to how seamlessly and unobtrusively the technology works. Essentially, if you notice the artifacts or have to restart the player in order to watch something, the quality of experience is poor. For streaming media to win widespread consumer acceptance, these glitches will not be acceptable. The industry will have to work to present streaming media in a slick, professional way that is at least as good as television, if not better.

Quality of Network Service

Closely related to quality of experience and quality of video performance is quality of network service. The truth about packet switched networks such as the Internet is that, unless you add intelligence to guarantee quality of service, there are no guarantees. With a switched-circuit technology, once you established a connection, it was there. You only lost it if the connection subsequently broke. With IP transfers, each packet finds its own route. The problem is that parts of the network can clog and individual packets can get caught in router queues. The more hops the packet takes from source to consumer, the more queues the packet can get stuck in.

Another factor that adversely affects quality of service on the Internet is the last mile of copper cabling that the phone company uses to deliver data. These local loops are often very old and not well maintained, so the connections can be electrically noisy, producing data packet loss. When most of the problem is due to the quality of the wiring from the phone company's exchange to the home, as is sometimes the case, even edge delivery networks cannot guarantee flawless streaming.

Ideally, the connection to the home would have far more bandwidth available than is required to maintain a media stream. If the peak transfer bandwidth is higher than the sustained bandwidth, it is possible to buffer rapidly, when the user selects a particular stream, cramming as much of the stream's data as possible into the streaming media player's buffer in the shortest possible time. Companies like burst.com have pioneered this intelligent buffering approach. What this minimizes is startup delay. This reserve cache of ready-to-watch video also provides protection against packet loss. The result is a more reliable playback of the media. End-to-end streaming systems like Microsoft's Corona and RealNetworks' Real System iQ incorporate similar quality-of-service techniques.

Forward error correction (FEC) techniques are also now finding application in streaming, as a way of guaranteeing quality of service. FEC techniques have been used widely for years to guarantee the quality of satellite transmissions and in reading data from a compact disk. FEC works by spreading redundant data over a longer time or greater code length. The resultant stream has more bits than would strictly be required to carry the raw streaming media payload, but the additional data, spread out over time, allows the connection to be broken momentarily or permits data packets to be entirely lost, without affecting the reconstruction of the original payload at the player. In other words, some of the data packets can be lost, but an exact reconstruction of the data stream can be recovered entirely, through the use of redundant data. Viterbi codes, BCH (Bose-Chaudhuri-Hocquenghem) codes, Reed Solomon codes, and XOR (Exclusive OR) codes are popular FEC implementations. XOR codes have been used with the RTP, according to the Internet Engineering Task Force's Request for Comment document RFC 2733.

Though not widely deployed, RSVP (Resource ReSerVation Protocol) can also help guarantee quality of network service. RSVP works by reserving Internet resources ahead of the transfer. In essence, it is a way for packets of data to avoid router queues. RSVP reserves the bandwidth. It provides the highest level of IP quality of service available. However, the complexity and overhead involved make it unsuitable for streaming across low-bandwidth connections and unsuitable for backbone traffic.

The presence of computer viruses and malicious denial-of-service (DoS) attacks can also degrade the quality of the network to the extent that media streams are interrupted. DoS attacks can be launched by one company against another, but are often the result of hacker activity. They work by swamping the resources on the network, either by generating storms of needless traffic or by requesting database lookups that bring DNS (Domain Name System) services to a halt.

Hackers are hard to eradicate. There are so many ways to create a virus or denial-of-service attack. Antivirus companies can only signature and inoculate against viruses after the fact. The source code to create a virus or DoS attack is protected in the US (though not in the UK) as freedom of expression and this code is traded freely on virus exchange Web sites. People posting this code cannot be prosecuted, because it is difficult to prove malicious intent. Most hackers claim they are merely alerting the world to potential security flaws in commercial software. If a virus or attack is launched, the IP address of the originating computer can be eventually traced, but it is difficult to prove who was at the keyboard at the time, or indeed if there was anyone at the keyboard at all. Hackers are often minors who cannot be prosecuted under the law, even though the penalties are already quite harsh. For example, in the US, a hacker who does a mere $5000 of damage can go to jail for ten years. The recent Nimda virus attack cost one small company I know over three times that amount.

In some countries, there are no laws against computer viruses and denial-of-service attacks. Hackers who launch these attacks don't think they will ever be caught. Indeed, the more Draconian the penalties, the "cooler" some people think it is to launch an attack. Many young hackers don't realize the damage they can do through their activities. In fact, there are many cases where hackers who were shown the devastation they caused changed their ways and beliefs overnight. Society sends mixed signals to hackers. Whereas on the one hand they are condemned as senseless vandals, the press tends to describe them as lone geniuses and job offers often flood in, once a hacker is caught in the act.

There are undoubtedly some ideologues among the hacker fraternity who claim they are fighting corporate greed, capitalism, and people who don't want information to be free to everybody. What these ideologues rarely explain is who will provide and pay for the content and infrastructure in the utopian world they are fighting to create.

Edge networks can provide some protection against denial-of-service attacks, since they can dynamically route content requests away from congested servers to resources that are available. It is for this very reason that Microsoft recently moved its DNS servers to the edge of the Internet. Unfortunately, the affected servers must still be cleaned up, taking them out of action for some time.

So, until such time as network quality of service can be assured, even from malicious attack, streaming media will find it difficult to compete with the guaranteed service provided by digital television transmission networks (although even these are subject to occasional outages). Fortunately, techniques and defenses are being developed and deployed with great rapidity.

Receivers and Players

Today, streaming media can be watched on a PC, through some set-top boxes, and on some of the newer mobile handsets and PDAs. None of these is a wholly satisfactory device for watching streaming media. What is required is a purpose-built entertainment machine, designed specifically to receive and present streaming media.

Powerful game consoles, like Microsoft's Xbox might be the "Trojan horse" in the living room that first introduces the world to the joys of streaming media. However, even this platform and the current crop of DVD players have their limitations where presentation of streaming media is concerned. Consumer electronics designers have not yet really addressed the problem of designing a streaming media reception and presentation system. What is known is that the device will need to accommodate far-field and near-field viewing simultaneously, since some content will look best displayed on a large screen on the other side of the room and some will require closer inspection on a palmtop device, for example. What is important is that the one streaming media program may require both display devices.

The software that runs on these future streaming media devices will need to be more carefully designed, from a human interaction point of view, than today's downloadable desktop streaming media players. If the hardware is used optimally, it is possible to play video back flawlessly using only an old 486 CPU (many early nonlinear editing systems did just that). Yet even the latest 2GHz Pentium 4 machine can stutter and choke when playing back a simple AVI file. The problem is not in the power of the underlying hardware, but in the real-time behavior of the software application and the operating system upon which it runs. Unless attention is paid to building both of these for flawless video playback, glitches will be inevitable. Designers of player software have not yet understood that it is unacceptable to randomly to drop frames of video, if the underlying software tasks cannot complete their functions in time, due to processor scheduling conflicts or resource contentions. Rather, the software must be designed so that this never happens.

Content Providers—Where Are the Big Names?

Because of the cost of producing feature films, which can run into hundreds of millions of dollars, Hollywood studios are understandably nervous about delivering digital content over the Web. They have a large investment riding on the strength of any digital rights management systems used to protect their assets.

The film industry was just as nervous when VCRs were introduced, yet what actually happened was that they were given another distribution channel for their media assets. Because their content was more readily available, they actually wound up making more money. Streaming media distribution is also likely to turn out to be just another distribution channel. What the major studios need to manage is the time value of their assets. The older the film, the less it is worth to pirates. What digital rights management systems need to achieve is to make cracking the encryption and protection more costly than the value of the digital asset. With the cost of computing coming down according to Moore's law, this cost is a continually moving target.

In fact, the major studios and content owners are not so much afraid of streaming as they are of digital media, because copying is easy and perfect and worldwide distribution of digital media takes only seconds. Streaming is just the messenger.

Streaming media creates opportunities for individual production companies to bypass the television networks and the Hollywood studio system altogether, since they could conceivably host their own content. If the makers of hit comedy series served all their content on demand, for a fee, and also sold merchandise relating to the series online, they could keep a larger proportion of the revenue for themselves. The catch is that in order to have a hit series, you need a mass audience. At present, the Hollywood studios and the larger broadcast networks are masters at building and serving mass audiences.

Major content owners such as the big record labels, broadcast television networks, and Hollywood studios have begun to take streaming seriously. The RealOne service has content distribution agreements with ABC, Fox, *The Wall Street Journal,* and others. Sites like Disney's Movies.com and Moviefly.com, a venture involving Sony Pictures, MGM, Paramount Pictures, Warner Brothers, and Universal are already making feature films available on the Web. Unfortunately, not every site can stream every movie, or even give information about it. However, it is reasonable to expect that streaming versions of movies on those sites will follow today's downloadable video.

Fresh Searches

When somebody publishes something new to the World Wide Web it can take up to two months or more before the search engines notice. It is estimated that even the most thorough search engines miss an incredible proportion of what actually exists on the Web. There is some content that isn't indexed in any search engine.

Hence, search engines are not only woefully out of date, at any given time, but also have incomplete knowledge about what is actually available. For streaming media on demand, this is a disastrous state of affairs. What is more important to viewers of streaming media is currency, rather than complete coverage, though both matter. Breaking news has a time value. Today, the only way people can find streaming media content that was freshly published is to go to a trusted news portal and look for the "breaking news" links.

Another possible solution is to develop new distributed searching techniques, which could make use of software agents. In this distributed search, somebody requesting information would actually send out a multicast message to all the search agents on the streaming media hosts where new streaming items were hosted. Because each search agent had fresh knowledge about the presence of new media items on its own server, it could do a match against the search specification locally, matching against only those media items on the server. Results from the search request would be posted back asynchronously to the person issuing the search.

Results from machines closer to the user would return first, whereas those from distant or busy machines would return later. People reading the search results could continue to wait for results for as long as it took to find what they were looking for, or else abort the search once they had found something of interest. If the search were aborted, another multicast message would be sent to all the search agents to discontinue the search and discard results.

The advantage of agent-based searching is that locally available and current media clips are reported first. The disadvantage is that complete exhaustive searching would take a very long time. In this case, it is better to rely on one of the traditional search engines.

For individual media servers, the search load could be significant, but the data set is limited and the rate of change of that data set is low, so an index could be cached. Technologies like Microsoft's .NET platform could enable the development of such distributed search methods. Search requests could also be limited by geography or by metadata keys, for improved search performance. Using the executable Internet to create fresh searches could be an important impetus for widespread streaming media adoption.

Web Publishing Issues

Publishing streaming media content to the Web is still relatively complex. Embedding streams into pages was, until quite recently, something only the specialist Web designers knew how to do. Encoding content required that video be digitized into a computer and then converted to a streaming media format using an encoding application, like discreet's Cleaner 5.1. Synchronized multimedia authoring tools that work well with streaming audio and video have only recently become available. Embedding, encoding and synchronizing media elements are all separate authoring processes, which require importing of files into separate applications.

Many Web site hosting companies are not equipped to deliver streaming content, since they do not provide streaming media serving hardware. Additionally, media servers from some vendors are licensed on the basis of the number of concurrent streams served, so hosting streaming media incurs additional costs. Streaming uses more bandwidth, the cost of which must be usually be borne by the Web hosting company. Web site hosting companies also see streaming media as a "bandwidth hog," which will degrade the quality of response for other static Web sites that they host, simply because their connectivity is at risk of being swamped by media streams.

Specialized content delivery networks exist and there are companies such as Digital Island and Akamai that can help replicate content to the edge of the Internet. However, to use their services, content publishers must negotiate content delivery agreements. This is beyond the capability of the casual streaming media content publisher. There are few, if any, content delivery companies that allow a small publisher to do an online e-commerce transaction, in order to take their streaming content and deliver it. Almost all require registration and a business agreement to be in place before publishing can start. This used to be the case with static Web hosting too, but these days Web-hosting services are available online with a few clicks of the mouse.

Digital rights management tools have not been widely available to date, making it difficult for content owners to protect their streaming media content. As a publisher, you not only need an agreement with the rights management system vendor (Microsoft, for example), you also need an arrangement with a company that will act as third-party notary for the online license transactions, an online payments company, and a company that will administer licenses for media on your behalf. That represents a complex business structure, beyond the scope of most small content producers today.

Even if you succeed in getting a piece of streaming media up on the Web and served, getting the search engines to recognize its existence can take a further six to eight weeks, as mentioned above. Search engines work by trawling the Web, or else by indexing the Web by hand. There is a vast backlog of sites to be indexed at any given time. For this reason, getting the content online is one thing. Getting search engines to find it is quite another, particularly since most search engines are sensitive to text on the page and metatags, but not equipped to read metadata embedded in a media stream.

If a publisher wishes to publish streaming media content to a number of sites, he can either use expensive content syndication systems and services, or else ask each site owner taking the content to revise and republish pages. Automated syndication for small publishers is still a dream.

Peer-to-peer content replication is just beginning to be available again. Napster was very popular for digital music, but few peer-to-peer networks existed for streaming video. The problems were that rights management was nonexistent and most people who could participate in file sharing of video did not have sufficiently high bandwidth connectivity to make the system workable. Even with digital music, users with high-speed connections often had to wait several minutes, to download music files. Until broadband connectivity becomes the norm, peer-to-peer sharing of streaming video content will not be easy or fast.

Mobile Networks and Devices

With few notable exceptions, mobile networks, at the time of writing, have insufficient bandwidth to enable streaming media applications (other than simple voice communications). Third-generation networks are not here yet and those in trial have unearthed bandwidth-limiting issues that remain to be resolved. The promise has not yet been fulfilled.

In addition, handsets currently available for mobile applications have not been thought through specifically for streaming media communications. Beyond adding a cursory camera and painting video pictures to slow-response, poor contrast, and viewing angle LCD screens, little thought has been given to the design of the applications or the underlying streaming performance of the operating system upon which they must run. The next generation of handsets and PDAs will make better use of streaming media.

Most mobile operating systems available today have their roots in isolated "island of technology" designs. They were originally formulated to be self-contained picocomputers, running handy applications for making notes and remembering phone numbers and appointments. Connection to networks was added on, almost as an afterthought in most cases. Yet connectivity is a fundamental property of streaming media, which mobile operating systems struggle to support well. Synchronization to other machines and data sources is also a vitally important property of mobile operating systems. Support for synchronization varies markedly in sophistication from platform to platform.

In the future, we can expect to see devices optimized for streaming media communications, including streaming cameras, PDAs with fast response screens, and mobile streaming media software applications that are easy to use. We'll also begin to see wireless, but not necessarily mobile, devices designed for installation in the home, which will bring streaming media into our lives more and more. We'll also begin to see streaming media in the car. These devices are still on the drawing boards, or only available as prototypes.

Cost-Effective Content Production

Producing streaming media is a labor-intensive process. Storyboarding, scripting, shooting, editing, and post-production processes require time and skilled people. Even though the tools used in these processes keep getting cheaper and more productive, there is an ultimate limit to how much technology can do to lower the costs of production. Yet production costs must fall. Even before streaming media emerged, the fragmentation of broadcast television audiences was already taking place. People had more to choose from, so each program made had less money to play with in production. Budgets were already under pressure.

Addressing the craft-shop nature of media production is the only way to solve the problem. Production companies need to adopt a few production-line techniques. What counts more is not the artist's sensibility to choose precisely the correct shade of blue or getting the light in exactly the right place, after eight hours of experimentation, but how productively an artist can produce quality work. Desktop tools are an obvious necessity and many improvements can still be made to these tools, as standalone packages. However, the tools for workgroup media production facilitation, like media management tools, file sharing systems, production planners, and organizers and automated sign-off and review procedures are still in their infancy. Most digital media production plants still battle with the need to transcode files from one format into another so that they can be passed from operator to operator and tool to tool. Much of the media passed around the production plant goes by foot. Catalogues of media assets are patchy and incomplete. Standard templates and ways of working are not well codified. Media production is not unlike software production. Tools, which have been developed for configuration management and version control in software production, can be readily adapted to media production.

Most streaming media production today takes place downstream from regular and expensive television production. This means that the streaming media presentation mimics television programming, thereby missing out on the more exciting opportunities for creating more engaging programming offered by the streaming medium. It also means that the production process cannot be optimized for streaming, to take advantage of cost savings available when one does not need to produce for television. For example, in the case of live studio production, it is almost impossible to obtain digital production switchers and automated lighting and camera controllers, which make use of cheaper fluorescent lighting systems and consumer-grade low-light digital cameras. All the available equipment is designed for live television production. This means it produces images and sounds that are far too high in quality to be economical for streaming, with a price tag to match. The available equipment is overdesigned and is overkill for streaming media production.

Multi-skilling is another necessary step in lowering the costs of media production. There are often problems with asking a production crew member to do things not normally in his or her job description, both from a union point of view and because production staff ought to be paid fairly, not taken on as juniors and then asked to do more senior tasks. However, some flexibility is needed, so that production crews can do more with fewer people.

The more that people who aren't media professionals become involved in desktop media production, the more they will need to understand the essential fundamentals of audiovisual communications. Just as when creating a PowerPoint presentation, it is important to understand how to tell the story in an engaging and exciting way, with enough structure apparent to ensure the audience gets the point. It is even more important to understand the grammar of motion pictures and sound to make good streaming media presentations, cost effectively. If the person creating the programming takes fewer missteps and can work quickly, without the need for major revisions and iterations, this will also drive the cost of streaming media production down. Education in visual communications is therefore an essential element in enabling cost-effective streaming media production.

Streaming media can be and is successfully produced with nothing more than a camera and a laptop computer. In many cases, the footage can be downloaded into the laptop, edited, and encoded for streaming in the back of the car as the crew is returning from a shoot. This rapid turnaround time, coupled with low production costs, gives the production company an added advantage, since it can often scoop the major television networks.

A Killer Application?

The streaming media industry is still searching for a killer application. Arguments rage about whether it will be video on demand, video conference calling from wherever you happen to be, keeping track of your children, teenagers using it to flirt, shopping in virtual showrooms, or a plethora of other possibilities. Nobody knows what will cause sudden and widespread adoption of streaming media technology, or if in fact it will be one killer application or many. What is clear is that desktop killer applications will not necessarily be the same as mobile killer applications.

Many technologists believe that "pervasive computing" is inevitable and will necessarily include streaming media as a first-class data type. The factors that will drive the next killer application will be consumer desire for immediacy and a high level of individualization, both of which present significant challenges for system operators. Personal relevance is the issue. Designers need to observe the daily lives of their customers. In real life, people like devices that can easily do the most basic things. All some people want is a time management system that organizes information from incoming phone calls, media streams, text messages, etc. into a wireless schedule. To quote Nokia's manager of product launches, Scott Gaines, "The killer app is you."

When Will Streaming Media Be Ready for Prime Time?

This section attempts to extrapolate from the current state of streaming media technology to a time when the system will be ready for mainstream adoption by the majority of people. Like all exercises in crystal ball gazing, the number of variables is enormous.

The economic landscape was being radically redrawn immediately following the infamous "dot-com crash" of 2001. Many streaming media companies were badly affected by the fallout, as investor confidence in technology stocks evaporated almost overnight. Following the atrocities of September 11, 2001, economic conditions worsened markedly. The United States officially entered recession, with a general economic slowdown in the US significantly affecting the global economy. After years of propping up moribund industries, the Japanese economy also struggled with its almost bankrupt financial system and spectacular government debt. Intended to stimulate the Japanese economy, but failing to have any noticeable effect on general consumer confidence, government spending was directed mostly into construction projects to build "bridges to nowhere," but not into the development of a broadband network infrastructure. Toward the end of 2001, the South African economy also teetered precariously, under the weight of the unchecked AIDS epidemic throughout the African subcontinent and much political instability, leading to the near collapse of the economies of nations such as Zimbabwe, a key supplier to South Africa. Argentina defaulted on its foreign debts, making it the biggest bankruptcy in history. Enron, once the doyen of the markets and considered almost "blue chip," collapsed due to irregularities in audited accounts, the largest *corporate* bankruptcy in history. Most of Eastern Europe continued to struggle with the transition from communist rule to market economics. Southeast Asian economies, once considered "tigers," had failed to perform for the past several years. The only relative bright spot was the resilience of the European Economic Union. This is obviously not the most stable of backgrounds against which to make predictions about a technology such as streaming media.

However, there are some trends that make it possible to make educated guesses about when gating items may be obviated. Regardless of economic conditions, even the extraordinary conditions that prevailed at the time of writing of this book, innovation historically tends to proceed at a fairly constant rate. The emergence of ideas seems to continue unabated, unaffected by outside influences. We can, therefore, identify key technologies and related factors which have, to date, prevented streaming media from reaching mainstream acceptance and divine where the innovation vectors might be pointing.

Broadband Penetration

Most pundits agree that streaming media will only trickle until broadband networks are widely deployed. This means that DSL (Digital Subscriber Lines), third-generation cellular networks and optical fiber to the home will have to penetrate more than about 30 percent of households for the industry to begin to take off.

A report from the Yankee Group* concluded that streaming media tools would become a significant part of the delivery of Internet content. "Rich media applications are problematic for users to download, because of the large file sizes," explained Amy Prehn, an analyst at the Yankee Group. "Streaming media has become established as a viable delivery channel for these complex applications, saving the user from significant delays since the entire file does not need to be downloaded." However, Prehn warned that the files would only be effective once quality high-speed networks were in place. "Streaming media files are highly sensitive to quality

---

A Look At The Streaming Media Value Chain, The Yankee Group, Boston, May 2001.

degradation due to packet loss, latency, and network congestion, and they effectively require a robust network infrastructure," she said.

In the US today, only about 40 percent of Americans are connected to the Internet and of those that are, about 20 percent have a broadband connection. So, relative to the total number of Americans (most of whom already have access to television and radio), broadband Internet connection only serves about 8 percent of the potential audience. "By the year 2005 there will be just as many households without Internet access, as those using cable modems or DSL," says Daryl Schoolar, a senior analyst at Cahners In-Stat Group. Dial-up will overshadow broadband connection in the US through to 2005, according to Schoolar. We have a long way to go before broadband penetration in the US reaches the critical mass necessary to fuel the adoption of streaming media.

There are several notable stars in broadband connectivity. Scandinavia and South Korea are already relatively well connected, though these nations remain the exception. According to Intel Chief Executive Craig Barrett, while acknowledging the European Union's efforts in trying fully to liberalize the telecommunications sector throughout the fifteen nation bloc, the adoption of broadband in Europe had been slowed down by delays in the opening up to competition of the "local loop" of telephone lines as well as high charges and a current lack of flat-rate pricing. Barrett pointed to the example of South Korea, which managed to achieve the world's highest broadband penetration through a mixture of government incentives and open competition. "It's strange that in a developing economy (Korea) we should have such a high broadband penetration. That is the result of a mix of government policy and strong competition," Barrett said. "Hopefully the EU governments and the U.S. will look at that as a symbol of what can be done."

According to declarative data from the TNS Establishment Survey, consisting of Computer Assisted Telephone Random Digit Dialing carried out by Taylor Nelson Sofres during August 2001, the number of European households connecting via broadband did not exceed 14% in any country and was less than 1% in many. Sweden led the way with 13.8% of households connected followed by Denmark with 13.2%. Germany, France, Spain and Norway scored 7.8, 6.4, 6.2 and 5.1% respectively. A paltry 2.3% of UK households have broadband connectivity, with Italy scoring a risible 0.9%.

Lobbyists Broadband Stakeholder Group in Britain has concluded that the UK government's broadband target of being the most competitive broadband market in the G7 by 2005 is hopelessly unattainable without radical government intervention. The UK ranks 22nd in the world in broadband penetration, according to figures published by the OECD (Organization for Economic Cooperation and Development), or, stated another way, in last place among the leading economically developed nations. Even with intervention to meet the government's stated aims, broadband penetration will struggle to exceed 24 percent by 2005, but on current estimates, it is unlikely to reach more than 6%. Even at 24% penetration the UK would still lag behind the US and Germany. Yet the economic and social prosperity of any town depends as much on broadband connectivity as on the quality of its transport links, according to the South East England Development Agency. Connectivity is as vital as good roads. The $32 billion third-generation license auctions that took place in the UK in 2000 are being held partially to blame. The auctions are thought to have effectively robbed the UK of telecom company investment in broadband.

Mainstream broadband connectivity continues to vary from country to country, but by 2004 more than 90 million households worldwide will be taking advantage of a high-speed connection, according to the Broadband Report from eMarketer.

South Korea and Canada are leading the broadband charge, with about a 50% penetration in each country. Most other countries are hovering around 10%. Demand remains high in the US, and the research firm expects the country to reach 30% penetration in households by 2004.

Accessing a broadband connection via cable will remain the access technology of choice through 2005, but DSL will continue to gain ground. In the United States alone, DSL subscribers will reach 13 million households by 2004, up from 1.5 million last year. The market for alternative broadband technologies—fixed wireless, satellite, and fiber—will top 32 million subscribers globally, up from 7 million in 2000.

Widespread adoption of high-speed Internet access connections has been slow because "the majority of mainstream Internet users still are not convinced of the value of upgrading to broadband," says Ben Macklin, a senior analyst at eMarketer. This is a chicken and egg scenario, of course. If the broadband connectivity is not there, it is difficult for applications that would convince consumers of the value of upgrading, such as streaming media, even to exist.

Cost has been another deterrent to widespread adoption in some countries. In England, where penetration has been relatively low, the average cost of service is roughly $60 a month. In South Korea, by contrast, service fees are about $25 a month, and the adoption rate is about 50%.

While all of this is seemingly gloomy reading, from the point of view of streaming media, it does nevertheless appear that a significant market for broadband streaming media will begin to be established around the year 2005. This is less than three years away.

The Fight for Rights

Until the squabbles between the owners of copyrights, represented by bodies such as the RIAA (Recording Industry Association of America) and the Internet file sharing and Internet radio companies reach some form of conclusion, streaming media will be prevented from reaching mainstream adoption. There are positive signs of early resolution, however. There was news of out-of-court agreements toward the end of 2001, along with several high-profile purchases of file sharing companies like Napster by established record labels like Bertelsmann. Hollywood was beginning to mobilize to make its content available online, if not streaming. Rights management solutions were being proposed by many companies, including streaming media heavyweights Microsoft and RealNetworks. It is likely that rights management will come of age during the latter part of 2002, or else in early 2003.

What remains less clear is whether or not consumer acceptance of rights management will be forthcoming. Rights management solutions restrict freedoms currently enjoyed by consumers of digital media and may be abused by rights holders to milk their customer base for higher profits.

Rights management may settle down to a form of shareware marketing, where consumers pay what they are able and according to what they think the media asset is worth to them. However, don't expect to see any such resolution between consumers and content owners before the end of 2003.

Digital Rights Management

To get an idea of the state of flux and lack of agreed standards presently applying to digital rights management, this section will scratch the tip of the iceberg and catalog some of the more relevant proposals under development. To explain each proposal would fill another book (*Digital Rights Management Demystified?*). Until the dust settles on these competing standards, adoption of streaming media by a mainstream audience will be problematic. Standardization may take up to five and perhaps even as long as ten years, judging by previous standards efforts applying to streaming media.

The Internet Digital Rights Management organization (www.idrm.org) has a proposal, under the umbrella of the Internet Engineering Task Force (IETF), to standardize digital rights management on the Internet. Its work is at an early stage and the group is currently surveying and cataloguing the problems involved in digital rights management on the Internet. The IDRM solution explicitly will *not* prevent illegal use of information through technical protection measures. It will, however, seek to use work done by the MPEG, adapting other useful technologies to its needs.

The Secure Digital Music Initiative (www.sdmi.org) is a forum that brings together more than 180 companies and organizations representing information technology, consumer electronics, security technology, and the worldwide recording industry and Internet service providers. SDMI's charter is to develop open technology specifications that protect the playing, storing, and distributing of digital music so that a new market for digital music may emerge. The open technology specifications released by SDMI will ultimately provide consumers with convenient access to music online and, in emerging digital distribution systems, enable copyright protection for artists' works and promote the development of new music-related business and technologies. DMAT (Digital Music Access Technology) is the trademark for products that are compliant with SDMI specifications. SDMI Portable Device Specification Version 1.0 has already been published. This organization had completed consideration of its phase two screening proposals in May 2001, but has not met since then (as of December 2001).

The Extensible rights Markup Language (XrML) has been proposed as a possible technology for digital rights management. It originates from work done by Xerox PARC (then called DPRL) in 1994 and provides specifications of rights and associated conditions and obligations for distributing digital content. It is based on XML (Extensible Markup Language), and version 2.0 of XrML was released in late November 2001. Unlike some of the proprietary rights management solutions already available, it supports gifting, library loan, site licensing, rental, personal lending, and payment to multiple rights holders, among a host of other business models. Some typical rights management terms are shown in Figure 2.28. XrML has been proposed to the MPEG as a possible technology to satisfy the requirements of the MPEG-21 standard. XrML is currently the only rights language being used in working DRM (Digital Rights Management) solutions, including DRM solutions from Microsoft.

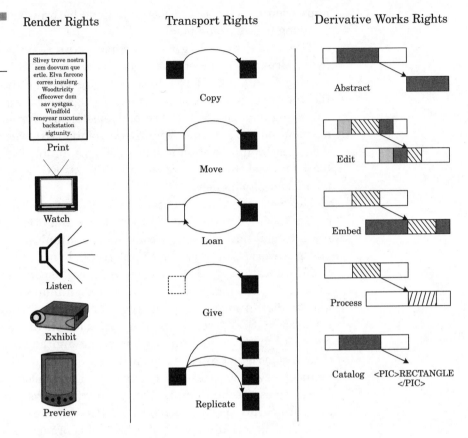

Figure 2.28
Some typical digital
rights.

The digital object identifier initiative (www.doi.org) has been around
since 1995. It was set up to provide a digital equivalent to the ISBN book
cataloguing system for digital media assets. The motivation for the technology can be summarized in a quote from *The Mystery of Capital: Why
Capitalism Succeeds in the West and Fails Everywhere Else* by Hernando
de Soto (2000): "Imagine a country where nobody can verify who owns
what, addresses cannot be easily verified, people cannot be made to pay
their debts, resources cannot conveniently be turned into money, ownership cannot be divided into shares, descriptions of assets are not standardized and cannot be easily compared, and the rules that govern property vary from neighborhood to neighborhood or even street to street. You
have just put yourself into the life of a developing country or former communist nation." Indeed, this quote accurately describes the current Internet as an environment for digital media commerce. To quote further: "One
of the most important things a formal property system does is transform

assets from a less accessible condition to a more accessible condition, so that they can do additional work. Unlike physical assets, representations are easily combined, divided, mobilized, and used to stimulate business deals. By uncoupling the economic features of an asset from their rigid, physical state, a representation makes the asset 'fungible'—able to be fashioned to suit practically any transaction." Of course, digital media assets are representations. The organization has issued its DOI handbook. The DOI is a system for identifying and exchanging intellectual property in the digital environment. It provides a framework for managing intellectual content, for linking customers with content suppliers and enabling automated copyright management of all types of media.

RealNetworks has been proposing another rights management solution in the form of XMCL (Extensible Media Commerce Language), included in the latest RealOne player and platform (www.xcml.org). It was set up as an initiative to promote an open-standard language to communicate digital media business rules, reasonably quickly, making maximum reuse of established specifications without trying to be too clever. An important feature of XMCL is that it tries to explain freedoms as well as restrictions. Like XrML, it is based on XML and includes digital signatures. It is a lighter-weight implementation than other systems proposed.

The Multimedia Rights Data Dictionary (2RDD) was begun in November 2001 under the umbrella of the DOI Foundation. This initiative includes the Motion Picture Association of America (MPAA), the Recording Industry Association of America (RIAA), and the International Federation of the Phonographic Industry (IFPI), among others. 2RDD will be a common dictionary or vocabulary for intellectual property rights, to enable the exchange of key information between content industries, and e-commerce trading of intellectual property rights. The dictionary will be submitted to the MPEG.

The Organization for Structured Information Standards (OASIS) is a non-profit, international consortium that creates interoperable industry specifications based on public standards such as XML. They have XACML (Extensible Access Control Markup Language), which is a specification, in XML, for expressing policies for information access over the Internet. They have a companion standards effort called SAML (Security Assertion Markup Language) for exchanging authentication and authorization information. OASIS is currently very active.

Another initiative is the Open Digital Rights Language (ORDL), which provides the semantics for a digital rights management expression language and data dictionary pertaining to all forms of digital content (www.ordl.net). The ORDL is a vocabulary for the expression of terms and conditions relating to digital content, including permissions, constraints, obligations, conditions, and agreements with rights holders. ORDL specification version 1.0 is available now. Current activity and future directions of this organization were hard to find.

Finally, we consider the initiatives for rights management that are being driven by the MPEG. MPEG refers to DRM as IPMP (Intellectual Property Management and Protection). Both MPEG-2 and MPEG-4 had provision for IP (Intellectual Property) datasets, using internationally recognized ID systems and provided hooks for proprietary protection systems. MPEG-7 includes identification of intellectual property, authentication of descriptions, and protection of those descriptions. However, the MPEG-7 IPMP system amounts to defaulting to internationally recognized ID systems. It only provides hooks to proprietary protection systems, just as MPEG-2 and MPEG-4 do. The current activity for MPEG is to extend MPEG-4's IPMP to enable interoperability. Thirteen submissions had been received, by October 2000. What rights management interoperability comes down to is that a rights language is nice, but not enough. Trust is essential. A content provider must trust the IPMP system. An IPMP system must trust the player. Players must trust the platform, and the content provider must trust the player and platform. Trust is not merely a technical issue. A Public Key Infrastructure will not suffice. There must be tamper-resistant implementations, but who will do the due diligence on players and who will check the platform?

MPEG-21 is putting all the IPMP elements together. In MPEG-21 all users have rights and interests and need to be able to express these. Many elements exist to build an infrastructure for the delivery and consumption of multimedia content. There is, however, no "big picture" to describe how the specification of these elements, either in existence or under development, relate to each other. The aim of MPEG-21 is to understand if and how these various elements fit together, to discuss which new standards may be required, if gaps in the infrastructure exist, and once those two things have been accomplished, actually to accomplish the integration of different standards.

To date, MPEG-21 has a working draft for a Digital Item Declaration—declaring the things that make up a digital item. They have proposals for Digital Item Identification and Description, with the capacity for unique identifiers and resolution. Management of personal and usage information by the end user is crucial, as is event reporting. They have called for requirements for a Rights Data Dictionary and Rights Expression Language.

In MPEG-21 parlance, there are Users. These include individuals, organizations, corporations, consortia, communities, governments, other standards bodies, etc. Users act in roles such as creators, consumers, rights holders, content providers, distributors, etc. User A and User B engage in a Transaction/Use/Relationship. They refer to a Digital Item. Under the transaction involving the Digital Item, there is an Authorization or Value Exchange. In order for this transaction to take place, six things need specification. These are:

- The Digital Item Declaration (e.g., a package, item or resource, including resources, metadata, and structure)
- The Content Representation (e.g., file format, codec, scalability, or header information)
- The Digital Item Identification and Description (e.g., unique identifiers, content descriptors, or resolution services)
- The Content Handling and Usage statement (e.g., storage management or content personalization)
- The Intellectual Property Management and Protection data (e.g., rights, permissions, encryption, and authentication)
- Terminals and Networks (e.g., event reporting, the e-commerce interface, performance metrics, or audit trail handles)

Basic MPEG-21 standards are scheduled for completion in 2002. This includes the DID (Digital Item Declaration) and DII&D (Digital Item Identification and Description). IPMP will take a while longer. For this reason, standards for digital rights management in streaming media will not exist until at least 2003. Streaming media will not be ready for prime time before then.

Mobile Media

Mobile media based on third-generation networks is another technology, essential for broad adoption of streaming media, which is suffering birth pains. There are precious few non-PC streaming media receivers designed for third-generation networks. Most of the network operators are struggling to get their infrastructure working to specification. Most will not be ready to handle streaming media content for some time to come.

In the home and in the car, standards for wireless or optical retransmission to portable devices are still in flux. Economical technologies for the mobile streaming media consumer are at least two years away and, more realistically, probably won't be widely available until 2005 at the earliest.

Appliances and Receivers

Broadband receiving equipment for the home still consists of either a PC or an advanced set-top box. There are few home media gateways under development. The one ray of hope is the influx of network-aware and DVD-capable game consoles that will be released widely during 2002. I was able to find very little information about in-car streaming devices, except those touted by telematics vendors for the far future. The problem with in-car designs is that they must last the life of the car, so it is important that standards be settled. Otherwise, the automobile manufacturers' wealthiest customers may find themselves with rapidly obsolete in-car entertainment systems, as standards and streaming technology continue to evolve.

Appliances and receivers for streaming media are not going to appear until after the broadband networks are in place. This means that streaming media will be an add-on to other devices, rather than commanding purpose-built appliances. We can expect devices to begin to see the light of day some time after 2005. Today's streaming media-enabled PDAs and mobile phone handsets will have to suffice until then.

Finding a Killer Application

Killer applications, in some senses, cannot be predicted and planned. They result from overwhelming consumer demand. That means that killer applications will not emerge until the technology is capable of supporting them. SMS messaging was a surprise success. The original intention of providing cellular phones to children was to enable them to stay in touch with their parents. Who could have known that they would subvert their use to flirt and date, or form little communities? SMS became a success because the technology was there and there were compelling reasons for a large user base to exchange low-cost, discreet (secret) messages.

What seems certain is that a killer application for streaming media has not yet been deployed. My guess is that the ability to schedule all incoming digital messages and media and organize those from whatever place and receiver you happen to have will be the most likely candidate.

The ability to effectively be in two places at once, interacting with people at remote locations as if you were really there, is the likely killer driver. However, until these applications are developed and deployed, nobody really knows which application will drive the mass widespread uptake of streaming media technology.

When Standards Prevail

We have alluded to the importance of standards in previous sections. Besides rights management standards, codec and player standards will have to settle better than they have to date. MPEG-4 looks like the technology of choice for compression and media descriptions. Several heavyweight industry players such as Apple and RealNetworks have aligned themselves around MPEG-4. Even Microsoft has some MPEG-4 support in an otherwise proprietary system.

Other standards that will have to settle include those for content syndication, streaming media transport, edge network caching, quality of service, and quality of experience. Today, there are competing standards, often proprietary, in each of these areas. Interoperability between products from different vendors is almost nonexistent. Imagine if you needed a different television set to watch different television networks. In a sense, that is what has happened with digital television. Different networks require you to use their digital set-top boxes and conditional access technology. If streaming media fails to standardize, it will hamper growth of the medium and prevent mass consumer uptake perhaps indefinitely.

Of all the factors that can prevent widespread consumer acceptance of streaming media, it seems to me that the glacial progress of standards is the biggest obstacle and the one with no definite finish date.

Sound Business Models

At the time of writing, very few companies have successfully found a way to make streaming media pay. There is some hope and indication that distance learning may be the first business to make streaming media profitable. The real reason that sound business models have not emerged is that the industry is only serving a population of technophiles and early adopters. There isn't yet enough of a market to milk for profits.

Sound business models eventually emerge once the technology settles to a standard and a majority of consumers adopt the technology. Early

DVD player vendors must have struggled to make the business pay. Every communications medium so far introduced has found a way to make money. I feel confident that as audience numbers grow and the technical solutions converge, there will be profitable and sound business models. However, this necessarily means that making money will be hard work until at least 2005.

Media Search Engines

Media search engines that make good use of metadata in order to index and find digital media items exist today, but are in their developmental infancy. These engines tend to focus on the text included with the media, but a few also do image and sound analysis in order to index media assets. These systems will undoubtedly develop in sophistication and usability. Virage is a vendor with a possible technical solution. Grass Valley Group's ContentShare also has potential in this area. However, there doesn't seem to be a widely deployed, agent-based, distributed search engine which can find content only recently published.

Several media search companies, among them Advanced Broadcast Technology in the UK, are developing hardware-based searching engines, to accelerate the process of trawling through billions of video fingerprints and signatures to find matching keys. Singingfish.com is a publicly available media search engine that can search for media-specific items. Search engine Google now allows searches on images as well as text. However, both of these are searching through textual descriptions. Taalee developed a more sophisticated version of the metadata search engine (see www.taalee.com), but Voquette has since acquired this.

Media search engines are a necessary component of the widespread adoption of streaming media, since without these, nobody will be able to find out what there is to watch or be alerted to newly published streaming media items. Text-based searches are better than nothing, but the only way to find fresh content today is to visit a known portal and notice what's new. These methods will suffice, but will ultimately be limiting. There is no accurate way to predict how long it will be before specialized media search engines will be widely available. It is likely, however, that demand will stimulate supply. When broadband critical mass is achieved, media search engines will rapidly follow. Thus, these search engines are likely to mature around 2005 or 2006.

Fast Seeking Support

One of the key differentiators between streaming media and television is the ability of streaming media players to skip through the boring bits. The ability to seek and fast forward and rewind is key. However, all the implementations of streaming media systems that I have seen to date fall short of expectations in this crucial area. Nearly all perform with unacceptable delay when jumping from point to point in the streaming media presentation. This need not be so.

All digital nonlinear audio and video editing systems have had to face this problem, in one form or another. Some of the better solutions to being able to jump around a piece of digital media were developed by Digital Audio Research and Lightworks in the UK. These companies found ways to ensure that no matter what the operator did, by way of jumping around a piece of digital media, the machines always provided near instantaneous response. In fact, both the DAR machines and the Lightworks nonlinear video editors gave users the impression of manipulating audiotape or film. It was possible to shuttle or instantly locate anywhere within the media, without noticing that the media was digital.

The key to this was various schemes of multilevel caching, with look-ahead prediction of media delivery, coupled with nested feedback. In other words, there was software on the machines that knew which part of the media the user was currently accessing. It then pulled media from ahead of and just behind the current play point. As the machine played through the media in a particular direction, the prebuffering was skewed to provide a larger buffer in the direction of playback. These systems also preloaded key frames of media, corresponding to scene changes or other significant marked points in the media. This allowed the user to jump instantaneously to these points, yet gave the multilevel caching software enough time to flush existing media requests and create new ones around the new play pointer.

The multilevel cache had to deal with several latencies, such as disk drive rotation and memory contention, but worked because the peak bandwidth available from source to player was several times that required for sustained playback. Another essential feature was the ability to send feedback to the source cache. The source media fetching task would respond in an intelligent way, discarding media requests no longer required. To date, there isn't a streaming media system that I am aware of which comes close to emulating this functionality. Connectionless models cannot, by definition, emulate the feedback required.

Schemes that do not burst load caches also cannot achieve the necessary prebuffering. I am not aware of any streaming media system that adjusts its prebuffering strategy in response to play speed and direction. I also do not know of a system that preloads key frames, to allow instantaneous jumps through the media at the player, without having to wait for the server to respond.

Without snappy and crisp media control at the player, streaming media will fail to meet consumer expectations and won't achieve widespread adoption. It is impossible to say when crisp and instantaneous media control will be available on streaming media players, since I am not aware of even a single solution in development.

The Audience

Who Will Watch?

For streaming media to succeed, it's not enough to have great technology. The audience must want to watch. For individuals to bother to make the switch, the medium must have characteristics that match consumer dispositions, desires, and aspirations more closely than what is currently available. Knowing where the audience will come from allows streaming media vendors to better understand and serve those people who will one day create their revenues.

This part of the book will characterize the audience in different ways, to reveal reasons why various groups might want to receive streaming media and what its appeal to them is.

Society has undoubtedly changed since television was first introduced. The people who embraced that technology are not the people who will embrace streaming media. Today's audience has different values and interests. The model of what a communications medium should be is very different from the view held by television's first audiences. They are ready to strike a different deal with providers of news, entertainment, and information.

Demographics

Children born since the Internet and home computing existed, the "digital generation," will be a more influential and powerful demographic group than the baby boomers. They already are. They are more numerous and have greater spending power than the baby boom generation born after the World War II. These children did not learn about digital media, they assimilated it, as a part of the environment in which they grew up. Games, videos, and the Internet are just a part of the landscape. Don Tapscott, in his book *Growing Up Digital*, has a very good analysis of this demographic group.

Children who grew up with video recorders have a great deal of difficulty understanding why all programs aren't available on demand. One of the hardest things to explain to my four-year-old is that if he misses something he wants to watch on television, which we don't have on video, there isn't a way to bring it back.

Streaming media appeals to this demographic group because of its on-demand and rich media offerings, but this group is also highly critical of technology that does not work well. Members of the digital generation are easily bored if the content is not engaging. With so much to see and do and so many things vying for their attention, they have little reason to persevere with uninteresting programming or with immature technology that does not give a gratifying user experience.

The Multitasking Viewer

Research has revealed that viewers today are quite comfortable with multitasking. Whereas the usual mode of watching television was to sit and watch, doing little else, today many people have the TV on as an audiovisual background, while they engage in other hobbies, work, write, cook, read, or play games on handheld machines. An astonishing number of people already surf the Web or chat online while watching television. In fact, many broadcast programs exploit this tendency, offering not only interactive content on the digital television set-top box, but also supplementing the program with a Web site, including live chat with key program makers and experts. Increasingly, program commissioning involves not only the television program, but also the Web site, the interactive features, a CD-ROM, a book, videos, DVDs, and other merchandising.

Streaming media is ideally placed to deliver this rich media experience via a single receiving platform. It makes little sense to have as little synchronization between these multimedia elements as exists today. If makers of receivers and IP delivery services could unify the delivery platform so that all these rich media offerings could be received in a more convenient way, with the platform doing everything that the television with set-top box ever did, what reason would there be not to watch streaming media?

This is not to say that the receiver needs to look exactly like a television set with its familiar handheld remote control. Rather, the streaming media receiving platform ought to be more like a home media gateway, redistributing synchronized multimedia elements to various display and interaction devices around the home. You should be able to view video in the shower, using a waterproof keypad for interactions, then continue to watch as you move to the kitchen, where you can access the Web-based text content on a high-resolution display. After making a cup of tea, you should be able to continue watching while sitting down in the living room, viewing the program on a large screen, high-resolution home cinema system, interacting with the other streaming content via a laptop computer connected wirelessly. Indeed, you ought to be able to continue an online chat (or many simultaneous cyber conversations) as you move from location to location. Table 3.1 lists some common stream profiles, applicable to a variety of devices that the viewer will want to use, perhaps at the same time.

TABLE 3.1

Common Stream Profiles

Profile	Use	Bandwidth	Window	Frame Rate
High quality full motion video	Near DVD quality	750 kbps	640 × 480	30 fps
Presentation quality broadband	Talking heads	384 kbps	320 × 240	24 fps
Presentation quality low bandwidth	Talking heads	128 kbps	240 × 160	18 fps
Presentation quality dial up	Talking heads	40 kbps	160 × 140	15 fps
High motion broadband	High motion video	512 kbps	320 × 240	30 fps
Screen capture broadband	Screen capture presentations	90 kbps	1024 × 768	10 fps
Screen capture dial up	Screen capture presentations	20 kbps	640 × 480	3 fps
Audio only	Restricted bandwidth applications	20–64 kbps		

Values

In contrast to the audiences that welcomed television and radio into their homes when those media were first introduced, the values of today's audience are different. Although perhaps a little more cynical, viewers today are less willing to believe the media and expect large corporations routinely to "spin-doctor" their message. The rise of the public relations industry has sometimes left people having to swallow messages like "toxic sludge is good for your children." Faced with that kind of media manipulation, people have grown less welcoming to viewpoints and less accepting of the truth behind news stories, when there could possibly be unstated vested interests.

What the audience wants is to be re-enfranchised. People expect choice and freedom and they want to have their voices heard. However, while they are not overly keen on Web regulation or censorship, they are concerned about protecting their children and society as a whole from "nasties" like pedophiles. They have concerns about the erosion of their privacy that is already happening, but are also concerned about their security. So they want safety, but don't necessarily want a Big Brother society. There is a growing feeling that too much television is not a healthy thing, as children need to run and jump and learn to socialize. They have also come to realize that television was designed to be a huge timewaster, demanding the viewer's exclusive attention, for long stretches of time. So paradoxically, they want to be better informed, entertained, and educated than ever before, yet in less time, so that they can live the rest of their lives to the full, away from the box.

People are no longer willing to passively accept what the media wants to say to them. They want to express opinions and ask questions. They want to be able to challenge established opinion and people in authority. They want to engage in debate as equals. On the whole, they are less in awe of authority figures and less naïve about the purpose of the media than their parents were. The media exists to sell things to you, not to guard the truth. If the two are coincident, then the media as we know it function satisfactorily; however there are frequent conflicts of interest.

Streaming media technology is ideally suited to allowing an audience to question assumptions and reveal special interests, as well as to public participation in issues of the day. Old media already tries to enlist the public through phone-ins, online votes, postings on Web sites, and so on, but they exercise editorial control. Streaming media allows a less regulated and controlled participation.

Expectations of the Media

Today's media audience wants analysis and informed comment, but also to be able to verify the facts for themselves. The ability to ask searching questions is accompanied by an expectation of greater transparency and honesty, with all possible conflicts of interest declared. It frustrates people when the media are complicit in missing key stories and party to manipulation by vested interests. Leading academic Noam Chomsky has made a career out of citing such examples. In short, people long for greater media integrity.

People want the media to talk to them at their level, not in a patronizing way, as is often the case with mass-audience programming, which must cater to the lowest common denominator. Hence the expectations people have of the media are that it provide trustworthy opinion and analysis, in an intelligent way. Streaming media, with its ability to take its time explaining technicalities and to allow an audience to participate and trace back to sources, provides a unique opportunity to fulfill these expectations.

Community Spirit

People like to feel they belong to something, especially since work, social and housing patterns have eroded traditional communities and increased the tendency for people to feel isolated and lonely. Online, people spontaneously form communities with astonishing facility. Groups of people readily identify with each other on the basis of common interests and gravitate to people they meet who hold similar outlooks and values. Streaming media is already facilitating this community formation, through streaming audio and video chat services, such as Yahoo! and PalTalk.

The Need for Speed

Who has time anymore to wait or to have time wasted? There are so many other demands on people's attention and leisure time. With companies working on "Internet Time" and requiring greater productivity, year after year, more is expected of everyone at work. Time is so precious, because nobody has more than a lifetime to give.

Streaming media allows people to get as much or as little information as they require; different levels of detail and the ability to control the pace of information delivery.

Expectations of Search Relevance

Search relevance is an important audience expectation. Nobody has the time or interest to wade through irrelevant or out-of-date search results. They can't possibly hope to be able to trawl through listings for thousands of on-demand channels. When the user specifies the search key, the results have to pertain to it. With metadata-enabled searching and some of the newer content management, publication, syndication, and replication systems being developed for streaming media applications, this is one technology that ought to be able to deliver fast and relevant pointers to material of interest.

The Need to Contribute and Interact

Whereas watching old media was like taking dictation, people today want not only to shape the editorial agenda, but also to create their own discussion groups and programming, to promote their own ideas and views. Media should not be for the elite. There is an expectation that anybody should be able to publish audiovisual content, just as on static Web sites today. With streaming media technology, it is possible to kick ideas around with peers and even collaborate on digital media production online. Production costs are not prohibitive, as they are with television production. The desktop content producer is a realistic prospect with streaming media.

Respect for Digital Rights

Most people will happily buy a piece of media, especially when it is easier or more cost effective to do so than to make copies. Many people with disposable income don't have time to fiddle with the process of making pirate copies. Those who have time, but no money, aren't potential customers anyway. Their accessing a piece of content serves only to spread the work to a wider audience and perhaps to create some word-of-mouth recommendation value to the content owner. Sentient beings realize that if favorite artists aren't paid they, like everybody else who isn't paid, go out of business.

Some activities of global media concerns have not won them any friends. It is widely known that digital content of all types is priced differentially across the globe. It can cost less to buy a CD pressed in the UK in a foreign country than it does in its country of origin, for example. Local taxes not withstanding, people do not like to feel that they are being taken advantage of.

The calculus for digital music downloaded or streamed from the Web might go something like this. If a CD with ten tracks on it costs $20 and if a person expects to play that CD a total of 50 times, then the value the consumer attaches to listening to a single track just once is around four cents. What this means is that an online music service cannot expect to charge the retail price of a single to access that track as a stream, especially when you consider that some consumers still have to pay connection charges by the minute and also pay for the storage, if they save the track for later replay. Fortunately, many digital music services have now emerged that charge a flat monthly rate for access.

In essence, people will resist paying more to stream digital music and video, at quality levels that rival CD and DVD, than it costs to deliver actual physical disks, pressed at a plant, packaged, and transported to warehouses and sold via a retail outlet. The amount paid for the bandwidth plus the contribution to the content owner cannot exceed the price of the DVD or CD without consumers feeling that the situation is absurd. Consumers know that the incremental cost to the content owner of shipping additional bits, once the initial production costs have been amortized, is effectively zero.

Peer-to-Peer (P2P) Streaming Networks

People who used Napster to access digital music online were branded thieves by the RIAA and the music industry in general. Perhaps somewhat controversially, I would argue that the music industry didn't provide a method as easy or convenient to buy their products. It isn't fair to brand somebody a thief when the industry has done nothing to make it possible to comply, short of engaging in a very much more complex and less immediate transaction. There are limits to this line of argument, of course. I would in no way condone shoplifting merely because the process of paying for goods in a store involves a more complex and less immediate transaction. The point is that when Napster arrived, most major-label music was not even available online, unless you ordered a CD and waited for it to come through the mail. There wasn't (and still isn't) a widely available, quasi-standardized, micropayments currency, other than credit cards (which impose practical limits on the minimum transaction), available to online consumers of digital music either.

As a method of distribution and as a rival to radio airplay, Napster and other file-sharing utilities were incredibly cost effective and efficient. Users paid for the storage and for the bandwidth to transport music from one person to another. Claims that CD sales were adversely affected were hard to substantiate, since Napster undoubtedly increased the exposure and hence sales of music by lesser-known artists who would otherwise have had a difficult time attracting the spotlight. The losers were the popular artists who already enjoyed saturation-marketing support from the existing music industry. Their audiences were somewhat diluted as people found alternatives online. Their monopoly of the airwaves counted for less. Consumers didn't see their use of peer-to-peer networks as theft, but more like a way to sample new things free. What was missing was a way for content owners to get paid.

Peer-to-peer file-sharing utilities had the added advantage of letting those knowledgeable about music gain additional kudos and respect by donating that knowledge in the form of recommendations to other users. The music was not the only valuable content available. Somebody looking to broaden musical tastes could save himself or herself a lot of time and money through having something recommended by someone whose taste he or she trusted. In economic terms, the recommendations short-circuited the search process and hence navigated the user to a satisfactory piece of music for very little investment in time.

There has been much talk about digital rights management as a way to allow content owners to get paid and not have their works stolen. If these systems are introduced clumsily, so that consumers now have a very much more complex transaction to complete in order to access digital media, consumers are likely not to bother. Digital rights management systems that threaten to blow up your stereo, or call the FBI whenever the digital license gets corrupted, or cannot be read will rightly encounter harsh consumer resistance. On the other hand, systems that make it easy to get digital works fairly will probably succeed.

Peer-to-peer file-sharing applications are here to stay. Companies like Kontiki and Red Swoosh are already working on next generation peer-to-peer solutions. No rights management system will ever be able to prevent all unauthorized access or copying on a peer-to-peer network. People determined to share their media with their friends will always find a way to do so. When Napster started removing copyright works from its database, users simply renamed those works with deliberate and obvious misspellings, or even in pig-Latin. In the end, there is no way to secure sound waves traveling through air or photons heading toward eyeballs in electromagnetic waves.

Recently, record companies have begun to buy online properties like Napster, Duet, MusicNet, and mp3.com. However, without content from all the major labels on the now-proprietary search engines, consumers are bound to be unhappy. One of the great things about Napster was that it was a one-stop shop; a kind of "celestial jukebox." If consumers must now access a number of Web sites and utilities to find music online, they have taken a giant step backward.

What was possible for audio will need to be possible for video content as well, if consumer expectations are to be met.

Protection from Perversion

Everyone online expects to be able to protect children from being preyed upon by cyber-perverts. People want to be able to access their entertainment and information without being deluged by unwanted and highly obtrusive advertisements for pornographic Web sites (does anybody really decide to buy something on the basis of some unsolicited electronic junk mail or other spam?). For streaming media to succeed, these issues will need to be taken very seriously and addressed. Privacy preferences and ratings systems are all steps in the right direction, but I know of few people who are free from unsolicited and sometimes highly offensive, if not wholly questionable, e-mail.

Audience members expect freedom to access any material they choose to access, but that it should not be easy to stumble accidentally upon something they don't wish to see (or don't wish to let their children see). Parents would ideally like to filter content and keep audit trails to manage their children's online excursions, but the truth is that the problem is one of curiosity. Technological solutions have never been good at teaching right from wrong. If a moral code is to be adhered to, parental guidance must do that work. Children need to be educated on why accessing certain kinds of material might not be such a healthy endeavor, rather than having the problem handed off to a piece of software or a regulatory body. As with anything dangerous or detrimental to children, working on helping them choose not to demand it is often more effective than blanket prohibition.

That said, there is an expectation that data about individuals held in corporate and government databases are protected against unauthorized access and inaccuracy. Data protection laws, such as those that apply in the UK, partly achieve that aim, though policing and enforcement is notoriously difficult and painstaking. With everybody able to publish audiovisual material on a streaming media system, the public expects some way to protect against the intrusion of grossly indecent or harmful material. However, since this is largely a matter of taste and depends on particular societies' attitudes to various taboos, finding a global solution is going to be challenging. Perhaps a better question for society to address is why some people feel the need to produce and consume what most people would regard as questionable material.

Silver Surfers

A significant group of people who might be a potential audience for streaming media is retired people who have the time to spend online and have the means to access the Internet. "Silver surfers," as they are colloquially termed, often use the Internet to monitor their pensions and investments, talk to their grandchildren with instant messaging services, and take and distribute digital family photographs. Silver surfers are often active genealogists and creative writers. Streaming media's main appeal to this group would be how it facilitates "grandchild-o-vision" or provides content related to their interests. Indeed, video tutorials teaching how to use particular computer applications or providing computer skill training are likely to appeal to this age group as well.

Streaming media is likely to be an ideal solution for silver surfers, since subjects can be presented in non-confounding ways, through the use of interactive audiovisual programs.

Serious Business

Another group that will significantly benefit from streaming media, whose characteristics need to be understood and catered to, are businesses. Small, medium, and large enterprises can all reduce costs and create larger profits using streaming media to enhance training materials, product demonstrations, sales pitches, corporate communications, investor relations bulletins, e-commerce offerings, marketing communications, tactical sales communications, teleworking, virtual meetings, and video conferencing. Microsoft's popular PowerPoint application already has the capability to create presentations using streaming media. Figure 3.1 shows a typical large enterprise intracast network topology, which might be used to deliver their internal streaming media communications.

This group of users of streaming media is likely to be one of the biggest and most important. Actual savings in use are relatively straightforward to demonstrate. In addition, most enterprises are already wired for broadband streaming. The only resistance tends to be

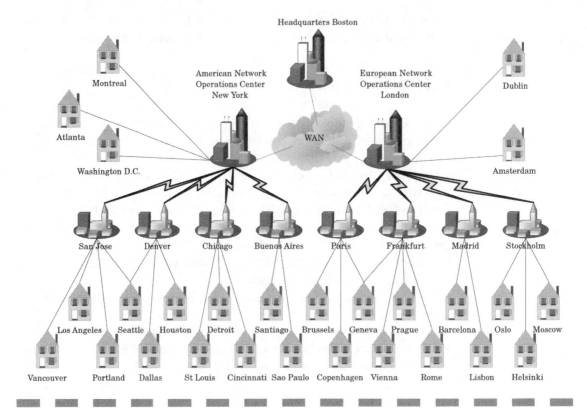

Figure 3.1 Large enterprise intracast network topology.

from the IT people who have to make this work and keep it working. The needs of these key stakeholders must also be met by the streaming media industry.

Learners

Streaming will appeal to those who want to learn from online experts. Imagine being able to watch Nobel Prize winners like Richard Feynman lecturing on their particular subjects. In the future, streaming media will make it possible to see both live lectures and archives of streaming learning material made by the best scientists in the world, for example. Over time, this will become an extremely valuable resource, since a subject explained by an expert with particularly lucid insight is always more educational than being taught by someone with less of a grasp of the subject. It's one thing to read a book on relativity, but quite another to listen to someone of Einstein's caliber telling about it.

I was fortunate enough to experience this phenomenon twice, while still an undergraduate. We were granted a guest lecture by a world authority and leading innovator in the field of full-color holography and later when addressed by a world expert on digital automatic control. The experience was life changing. Suddenly, complex and abstruse subjects become clear, simple, and accessible. If streaming media makes it possible for this experience to be replicated often and widely, it will, in my opinion, have justified the investment in the technology.

How Will We Watch Streaming Media?

Streaming media will be consumed in surprising ways, using surprising devices. The old assumptions the consumer electronics industry relies upon when designing its products won't hold. Streaming media receivers are not "point products" that work in isolation. Rather, they are intimately connected to each other and to the wider world. They will work in concert with other machines. Product designers have yet to stretch their imaginations to accommodate the myriad possibilities for streaming media reception.

Streaming media will be something that we watch everywhere. The trend toward pervasive computing will be significant to streaming media, since digital media will provide the interface from device to user. The PC will not be the only device we use to access streaming media, but most receivers will have significant embedded computing power.

Today there are few technical constraints to the creation of these streaming media appliances. The reason there isn't a receiver on your fridge is more related to the lack of a broadband streaming media infrastructure than to unrealized technological breakthroughs or cost. According to research done by Streaming Media Inc., in association with The Carmel Group, non-PC streaming media devices will number 48.7 million units by 2007. They estimate there will be 7 million users of non-PC streaming devices by that time. Streaming to devices other than PCs will remain a novelty at least until 2005. Figure 3.2 illustrates the projected trends in non-PC streaming devices, while Figure 3.3 shows the corresponding number of consumers. This section is about the devices and situations that will define how we watch streaming media.

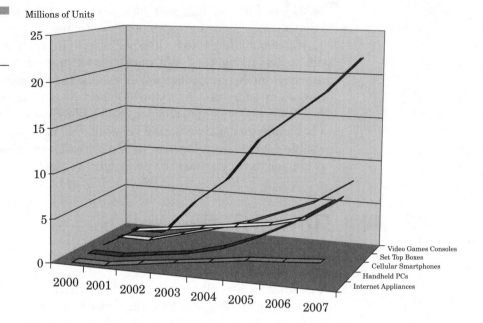

Millions of Units

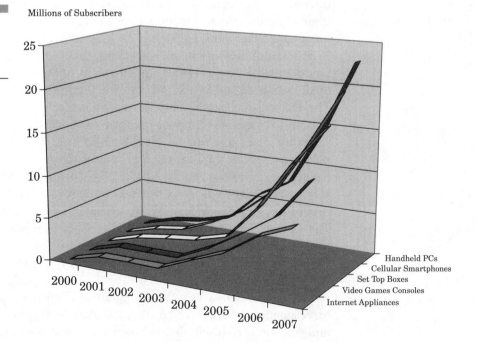

Millions of Subscribers

The PC Platform

Today, the desktop PC is the only really practical, cost-effective, and widely available device with which to receive streaming media. Machines have enough processing power, memory, graphics performance, storage, and throughput to do the job. The operating system de facto standards, Microsoft's Windows and Apple's Macintosh operating system, make possible wide deployment of streaming media player software. Operating systems encountered less frequently in the home, such as Linux, also have player software support. Network connectivity exists and is getting faster and cheaper, as people desire to access the World Wide Web with less waiting. People are able to get almost satisfactory streaming performance as a side effect of improving their general Web-surfing ability.

The PC will be the most significant receiver for streaming media for the foreseeable future, but consumers will demand more reliable playback, better player performance, and higher resolution. The PC will be suitable for self-study applications or acts of individual viewing, but poor for entertainment applications, where a large screen at the other end of the living room is a better choice. Anybody who has ever tried to watch a feature-length movie on a PC DVD player will attest to that fact. The other factor that limits the utility of the PC for streaming entertainment is that, unlike a television set, it doesn't come on instantly. PCs take minutes to boot. There is no technical reason why this should be so, because many other computing devices have that capability today, but PCs are tied to legacy design and won't change overnight. A glimmer of hope is Microsoft's XP fast-booting feature.

Set-Top Boxes and Beyond

Many set-top boxes are just PCs in disguise, because the cost of the parts needed to make a set-top box came to be about the same as the cost of the parts needed to make a PC. Simpler set-top box designs lost their cost advantage and certainly were no more reliable. Coupled to this was the increasing demand put on the set-top box's hardware to deliver interactive television content and the fact that writing software for a set-top box was not as easy or cheap as writing software for a PC platform. Hence, the latest generation of set-top boxes is very close in design to desktop PCs. This is a positive development for streaming media, since streaming media players require a good deal of processing power and memory. Streaming-capable set-top boxes for satellite and cable are predicted to arrive some time in 2002, but legacy infrastructure remains an inhibitor to innovation and consumer uptake.

Web TV-compliant set-top boxes were an early attempt to converge Internet and broadcast television technology. Ultimately, those set-top boxes did not find widespread favor with consumers, because of the limitations of the television for displaying Web content (it was all too small, indistinct, and too far away) and because the set-top box's hardware was exercised to its limits. In addition, the back channel was a simple telephone line, so Internet access from a Web TV set-top box involved more waiting than would be experienced on a desktop. In other words, it was too slow.

Game consoles, such as Microsoft's Xbox, are thought to be likely contenders for bringing streaming media to a mass market. These boxes are powerful and relatively cheap and can produce crisp, non-interlaced video. Television display technology may soon change too, adding the ability to display non-interlaced, progressive scan video, either using back-projection technology, flat-screen TFT (Thin Film Transistor) technology or light-emitting polymer technology, instead of today's bulky cathode-ray technology. Digital video recorders are also expected to allow streaming media to be received and presented to viewers. Indeed, the optimal streaming media reception environment might be a combination of devices, with a home media gateway taking the streams from the incoming data pipes and redistributing them simultaneously on a home video network (wired or wireless) to a large screen at the other end of the living room and a laptop computer on the armchair. It is estimated that one million users will be streaming via gaming console by the year 2007. By that time, streaming set-top boxes will account for 18% of the non-PC streaming media receivers, serving 21% of the non-PC streaming media users. Video game consoles will represent 45% of the non-PC receivers and attract 15% of non-PC streaming media viewers.

Mobile PDA and Web Tablets

Streaming media can already be received by the latest generation of cell phones, palm tops and PDAs. Web tablets are also in design, with improved video performance and these will be ideal streaming media receivers for applications like mobile on-demand video rental. The use of these devices for streaming media is entirely dependent on the rollout of third-generation cellular networks.

From a usability point of view, there are some questions about whether or not video streaming on mobile phone handsets will be a killer application. In order to videoconference with somebody or watch some video, you have to be still and pay attention, or else risk walking into things. This is not the case with voice communications, where you can quite happily talk on the move. Commuters may watch while on trains or in cars, if they are passengers, so this might be the main use of mobile streaming media. Skeptics note that portable televisions have been available for a long time, yet they never became as popular as DVD players. In fact, portable DVD players have also been available for some time, but we don't see people watching movies wherever we go. Perhaps selling mobile video to consumers will prove to be harder than we think.

According to the research we have been quoting throughout this chapter, handheld PCs will account for 16% of all non-PC streaming media receivers by 2007, reaching 28% of all non-PC streaming media users. Cellular devices will represent 17% of the volume, reaching 26% of users.

In the Car

Vendors of in-car telematics systems are already working on technology to allow people to receive Internet radio, on-demand streaming video, games, and quizzes in the car. The technology to do this with terrestrial digital television has existed for some time, but there were some issues with receiving digital video broadcast signals while the receiver was in motion. Also, the market for in-car electronics and computing was not sufficiently mature. Third-generation cellular networks promise to make reception of IP-based streams reliable, even if the vehicle is moving, because of the technology, built into the standard, to allow transition from one cell to another.

In Public Places

Outdoor display technology to make video billboards exists. Large outdoor digital projection screens are also within our current technological grasp. Today, those outdoor displays have small, dedicated video servers attached to them, to supply the video. In the future, they could be connected to streaming media receivers instead. This would allow the content displayed on the screen to be changed remotely and often. Applications could include shopping mall TV, where retailers screen advertising to passing shoppers using streaming technology. Manufacturers could buy advertising space on screens at the end of every aisle or at the point of sale, in the same way that they buy premium shelf space in supermarkets. Road signs could even include streaming video of breakdowns or traffic congestion up ahead, allowing drivers to take alternative routes.

Streaming media is likely to be used in these applications, because it is the most cost effective means yet devised of transmitting localized video content. Although not yet available, rugged, low-cost, simple, and reliable streaming media receivers, suitable for outdoor installation, will be the key enabling technology. Once these are available, it will be possible to screen highlights of play at cricket matches, baseball games, football matches, and ice hockey plays using wireless links. Spectators could view the content on the large screen, or else receive and navigate that content on mobile, handheld, streaming media receivers. In the latter case, it would be possible to stream the output of all the cameras covering the event on individual streaming channels. This would allow spectators not only to watch the live action, but also to get unique perspectives on the action, via streaming media receivers. For example, while watching a motor race, trackside spectators could also view their favorite driver's in-car camera. We will undoubtedly see streaming media in public places.

At the D-Cinema

One of the ways in which we will watch streaming media is sitting comfortably in a chair, in a cavernous darkened room, eating popcorn, surrounded by strangers. Distribution of feature films, using digital media, is already happening. Streaming those movies, via high-speed optical networks or satellite, is the next step. Although the lack of rights management and adequate security are issues to be addressed before the adoption of streaming technology, most digital cinema applications require 45 megabits per second of bandwidth and use MPEG-2 compression. There are claims that MPEG-4 compression can do the same job with around 3 megabits per second of bandwidth.

If cinema-quality video can be projected to an auditorium using only 3 megabits per second of bandwidth, it becomes economical to stream live rock concerts to several venues at once and to change the film on show far more frequently. It is also possible to support smaller audience numbers profitably. It is even possible to create battery-powered screening equipment that can present streaming digital cinema presentations, via satellite, to the remotest audiences.

Streaming media will undoubtedly be the way films are delivered to movie theatres, once concerns over copyright protection and security are addressed and once exhibitors and studios agree on who should pay for the digital projection equipment.

On the Fridge

It seems absurd to imagine a refrigerator as a streaming media receiver, but in reality, it is not so outlandish. What could be better than watching a cooking show, or interactive instructional cooking training program, or even interacting with a real chef, from the comfort of your kitchen, while you cook?

Many kitchens already have televisions in them, with people watching soaps or daytime chat shows, as they carry on with their meal preparations. Replacing that television with a streaming media terminal is not such a stretch of the imagination. If agreeing to receive advertising on your fridge meant that you got that fridge free, for example, many people would gladly welcome such a streaming media appliance into their homes.

The streaming media receiver would add capabilities that television couldn't match. As examples, it could allow you to order ingredients for home delivery online, but also let you see the goods you were buying. Local stores could economically advertise fresh stock to people living within driving distance, as soon as it arrives, via a local advertising insertion operator. They could do more than just announce it; they could actually show it. It would be possible to view the lines at nearby supermarkets before deciding to venture out. You could interrogate a supermarket's stock list or even look at its shelves, to see if what you need is in stock. Why drive all the way to the store to pick up bread, if there is no bread? Skeptics might even install streaming cameras inside their fridges, just to satisfy themselves that the light really does go out.

Streaming media appliances for the fridge have been prototyped, but are not yet widely available. However, the case for such specialist receivers is intriguing.

Around the House

We alluded to the presence of home video networks earlier in this chapter. It is highly likely that the home of the future will have a gateway to marshal and catalog the streaming media coming into the home via a multitude of carriers. The edge of the Internet will be in the cupboard. These media gateways will rebroadcast the media to streaming media appliances throughout the home, using either high-speed copper or optical cabling, infrared transmission, or wireless networks, such as Bluetooth (now encompassed within 802.15 WiMedia).

Some researchers are proposing the use of proximity-detection technology to allow a streaming media program you are watching to follow you from screen to screen, portable device to portable device, as you walk around the house. Home media gateways often include the ability to video whoever is at your front door and facilities to monitor your home, rebroadcasting a secure video stream that you can log onto and view when you are away. In fact, video originating from inside the house could be relayed to other rooms, allowing mothers to watch children sleeping in the nursery from the kitchen, bathroom, or even the garden.

Today, home media gateways exist only on the drawing boards of consumer electronics companies. However, they are likely to be important appliances in the homes of the future and purpose-built to receive and redistribute streaming media. Internet appliances will represent 4% of the non-PC streaming media appliances, by 2007, reaching 11% of non-PC streaming media viewers.

Surveillance Centers

Love it or loathe it, streaming media will make it possible for surveillance systems to obtain very much better quality images than can be obtained today, transmitting these back to an operations center using very little bandwidth. These higher-resolution images, coupled with the zoomable image compression techniques coming soon, will make it possible to identify a culprit's face in crisp detail, without the blockiness or blur associated with today's CCTV (Closed Circuit TeleVision) systems. Streaming video is also very easy and economical to store on mass-storage devices like hard disks. So, streaming media will undoubtedly be watched by security personal. Not only will there be many ways for us to watch streaming media, but many ways for streaming media to watch us.

When Will We Watch?

We live in an attention economy. The most precious thing that people have is time. There are myriad competing demands on a person's time. For companies to be successful, they must successfully attract people's attention, for long enough to do business with them. Grabbing the viewer's attention is becoming increasingly difficult and wasting a person's time is punishable by the viewer's defection to other pastimes. With so much information to deal with, so many distractions, so many options for how to spend our lifetimes, yet with every consumer transaction seemingly designed expressly to waste our time, will there be any time left over to watch streaming media? Indeed, will streaming media become a tool for absorbing condensed information much more quickly and in a more ordered way than anything we have been offered so far? Will enough people spend enough of their lifetimes watching streaming content to make it an important communications medium? When will we watch?

One thing is certain. Broadcast television audiences are in decline, as people find themselves increasingly unwilling to devote the hours they once did to watching the box. For streaming media to succeed, the medium must find other times in people's days when they will willingly watch.

Clearly, the broadcast television paradigm has run its course. Sitting passively and watching, as the information is rationed out, is no longer a satisfactory experience. For some idea about what streaming media might do to change the paradigm, consider the PointCast network (now infogate), which was once the doyen of the media industry. Using proprietary technology and networks, PointCast delivered a personalized newspaper to the desktop. Unlike a newspaper, however, the visual experience was similar to one of the cable infochannels, like MSNBC. There was moving video, with ticker tape scrolling banners and side bar graphics. Although horrendously expensive, in terms of the percentage of corporate bandwidth the system required to operate, when bandwidth was not plentiful anyway, PointCast nevertheless showed the way forward for the streaming media industry. In many respects, it was the inspiration for much of the subsequent streaming media technology development.

The key to PointCast was personalization and localization. You could receive active alerts on breaking news of interest to you, by keyword. For example, you could set the system up to alert you with news of your

IBM stock's going above $120, or with a storm alert for your home town. Information was drawn from reputable and leading content providers, including some of the more respected newspapers and television news networks. The beauty of the system was that the information could follow you, passing your alerts and news via your cell phone, pager, e-mail, or ICQ* account, if you were away from the desktop. Infogate, the successor to PointCast, is based on open Internet standards, unlike its ancestor. With seven years of development behind it and with three million downloads, the new infogate system is worthy of examination, as a taste of another way of presenting information to busy people who don't have time to sit around and watch and wait.

This chapter will show that streaming media will be watched more than television and for many more compelling reasons. It will form a pervasive part of our everyday living and help us conquer the information glut we all must face.

The Competition for Attention

We all know somebody like this: a senior executive in a fast-moving company, who spends virtually every minute of his day working or catching up on family business. He has no time left for his "hobbies," which may be fondly remembered as a somewhat quaint self-indulgence. At home, he's devoted to his wife and school-age children, but even so he feels regularly compelled to check voice mail and e-mail, almost compulsively and obsessively. At work, he has a team of 40 people working for him. All of them feel the need for more of his attention. They and his peers in the company ambush him on his way to the men's room or in the corridors on the way to his next meeting. Sometimes, the best he can do for somebody who needs a meeting with him is to offer a shared wait in the lunchtime cafeteria line. His plush offices are located within walking distance of some intriguing and wonderful restaurants, but he rarely has enough spare time to use them. He spends the great majority of his working day in meetings, and the balance of his time answering e-mails and voice mails in between. His colleagues often resort to instant messages, because regular e-mails aren't attention-grabbing enough to get answers from him when they need them. Even on the way to and from work, his journey is consumed by cell-phone conversations or voice

---

*ICQ is a popular Web-based instant messaging service. The name is pronounced "I Seek You".

mail. As the information assault persists, he worries that he is not giving his family enough attention. As a manager, he feels he is short-
changing his employees in denying them all the attention they deserve.
He has no time for quiet reflection. This man lives in information-over-
load hell. Maybe this is you.

Streaming media must compete with all these other distractions and
activities. At home and at work, streaming will find a role in people's
daily lives for a single significant reason. You can access the information
you want, while doing something else and easily skip over the boring
bits. Because the information is portable, from device to device, yet the
rate of access can be controlled, streaming media has the potential to
pass control back to the executive. Reading e-mail is not as fast or rich
as being able to skim through a video making the same point. Voice mail
is a tyranny, unless streaming media technology makes it possible to
skip ahead. Streaming media makes it possible to appear to be in two
places at once. Rather than spending large parts of the day sitting in
meetings, or on the way to and from them, having several meetings in
progress at once on the desktop, using streaming video technology to
host the virtual meetings is a way to skip in and out of discussions,
without having to leave and return.

Time Is Precious

Streaming media will be readied for the mass market in an environment
when money is relatively plentiful, compared to lifetimes. There will be
enough bandwidth and there will be an ever-worsening information
glut. Streaming media that works seamlessly with other messaging systems will strike a blow for fighting the information deluge by allowing
you to search effectively and sort all the media directed toward you by
metadata keywords. Streaming media allows you to preview the longer
message, especially if creating an executive summary attached to the
longer streaming media piece becomes the normal practice. Streaming
media can be presented as synchronized multimedia, so many more of
the senses can be addressed at once, allowing a person's attention to be
parceled out in small doses, so that a person can pay attention to all the
incoming information in enough detail to effectively "get" the message.

Because streaming media is a richer form of communication than
simple e-mail or voicemail, it is possible to use stronger techniques for
automatically sorting and filtering streaming media messages. There
are many more clues and much more contextual information to work

with. Streaming also allows you to unify all your messages and information, so that you can schedule reading and viewing all those messages to a time that suits you. Because you can access your media from mobile devices as well as your desktop, you will be able to create catalogs of information of interest to you, grouped by metadata tags. You won't need to carry these assets with you everywhere, or even carry the catalog. The point with streaming media is that when you need to access anything of personal interest, you first access your catalog (which might reside on your company's server or your home media gateway) and just stream the content to your device, wherever and whenever you need to. Being able to access all your media from wherever you are, as a stream, enables you to complete more tasks on the spot, rather than filing them as pending, until such time as you have at hand the information you need to complete the task, as is the norm today.

Saving Time

Saving time by being able to skip irrelevant items or familiar detail is one of streaming media's greatest strengths. You can get to the nub of the information in a shorter time. However, streaming media can save time in less obvious ways. If you can view the lines ahead via streaming media technology, you can make alternative choices instead of getting stuck in the traffic. Indeed, the ability to telecommute and attend virtual meetings will have a tremendous impact on giving time back to working individuals. Traveling time is often long and wasteful, adding to the wear and tear on an individual's body, particularly if jet lag is a factor. The business traveler is effectively "trapped in transit" for several hours at a time, and can neither work productively, nor attend to family business.

Another way in which streaming media will save time is with just-in-time learning, where information can be imparted rapidly and richly with streaming media training materials. Another time saving in training is with a virtual mentor: one who is physically located elsewhere, but is in direct and instant video contact with the apprentice at all times.

Time savings also accumulate because of streaming media's ability to unify all incoming information and place it in a single streaming media in-basket. If all the information assaulting you can be directed to this single marshalling pool, you have only one decision to make for each item in the in-basket: does it require action? If the item does not, it can be discarded, or scheduled for review at a later date when it might need action, or else cataloged in a personal metadata filing system (a database of references to items of media of interest to the executive, grouped according to his personal schema). If the item requires action, on the other hand, the executive can do it, delegate it, or defer it. If he chooses to defer it, it can be scheduled to reappear in his in-box at some specified date in the future, or else as soon as he has time to attend to it. If he wishes to do the action immediately, he is confident that he can access any necessary information wherever he happens to be.

Streaming media technology, coupled with an intelligent cataloging and scheduling application like this, which worked on the executable Internet as a distributed suite of software services, would allow business users to save incredible amounts of time. This is because there would always be a way to take the next effective action, regardless of how the message was delivered and what media format was used. Just being able to unify voice mail, phone calls, video conferences, e-mails, Web pages, video material, and even text documents and then deal with each item decisively, using technology to delegate or defer items and streaming to access information necessary to complete tasks, would save industry billions of dollars. This solution is not yet available, but streaming media technology makes it possible. The technology exists to realize this application, which goes beyond the vision of today's PDA and smart-phone vendors.

Every Business Is in Show Business

To get consumers to pay attention and to build brand awareness, businesses are finding themselves forced into show business. Their content must be as good as any other content available to consumers, or else consumers will defect to more interesting content or pursuits. If they want to sell to consumers or influence them, they have to do it in an entertaining way. A company's competitor for attention is not the other company making and selling similar products; it is the Disney Corporation!

In fact, businesses that are the most profligate with bandwidth, providing high-quality information, which looks and sounds great and which redirects consumers to other company information that is equally interesting, will succeed at attracting the attention of consumers over those businesses that use bandwidth more conservatively, as if it were rare. Hence, one of the significant times we will watch streaming media will be when we are planning to buy something or transacting business with a company. Figure 3.4 shows the business potential for streaming media.

We'll also spend time watching streaming media when we work for companies that use streaming media for corporate communications and

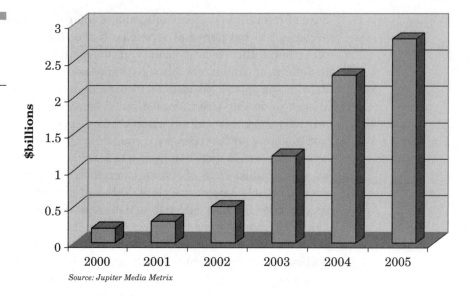

Source: Jupiter Media Metrix

training. If the internal communications are entertaining and engaging, as well as brief and to the point, the company will act like a more intelligent and informed whole. See Figure 3.5 for the trends in business streaming applications. In the language of Nobel Prize-winning Ronald H. Coase, who first described why companies exist, the result is that internal transaction costs are further minimized. The effect of this is to improve the firm's competitiveness, which translates into higher earn-

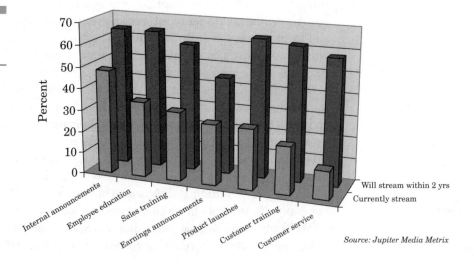

Source: Jupiter Media Metrix

ings and profits. Although a full description of the economics of the lowering of internal transaction costs due to the increased richness and reach of internal streaming media communications is beyond the scope of this book, we can say with confidence that the effective use of streaming media can positively affect the bottom line. We will, therefore, watch streaming media when it makes us more money.

Getting What You Need

Streaming media's ability to help you find exactly what you need, without having to call or visit places in person before making a decision, means that search costs are minimized. In other words, because you can see the goods you want to buy or can videoconference with the service providers you are considering, you save both time and money. Conducting your survey of what's available can be done through the use of streaming digital media. Educating yourself about the details of the subject at hand can take place quite quickly, through the use of streaming multimedia guides. So, we will watch streaming media whenever we want to get what we need as cheaply and as quickly as possible.

Sometimes getting what we need requires some privacy. Because streaming media is not censored or regulated, we will also watch streaming media when we need to get things that require some discretion (bearing in mind that information about what we view is relatively easily collected).

When You Don't Know

One of the best times to access streaming media is when you want to learn some fact quickly. Because streaming media is so rich, you can access knowledge in a variety of ways. Not only could you read about something of interest or watch a video about it, the article could be read to you while you are driving, for example. A helpful assistant, as another example, could whisper pertinent information into your ear at a live meeting. Streaming media is both cheap enough and simple enough to make these applications easy to realize.

If you visit a building you have never been to before, streaming media can deliver current views of the front of it, along with a helpful aerial photograph and street map, with your current location superimposed on them. Even with street maps and electronic guidance to your destination, an actual video view of what you are looking for is sometimes essential, especially in bustling inner cities. Streaming is the technology that enables such applications.

If looking for somewhere to eat, you might log into live video suggestions that are offered based on your current location, which not only show video of the dishes and how crowded the restaurant currently is, but might also show you the state of the kitchens! There are some restaurants that suddenly become deserted, after a period of great popularity. The reason that loyal customers, who have no complaint with the restaurant, suddenly desert is that they all simultaneously and independently make the decision that the place is best avoided, because it is so popular and always crowded. This results in the paradox: the place is so popular that nobody ever goes there. This apparent enigma is an emergent behavior studied by mathematicians interested in complexity theory. A streaming camera of the restaurant, broadcast live to the Internet, would allow potential diners to verify whether or not their view of how crowded the restaurant is true or not. Restaurateurs who wish to keep people coming to their restaurant need to provide streaming views of their dining rooms. Diners, similarly, will watch streaming media whenever they aren't sure and need to know.

When You Want to Know More

One of the great features of streaming media is the ability to hyperlink media items to each other. It allows digital media to provide sidebars to the main program, allowing consumers to bolster their knowledge on an as-needed basis. As an example, news reporting with streaming media can allow viewers to hyperlink to analysis or background to the news story. How often have you watched broadcast television news only to be frustrated at the brevity of the coverage of some subject of interest to you, or else because you wanted to know more, but couldn't ask for more? Hyperlinked streaming media provides opportunities to find more information. Indeed, because streaming media lends itself well to audience participation at a very direct level, it would be possible for some news services to answer your questions about news items as you pose them. If there isn't enough analysis or detail, you can simply ask for more.

This kind of news service could never exist on a mass medium like broadcast television, because the editors of mass media news need to avoid losing viewers through insufficient brevity or through a barrage of detailed information that they are not interested in. A good streaming media content provider will find ways to leave the detail as an on-demand item. Those interested will access it; those who aren't will stay with the main narrative flow. We will therefore watch streaming media whenever we need or want to know more.

Anywhere

With small, power-efficient mobile streaming media receivers, wireless home networks, third-generation cellular networks, and Internet-in-the-sky satellite networks, streaming media will be accessible virtually anywhere. Taking interactive language lessons as you drive to work in your car or on a train will be possible. Instead of listening to the radio in the morning, a personalized summary of what is happening in your sphere of interest might be the program that you listen to. Streaming media programs that you start playing at one location, on one device, could follow you around the house, moving from device to device in response to your physical location.

Even places where cell phone coverage is difficult today could one day be connected, through local wireless networks or repeaters. For example, it isn't possible to make a cell phone call on an underground train. However, with repeaters and trackside wireless network access points, or even transmission through collector rails, the problem could be solved. Consumer demand will drive the installation of the infrastructure. In remote locations, beyond the reach of wired and wireless networks, satellite coverage will suffice.

The only limitations to when we will access streaming media will be when it is considered impolite or antisocial to do so. For example, most people agree that people talking on mobile phones in theatres and restaurants, or on a crowded train, are a nuisance. So, too, there will be societal limits on when we watch streaming media based on a tacit etiquette.

Anytime

Even streaming media programs that are initially broadcast according to a schedule can be stored for on-demand viewing at a later time. Time shifting of digital media is a simple and routine thing. Although most desktop streaming media players do not allow you to record the stream to disk, the stream itself can be accessed directly from a media server later, if the content provider makes this facility available.

Most players do not allow recording of the stream today as a very crude copy protection scheme, preventing users from making unauthorized copies of media for redistribution and preventing virus attack on consumer PCs. With digital rights management and improved virus scrutiny, there will be no good reason to prevent copying of a stream to disk.

Streaming media servers can potentially alert viewers when a scheduled streaming program is about to commence, using instant messaging or even e-mail. Deep archives of historical media can be repurposed for streaming automatically. The availability of deep archives, online, lets you effectively view anything that ever went to air, anytime you choose. The availability of such a broad range of choice coupled with the convenience of on-demand viewing means that we will consume more hours of streaming media, since whenever we feel inclined to watch something, we are almost always guaranteed to find something interesting to watch.

The Simulcast Experience

If the streaming media industry worked to adopt a simulcast standard, it would be possible to support scenarios such as watching news reports at breakfast, which then continue audibly, while you drive your car to the station. While you wait for your train on the platform, the news program could continue on your mobile phone, as audio, into a personal earpiece. On the train, the same program could be picked up on your pocket PC. When you reach your desktop at work, the news could continue uninterrupted, or as alerts and bulletins, where news items you care about are announced automatically.

With simulcast technology, the same program, authored only once, would be delivered via different networks, to different devices, yet it would be possible to pass seamlessly from location to location and device to device, with continuity unbroken. Some parts of the MPEG-4 standard hold promise for multipath delivery.

Proximity sensors are being used by researchers to effect the smooth handover of streaming program from device to device and room to room, in simulations of digital homes of the future. Personal identification remains problematic at present, but in principle, the home network's proximity sensors could track individual family members, chasing them with the program they are watching and resolving conflicts between family members in the same location as they arise. The home network could put a picture in a picture or split the screen, or else pause the playback of one program, in favor of the other. Streaming media technology coupled with user tracking could act to encourage us to watch streaming programs more of the time, as we go about our daily routines.

Personal Streaming Universes

If streaming programs can follow the viewer, then the viewer's own preferences and favorites can also follow. This would allow a consumer to access favorite music, videos or other digital media from whatever player was in use, adjusting the program stream delivered according to the current device's capabilities. If you have purchased digital rights to a favorite movie, then if the streaming media system is able to authenticate your presence at a streaming media player other than your licensed player, the license permissions could automatically track you. You may need to carry some form of radio frequency ID key fob, or some equivalent identifier that is machine readable, but this would be a far more satisfactory rights-management solution for consumers than systems that license players, not consumers.

The advantage to consumers of being able to access their paid-for media, regardless of where they are and what machine they are using, is obvious. They need not store a single physical copy of their digital media, streaming it from source every time they wish to play it. The capacity of searching and indexing your favorites in a number of different ways, simultaneously, adds additional flexibility and utility. Indeed, such a database could be built for a consumer automatically, upon payment for licenses to access particular media properties.

Interestingly, the list of your favorites is, itself, a piece of information that has a potential commercial value. The ability to have all your favorite media available to you everywhere will mean that you spend more time accessing that media. We will consume more streaming media because it will be easier and more compelling to do so.

Why Watch Streaming Media?

Streaming media has unique characteristics highly consonant with human needs and lifestyles. With the cost and convenience of streaming becoming more in keeping with consumer expectations, consumers may soon find there are compelling reasons to become streaming media adopters. Once streaming media achieves a level of maturity that makes it satisfy these human needs, there will be strong reasons for consumers to bother making the switch from broadcast mass media.

In the late 1960s Abraham Maslow outlined his hierarchy of human needs. The underlying idea was that until the lower needs are satisfied, the higher ones are not pursued. According to the theory, society improves by satisfying lower order needs universally, gradually moving toward the top of the needs pyramid. The needs that Maslow identified, from lowest to highest are: physiological, safety, love, esteem, and self-actualization.

Physiological needs are the most basic, satisfied by things like eating, sleeping and breathing. We satisfy our physiological needs by using information and skills to help us cope. The need for safety is satisfied by things like having a home, the legal system, a police force, etc. These needs express the desire to find information and schemes to make us feel helped, rather than helpless. Societal organizations such as family, community, partnership, and marriage solve the need for love. When we seek to satisfy our need for love, we are seeking enlightenment, according to the theory. The need for esteem describes our desire to feel satisfaction from achieving personal goals and also from the attention and recognition we attract from others by our achievements. The solutions are said to be empowering. The need for self-actualization is the quest for edification, including higher aesthetic achievements. Musicians who compose, artists who paint, and poets who write are all seeking to solve their need to self-actualize.

Streaming media has the power to deliver information that helps us cope. We satisfy some of our physiological needs by finding out information about our health, for example. We use the medium to help ourselves survive by accessing information that helps us cope with the world and our lives. The ability to keep in touch with our children and to monitor our property with streaming media communications contributes to the satisfaction of our need for safety. Streaming media technology also facilitates the formation of communities, interest groups, and spontaneous lobby groups. The need for esteem is solved by streaming media technology through each individual's ability to contribute to debate, through streaming media virtual meeting technology that allows people to complete complex projects collaboratively and to gain kudos from colleagues by making contributions to collective activities. Finally, the need to self-actualize is fostered since everyone has the freedom to contribute his or her own artistic works to the open digital media marketplace enabled by streaming media technology. People can also use distance-learning materials to master skills once beyond their reach. Hence, on every level of Maslow's hierarchy of needs, streaming media enables satisfaction of those needs.

Better Than Books?

It is a very bold assertion to suggest that streaming media might be a better form of information delivery than books. However, there are some significant characteristics of streaming digital media that books cannot match. With streaming media, you can access vast libraries of information from wherever you happen to be, without having to travel to a library or carry the library with you. There is no need for consumers to store the physical media assets or even their catalog of items of interest. These can be streamed to whatever streaming media playback device you happen to have to hand.

Streaming media adds graphical animations and vivid moving images to illustrate the points made in a piece of text, for example. Books are unable to provide such animations to explain points made in the body of the book. Video can also aid imagination, creativity, and understanding in ways that books cannot. Streaming media can also be updated and edited much more readily than a printed edition of a book. If errors are discovered in a printed book, errata must be issued, or else a revised edition printed and published. Streaming media can keep works more up to date and current.

As with any book, random access to sections within the body of the work is made possible with streaming media technology. However, with electronic indexing and hyperlinks, navigation within an individual streaming presentation is enhanced. Another advantage of streaming media over books allows hiding of detail, in the main flow of the narrative, with links to sidebar information for those requiring deeper treatment of the subject. The ability to navigate and search within a streaming media presentation and across multiple presentations, simultaneously, could be the killer application that streaming media is looking for.

In contrast to reading a printed book, you can edit pieces of streaming media (assuming you have digital rights to save copies and abstract sections) into your own version of the presentation, with the narrative paced to your own particular taste and the subject matter categorized according to your own schema. You can even capture sections for your own edit on the fly, as you watch the streaming media piece for the first time. Like a summary or précis, your own slant on a piece of streaming media is itself a piece of media worthy of publication; in the same way that literary critique is valued. You may even add a voiceover track to the main presentation, perhaps with a headshot of yourself as you speak, giving responses and insight into the main program flow. Like margin notes that one would make in a printed book, links to explanatory detail can be embedded in a piece of streaming media, so that the next person viewing the media can discover more. Indeed, total strangers, merely by adding to the work, can create streaming media presentations collaboratively. In the same way that open-source software is created, so too streaming media treatments of complex subjects can grow organically, with contributions by a large number of authors.

One of the more famous experiments into this form of authoring was the www.h2g2.com site set up by the late science fiction author and humorist Douglas Adams, renowned for his "Hitchhiker's Guide to the Galaxy" series. Like the fictional Encyclopedia Galactica from his books, the site grows into a reference work on practically everything. All the text comes from individual contributions, some of which are undoubtedly dubious and not meant to be taken seriously; others are deeply insightful and lucid. It is a small leap of the imagination to go from collaborative text to collaborative multimedia.

Fast Variety

Searching for a piece of streaming media that catches your interest is much faster than channel surfing on a television, because search engines that index on metadata can get to the media you desire more quickly than a random walk through the channels currently broadcasting to your television set. Streaming media receivers can be designed to offer several streaming windows at once or even display pictures in a picture. Many digital television broadcasters are currently experimenting with multi-view presentation, where several camera angles are displayed at once, for interactive sports coverage. Streaming media already lends itself to multiple views at once, bandwidth permitting. It is entirely possible to be watching one piece of streaming media while listening to another.

Previewing of longer presentations is another area where streaming media can outpace television or videotape. With streaming media, it is possible either to shuttle through the presentation rapidly (assuming the player software can do this), or to jump from point to point. Content producers may also produce summary capsules of their longer work, like Hollywood trailers for feature films. Sophisticated media publication systems may be developed to publish a series of thumbnails based on scene changes, packaged as a piece of streaming media, automatically. This would then allow any potential viewer to scan the presentation in almost a storyboard presentation before deciding to watch the piece in full-motion resolution. Indeed, an automatically generated thumbnail storyboard may be the ideal interface for navigation within the piece of streaming media. Rather than using the fast-forward, play, and rewind paradigm, viewers might actually prefer to point to images within a storyboard in order to seek to some midpoint in the presentation. A storyboard presented as an HTML page could be embedded with every piece of streaming media published as metadata, enabling richer search engine indexes to be created, without the need for the search engine's Web crawler to access every frame of video in the presentation.

Whenever You Want

Streaming media, by its very nature, makes programs available on demand via wireline or wireless networks, both fixed and mobile, 24 hours a day, 7 days a week, 365 days a year. Unlike Personal Video Recorders (PVRs), which enable time shifting of broadcast television programs to a time that suits the viewer, time shifting is a fundamental property of on-demand streaming media. For streaming programs scheduled to run at particular times as a multicast program, time shifting is still required, but many streaming media providers who multicast would rather archive the presentation for on-demand viewing at a later time than risk losing potential viewers. Programs which are streamed and never archived for on-demand viewing are likely to be the exception. Even where the provider does not make an on-demand version available, a personal video recorder is relatively simple to emulate with streaming media technology, compared to performing the same feat with digital video broadcasts. This is because the metadata in streaming media can be richer, the video and audio is more highly compressed and so places fewer demands on hard-disk performance, there can be closer dialogs between the program scheduling software that will play the piece at a certain time and the player software that will try to detect it and record it, and digital rights management can circumvent fears about unauthorized rebroadcast of time-shifted content.

Humans have a fundamental need to find information to help them when they are under threat. Because streaming media can deliver rapid access to answers wherever you are, whenever you need them, it satisfies our need to be safe. It also satisfies our need to seek feelings of belonging and esteem, because it can be always on. Whenever we need to connect with other humans who will love and value us, we can do so through streaming technology. Cellular phones are the first demonstration of this fact. This form of audio-only streaming, from person to person, shows how we value the ability to call for help or talk to loved ones whenever the need arises, no matter where we are. A more poignant and stark example of this phenomenon than the calls made on cellular telephones from the doomed flight 93 over Pennsylvania on the morning of September 11, 2001 would be difficult to find. Had streaming video been available, it is entirely possible that those would have been face-to-face as well as voice-to-voice calls.

A Personal Information Shadow

Every person living in the developed world has a data shadow of some description. Your data shadow consists of the computer record of your transactions with companies, your government-held tax and social security details, the Web sites you have visited, information pertaining to your credit rating and so on. Nobody's data shadow exists in one place on one computer, but in principle, somebody with the right motivations could pinpoint your movements and the private details of your life with astonishing accuracy, if the data already available were mined.

A part of your data shadow consists of your tastes and preferences. Amazon.com already holds preference data for millions of book readers. It uses this to direct them to other books that people of like mind have already enjoyed. This has obvious benefits to consumers. A list of favorite Web sites and the sites visited is another part of that shadow already stored, though often held privately.

It can be very useful for your list of favorite streaming media presentations to be available for access, no matter where you are or which device you are accessing from. This would let you watch your favorite movie, for example, wherever you happen to be. Indeed, what you watch, where, and when is also a part of your data shadow, which may have real commercial value and may be sold by consumers who wish to do so, in exchange for cheaper access to media, for example. More advanced rights management systems could make use of this data shadow to provide more flexible licensing than the currently touted per-PC licenses. The richness of a streaming media shadow could provide tangible benefits to consumers (as well as some dangers for civil liberties and privacy). On balance, as with the Web, many consumers will consider that the benefits of having such a rich data shadow outweigh the disadvantages. Indeed, rich data shadows are likely to be inevitable as more and more surveillance cameras, driven by streaming media technology, record our faces wherever we go.

Video Beats Text

We absorb information presented audiovisually much more rapidly than we do when reading text. Evolution has meant that we are all adapted this way. Moving pictures and the right accompanying sounds can illustrate concepts more clearly and concretely than can words alone. Although reading and listening both allow the person engaged in those activities to use imagination to fill in the missing detail, to communicate a concept clearly often requires moving pictures.

Many people have found it impossible to follow recipes, for example, because the text didn't make clear what the cook ought to expect. Watching somebody bake a cake, step by step, however, fills in the missing details. You can see the color and texture of the mixture before it goes into the cake tin. You can see the color of the baked cake. Without this information, many text-based recipes are inadequate instructions on how to reproduce the author's intended confection.

Commercial messages are also more attractive and effective when presented as audiovisuals. Audiovisual material is particularly attractive to people who wish to watch passively and have the story told to them. Streaming media technology makes it possible for more of the information we currently access from books to be presented as audiovisuals. In contrast to broadcast television, the on-demand and online nature of the information makes it accessible by anyone, anytime. Access of streaming media material is more akin to going to a library and finding a book than trawling through the program schedules in the hope of finding that what we want to know will be broadcast when we can view it. Streaming media marries the sheer richness and impact of audiovisual presentation with the convenience and access features of a library of books.

Informative and Interactive

Streaming media is a satisfying technology for exploring and navigating through information, since it marries the hyperlinked nature of the Web with the impact and richness of audiovisual media. The viewer can encounter live programs and archive material, without necessarily being aware of the difference. Because people can follow their own learning paths they can learn according to their own particular ways of learning. No two consumers will ever follow the same path through the streaming media available.

The advantage of streaming media over other forms of audiovisual media is the fact that search engines and metadata descriptions can help us match streaming media items to our own interests, just as we can now find text items to satisfy our own information demands on the Web, with the more comprehensive search engines. If digital television programming were available on demand and indexed with a search engine similar in power to www.google.com, with hyperlinks from program to program and within a program, it would be streaming media. Indeed, once most television programs are repurposed in streaming formats and indexed by media search engines, much of the streaming media we view from archive will be old television shows.

The Best Mentors

Streaming media will be a tool par excellence for self-actualization. It will be possible to access the very best teachers on whatever subject you choose; you may even interact with them. Some will demand a fee for access (using digital rights management) or else some may offer their knowledge free or in exchange for other knowledge. The medium may even foster information exchanges or clearinghouses for intellectual property as teachers find ways of imparting what they know.

With streaming media, companies may pay consultants to encourage and guide their staffs using telepresence. Management guru Tom Peters, for example, may never have to board a plane again. Companies could pay for his virtual presence and his mentoring online, as a rolling videoconference.

The Best Salesmen

Every company has one salesperson who sells the product with more passion and more conviction than his or her colleagues. These salespeople often turn in the best results as a consequence of their facility in selling to consumers. However, even the best salesperson has some days better than others. Some days, he or she just cannot reproduce the enthusiasm or verve, for whatever reason. With streaming media, the best salesperson's best sales pitch could be recorded and used either in direct sales situations, or else as training material.

Founders of startup companies often complain that the burden of speaking to yet another stony-faced and unreceptive audience about their passion for their enterprise greatly taxes them. Investor relations often drain the life from entrepreneurs wishing to spend their precious time on developing their ideas, rather than convincing the unconvinced. It is a truism that if an entrepreneur's idea is any good, he won't have to fear its being stolen, because he will almost literally have to ram that idea down the throats of all and sundry. Smarter entrepreneurs will have video made, capturing them in full flourish, for replay at later dates.

Streaming media will be a great conduit for people's passion and commitment. Seeing the most motivated person on screen, via streaming media, in a variety of business-related scenarios, could significantly improve the outside world's response to that person's ideas and this could translate to better business.

Body Language

So much of what we communicate is carried in vocal inflections, pace of delivery, and in the stances and gestures we make with our faces and bodies. Text chat and e-mail are prone to "flame wars," in which comments are taken out of context and lead to heated exchanges, because of the information lost in text-based communications. What we want to say is often not adequately conveyed by what we write. There are occasions where a double positive is actually a negative. For example, if you were to respond to somebody with "yeah, right," in the absence of gesture and body language, it wouldn't be possible to tell if you were wholeheartedly agreeing with or sarcastically mocking the person. Similarly, when apologizing to somebody, if they respond with "it doesn't matter," there is no way to tell if you are being forgiven or if they are further accusing you of gross insensitivity. The message is all in the nuance, which is lost in text communications.

Phone calls have always been less ambiguous means of communication than text, but on occasion, even these leave communications open for interpretation. Streaming video communication, on the other hand, allows smiles and gestures, as well as body posture, to accompany the spoken words. Streaming media is the only long distance communications medium that can do this economically. Previous attempts at video telephony largely failed because the compression was so severe that the gestural nuances that were the very reason to communicate visually in the first place were grossly distorted, delayed or, in extremis, totally lost. With streaming media's compression refinements and the advent of broadband networks, we might finally get reliable videophones that add to conversational richness rather than detract from it.

Intimate Connections

People have cyber sex with each other today using text-based chat or over the phone. In cyber sex, each consenting adult pleasures himself or herself while encouraging the other. As a form of sexual activity, it carries no risk of sexually transmitted disease and may have become popular for that very reason. Also, participants can claim that they have not done anything other than pleasured themselves; in many relationships, this is considered an acceptable way to explore sexuality. Many couples that must be apart for business reasons for weeks at a time also maintain continuity in their sexual relationships through cyber sex. It can actually support fidelity in marriage.

Cyber sex using streaming media technology has the potential to be much more evocative and erotic, yet offer just as much safety from sexually transmitted diseases as previous technologies. Live images of people experiencing cyber sex with each other are traveling across the Internet already, through video chat services. As video quality improves due to improvements in available bandwidth, this application of the streaming medium is likely to experience rapid growth.

Other uses of high-quality streaming will include being able to communicate with distant friends and family, as if you were in the same room with them. Streaming media applications can satisfy the basic human need for love and belonging in many ways. Previous technologies have not provided a satisfactory user experience at an acceptable price point and were not realistic solutions for those wanting to maintain intimate connections with people they love at a distance.

Natural Modes of Thought

We all investigate the world we live in, organize our projects, catalog our information, and express our creativity in often idiosyncratic ways. The choice of canvas for our natural modes of thought is important, if our thought processes are to be supported rather than hindered. As a blank medium for expression, streaming media ranks among the richest ever conceived. Few media have ever been as accessible, global, or adaptable. Though many media are cheaper, streaming media is far cheaper to access for the artist, than making a feature film or getting a DVD to market.

We don't plan our real-life projects and activities in the way that project-scheduling methods used in business tell us we ought to. I've never yet seen a Gantt or Pert chart on my refrigerator. Real-life project planning is more informal than that. What normally happens is that you brainstorm, writing down "important stuff you need to consider in order to get the project done." Then you organize that information, so that all the relevant notes and information, or pointers to information, exist somewhere and can be accessed by some organizing principle, like a loose-leaf folder with alphabetical tabs, for example. Next, you generally set up meetings needed to get the project to progress. Finally, you might need to gather other information, so you wander off looking for it in books, on the Web, or wherever you know it to exist. Project planning has random elements, an informal and constantly evolving list of things to do, and a collection of important related information, organized in an ad hoc way, appropriate to the scale and type of project being undertaken.

Many times, you'll think of something you don't want to forget when you're at a place that has nothing to do with the project. You're driving to the store, for example, and think of a great way that you might want to start the next staff meeting. Or, you're stirring the spaghetti sauce in the kitchen and it occurs to you that you might want to give out nice carry bags to participants in the upcoming conference. You might be watching something on television when you suddenly remember another key person you might want to include in an advisory body you're putting together. How do you capture these randomly occurring brainwaves and catalog them usefully in your project plan? I believe streaming media offers some compelling solutions.

With mobile streaming media becoming an omnipresent feature of our lives in the future, there will be a way to capture those thoughts with whatever streaming device we have to hand. If distributed software begins to allow us to access all our information with streaming receivers, regardless of where it is physically held, then the project plan really consists of nothing more than pointers to information resources we know we can access by streaming to the device we happen to be using. The plan itself is also something that can be streamed to whatever streaming media receiver we are accessing. In such a world, adding to the plan is as easy as making a phone call. Documents, hyperlinks, video material, voice notes, or just a video of yourself talking to camera, can be captured and cataloged in our virtual project plan, any time of the day or night. No current desktop project planning system can do that. No paper-based project planning system offers the same level of portability or integration of media types. Streaming media technology could therefore revolutionize the way projects are planned and executed, matching project-planning methods to our natural and instinctive ways of getting things done. The result would be greater effectiveness at moving plans along and hence greater productivity. This is one of the less-considered ways in which streaming media technology will one day make obvious returns on investment.

It should never be forgotten that video is an extremely potent messenger. Video can inspire creativity in others simply because it is video. This is the reason why advertising on television commands a premium. This is why instructional videos outsell paper-based versions, particularly in musical instruction, where a virtuoso can be enlisted to present the material, inspiring as well as demonstrating. The power of the visual image is the reason why Hollywood films can so profoundly affect the emotions of audiences worldwide. Streaming media, for the first time, brings together the power of audiovisual communications and a universal and relatively cheap distribution channel, putting the power of that medium into the hands of anyone with the means and will to use it. We will watch and use streaming media simply because other ways of communicating will seem impoverished and unsatisfactory by comparison.

What Will We Watch?

This chapter will talk about some of the unique forms of entertainment and other experiences that content providers will be able to create with streaming media. Obvious things include video on demand, movie rentals, digital music, television shows repurposed for the Web, corporate communications from businesses to consumers, streaming ads, home encyclopedias and Internet radio. All of these have already been done to some degree. However, the aim here is to talk about styles of streaming programming not yet widely available, which stretch the capabilities of the medium.

Streaming media is concerned with immersion, interactivity, personal relevance, information density, individualization, richness, and reach. More than any other medium yet invented, streaming media allows program makers to achieve those aims to a greater degree than was ever possible before.

Hyper News

With streaming media, news production can be revolutionized. Streaming news stories can contain sidebar links or even links embedded within the video itself, allowing the viewer to jump from story to story, perhaps following a series of successive reports back in time, to see how related events unfolded: news with hindsight. Background stories can also be traced, as can sidebar stories to explain particular issues of relevance to the main featured report. What many news sites, like http://news.bbc.co.uk already do with text and graphics, could readily be achieved with television-quality broadband streaming media content. In fact, viewers bent on checking the veracity of a report could potentially refer to the original camera reels from which various news items were edited, cross-matching these with other camera reels with the same timecode and geographic location metadata stamps. Media management systems, which are increasingly finding application in television news production, make this possible.

Taking news reporting one step further, it is possible to check the track record and credentials of a reporter, simply by pulling up the relevant resume and a list of previous reports. With this information, we can build up a list of reporters whose editorial commentary we trust. For example, I think the reporter John Pilger, famous or infamous for his exposé of what was occurring in East Timor, uses overly emotive language and imagery in his reporting, yet often reveals grains of uncomfortable and not widely reported truth. I only have this opinion because of background reading I have done on him and the subjects he reports on. With streaming media, this additional information could be presented as hyperlinks for me to follow.

Spokesmen who appear on camera could similarly have their potted history available as a sidebar for viewers, revealing their affiliations and financial interests, for example. In the UK, the BBC's flagship news magazine program, *Panorama*, recently posted a Web site, accompanying their report entitled "Under the Skin," listing the criminal records and links to organized football violence of senior members of a militant far right-wing political party. The party's spokespeople were trying to create a cleaner media image, in order to appeal to more moderate voters. However, the background information provided on the Web site stripped bare the deception, in greater detail than could be achieved with 50 minutes of video alone.

Just as it is possible for audience members in a live studio debate to pose questions directly to panelists, interviewees, and reporters, so it will be possible, with streaming media technology, to participate in debates from the comfort of your own home. Like a phone-in, this technique will use video technology instead, allowing people to see the person posing the question, while they contribute to the direction of the program. Indeed, streaming media technology makes it possible for audience members to simultaneously discuss the issues with each other, using live video chat. Once audience members are allowed to participate actively in this way, control over the agenda passes subtly from the hands of the news producer into the hands of the audience. News is no longer dictation. There are, however, some very real dangers to this approach. The potential for legal action against individual audience members, if what they say is in some way damaging to another party, must be borne in mind by every participant. The role of guarding against legal problems once fell to the news editor, but with greater participation by the audience, it becomes impossible for that editor to fulfill.

With the infinity of available channels on the streaming medium, the requirement to keep program lengths to their allotted time could vanish. It is possible never to be "out of time." With broadcast television news, politicians notoriously "sandbag," evading having to answer a difficult question until the time allotted for the interview has run out. With streaming media, the time allotted need not be so critically regulated. If a politician refuses to answer, this can be brought out clearly in the interview; this can be just as damaging to the interviewee as answering the question might have been.

Streaming media technology has advantages in the speed with which breaking news can be distributed to audiences around the globe. Reporters with little more than a satellite phone or laptop, third-generation mobile phone and camera can post instant reports from wherever they are. With content delivery networks replicating these media clips to caches at the edge of the Internet, an audience of millions can be watching events more or less as they happen. Broadcast networks struggle to compete with this turnaround time, using traditional broadcast news production equipment (ENG Electronic News Gathering kit, for example). In fact, many broadcast news networks have conceded defeat, using streaming satellite technology to relay news instantly from Afghanistan during the anti-Taliban war. With streaming media, there is no need to transcode from a streaming format to a broadcast television format, as there is with the major television news networks today. A streaming news report is already in the medium's native format from source.

Because streaming media uses efficient compression technology, a deep news archive, which could be accessed and searched by individual viewers, could be made available quite cost effectively. Researchers, in particular, could access this archive to obtain "first drafts" of historically significant events. Over time, this will be a highly valuable resource, accessible to many more researchers than current broadcast archives, with their bulky, unwieldy, and ever-degenerating videotapes, could ever be.

Because desktop streaming media production is realistically possible, it is possible for any member of the public to produce a news item. Community news or special lobby groups have access to the same news production resources, such as desktop production tools, library footage, and the means to distribute their finished reports, as the professional news organizations. For example, organizations like Greenpeace or Amnesty International, who already have media units to produce material suitable for inclusion in broadcast news, need not wait at the gate to the distribution channel, as they do today, in the hope that their item will be broadcast to an audience. They can set up their own streaming channel and distribute their video footage directly to audiences worldwide. Indeed, many of the regimes and companies on whom they report would find it difficult to stop citizens or investors from accessing those reports; something that they can still more or less prevent today. News organizations using parts of those video reports could, for example, refer viewers to Greenpeace's streaming media serving site (if there is one) to see the full version.

Streaming media fosters a greater diversity of opinion. An example of the power of this was the Web site set up by the Revolutionary Association of the Women of Afghanistan (www.rawa.org), which served media streams of public executions of women by the Taliban and other human rights abuses, caught on video. The content on this site was included in several broadcast news reports and served to harden public opinion against the Taliban. While streaming media technology doesn't make the cost of getting a message out to the public entirely free, it changes the economics of news provision to such a degree that more and more organizations with particular points of view will be able to speak directly to audiences.

Effective Education

CD-ROMs have been used for years to create interactive multimedia learning packages. Their advantages and shortcomings are well known. What streaming media adds to the mix is live interaction with teachers and fellow students, up-to-the-minute content and individualization.

These are important factors in making media-based educational materials more effective.

Streaming media can be used to create a rich educational experience, with aids to learning embedded in the main program stream and on-line review quizzes administered and instantly graded. Early indications from people interested in using streaming media have shown that distance-learning applications will be an important sector of the market.

Writing this kind of content remains a cottage industry requiring extensive handcrafting. Manufacturers of desktop production tools have some distance to go before they create the ultimate distance-learning streaming media creation solution. Live production equipment, suitable for educational program making, is also in its infancy. For these applications, the producer would ideally like to have a machine, making it possible for a single operator to use computer and robotics to control lights, camera, and sound. If the machine could also add synthetic elements and rolling or crawling on-screen titles, that would greatly reduce the cost of production of educational streaming media.

Players that render synchronized multimedia for viewing are also still in their infancy. They stall and stutter, and text panels flash obtrusively when changed. They need to be designed to protect visual continuity. With players needing to deal with simultaneous media streams from distributed sources, it just might be that a single network protocol stack isn't adequate for the task.

Nevertheless, streaming media programming that serves to educate is going to be commonly available and widely viewed.

Help at Hand

Some of the streaming media content that will be received won't be content, as such, at all. One of the applications of streaming media, particularly in business-to-consumer applications, will be to offer help and advice to customers. Customer-support personnel will be able to watch what you are doing and see what your problem is, guiding you to a solution, using streaming media. If your dishwasher breaks down, the first content you might access could be a short video showing what to check and where to look. If all those solutions fail or if you lack the confidence to follow the instructions in the video material, a human could then come online, asking you to point a mobile camera at places on the machine and talking you through the problem.

Products could even be designed to include wireless cameras, so that remote technicians could diagnose and suggest remedies without making a home visit. Cameras might be installed in inaccessible or inhospitable places, like under the hood of a car, for example, so that a trained technician could observe the problem as you drive. Third-generation wireless networks will make such applications possible and perhaps even commonplace.

When we buy a new product, we might view a stream showing how to set up the unit and make it work. Instead of reading an instruction manual, or more commonly just figuring out what to do without instructions, customers could be guided through the setup process by a streaming media program.

We might even access streaming media product demonstrations to have the demonstrator sell us on the virtues of a particular product we are unsure about buying, or to comparison shop. These demonstrations might be prerecorded or live. Much of the streaming media we watch in the future will be programs that help us get things done or help us make decisions.

Love Interactions

Just as people have found ways to use telephones, the Internet, on-line chat, and SMS messaging services to flirt and to have cyber sex, so a section of the population will undoubtedly use streaming media for those purposes too. A certain proportion of users will spend time accessing live pornography, where users direct the participants on camera, or else view each other in intimate and erotic ways. How popular this kind of programming will be can only be extrapolated from the success of online pornography Web sites. Indications are that this application will account for a significant amount of the time spent viewing streams, at least in some circles.

Immersive Entertainment

Streaming games, combining rich media types such as photorealistic 3D models with live video textures mapped onto them, surround sound, and peer-to-peer interactivity could be a popular streaming media programming style. The video gaming industry already earns more than

Hollywood movies. Much of the content streamed to households in the future is likely to be of this type.

Broadband connections and extremely capable real-time 3D graphics engines, like the one on Microsoft's gaming console Xbox, provide a platform for multiplayer, multiview, networked, photo-realistic entertainment.

D-Features

Independent digital filmmaking, using cheap DV cameras, is already a popular activity. These are called *D-Features* or Digital Feature Films. With a little ingenuity and a great deal of bravado, individuals can shoot, edit and post-produce theatrical feature films for a few thousand dollars. Compared to Hollywood budgets that run into the millions, this is staggering. Indeed, films that cost very little to make, like *The Blair Witch Project*, can sometimes go on to become box-office smash hits. Films made on a shoestring, which find large audiences, are very profitable indeed.

Independent filmmakers encoding their films to high-quality compressed streaming media formats are already finding an audience through sites like www.atomfilms.com. As the infrastructure for delivery of streaming media evolves, people are likely to watch more of this kind of programming, simply because of its variety value.

Specialist projection houses (microcinemas) may also spring up to serve film-quality streams to small, niche audiences. Feature-length films or short films made by underground filmmakers may one day provide real competition to Hollywood's output.

Video Instant Messaging and Mail

Much of the streaming media that will be watched in the future is already being watched now, but in very poor quality. Online video and voice chat services, like PalTalk and Yahoo! Chat, already allow people to communicate with each other using audiovisual media.

Streaming media technology has not yet been widely adopted in this application. Most video chat sites manage to transmit just a few stills per second. There isn't a full motion version yet available. Microsoft's NetMeeting does have Windows Media Technology under the hood, though what pieces of that technology are actually used is not clear to me. Streaming media cameras that encode directly to MPEG-4 are likely to become widely available soon, since several chip vendors are embedding the encoding hardware in their products. When these are everywhere, people will be able to communicate with each other, in full motion. This is the best way possible of staying in touch with distant relatives whom you rarely see face to face. It is also a good way to meet new people, since in video chat, it is harder than with text chat to fake your identity and tell lies about yourself. Much of the streaming media that people will watch will be pictures of each other!

Special-Interest Magazine Shows

Just as desktop publishing spawned a rash of magazine titles and fanzines, catering to every conceivable niche interest, so too will desktop streaming media production spawn streaming magazine shows catering to those same niche interests. The "streamzine" will be a logical outgrowth of the trend for *samizdat* (underground publication of manuscripts) on just about any subject imaginable.

In alternative bookstores across the US, it is possible to find "zines" (short for fanzines) with slick covers, color printing, tight binding, and sharp design that make them almost indistinguishable from mainstream magazines. "Ezines," Web sites that resemble their physical cousins, have already become a phenomenon. Costing virtually nothing to produce, an ezine can have an audience of thousands, with a potential audience reaching into the millions. Ezines are growing by astronomical proportions, with virtually no end in sight (see www.bestezines.com for example). Streamzines will be the next logical step. Looking like a well-produced television show, with synchronized multimedia elements, the streamzine will provide content to cater to all tastes and interests. Through the power of streaming media, people can meet like-minded individuals, engage in discussions on topics of similar interest, and create virtual communities.

Archives and Vaults

All the audio and video that ever existed in an archive, anywhere in the world, could one day be available as on-demand streaming media. Assets that were effectively "dead" could once again begin to earn revenue for their content owners. Conversion of old film and sound archives to streaming media would preserve our audiovisual heritage.

Applications of digital media to make precious and fragile archives more accessible have already been pioneered by the British Library, for example. The Library has digitized the pages of rare and precious first editions and created a computer simulation that allows online readers to leaf through those books virtually, reading the pages as if in real life. If those actual first editions had been read by as many people as have used the digital simulation, the books would have crumbled and disintegrated by now. With digitized content, the reader can see the book as it appears, margin notes and all, without destroying or degrading it.

For audio and video archives, this is also true. Archives of film stock and magnetic recording tape become extremely fragile with age. Magnetic tape, on which most of our recent video history is stored, lasts only about ten years, before it is necessary to dub to a new tape. Storage costs are enormous, as are the costs of periodic dubbing and restoration. The process of playing an old tape or showing an old film can significantly degrade the asset. If these could be digitized just once and then made available through streaming media servers, rarely seen or heard events from history would be available to all. It is anticipated that this will comprise a significant amount of the streaming media available in the future.

Streaming Auctions

If the law tells you "let the buyer beware" then whatever buyers can do to make themselves aware of what they are buying and from whom, will be like insurance for them. People buy things using online auctions sites like eBay every day. Yet the buyer could be anyone and the goods might not be as good as they appear in the still photographs posted on the auction site. Reports of fraudulent behavior by sellers using online auction sites to cheat unwary customers are a major source of consumer complaint.

Streaming media technology makes it possible to inspect an article you wish to buy virtually. The goods can be modeled in three dimensions. Using one of the many virtual-reality streaming technologies available, such as MPEG-4 3D add-on X3D (eXtensible 3D), the buyer could manipulate the model in three dimension and view it, as if inspecting in person. The technology allows buyers to avoid having to buy "pigs in pokes."

Fly-By, Walkthrough Streaming

Much of the streaming media content available for viewing is likely to be telepresence material that allows people to tour virtually holiday resorts, homes, or architectural layouts for new kitchens, before they buy. They may walk the route to a destination of interest in cyberspace before they ever attempt the journey in real life. Products may be demonstrated in closeup detail, allowing customers to see every aspect of the product before making a purchase. Indeed, retailers and trade-show organizers might use telepresence streaming to entice you to visit their store or exhibition.

Extreme Retailing and E-Commerce

Retailers are seeing shopping as more and more of an entertainment experience. "Extreme retailers" are making use of digital cinema presentation equipment to show free in-store content, in an immersive and impressive environment. Of course, interspersed with the content are the ads, which customers must sit through in order to complete the fun experience.

E-commerce sites can fight back aggressively by making their sites just as attractive and entertaining, with companies moving toward making online shopping fun, rather than presenting potential customers with a dry and boring online catalog, as many do today.

Streaming media underpins all this content production and distribution. It would not be surprising for a major label (Coca-Cola, for example) one day to provide some of the most entertaining content available anywhere. Retailers may become more prolific media producers than media companies.

Honey, I Shrunk the Children

One of the more popular applications of streaming media is likely to be keeping watch over those you love. Watching your children at school or play is relatively easy and cheap, using streaming media technology. Mobile Web cams, using third-generation wireless networks, are easy to deploy. As a parent, you can, from your office, watch your baby sleeping in the nursery, or else watch your children with their childcare worker.

Hong Kong-based company Satellite Devices showed an interesting product called Kid Finder at the Innovation Expo 2001, held in Hong Kong in December 2001. Using GPS (Global Positioning System) and GSM (Global System for Mobile communication) technology, it can monitor where a child is via the Internet. In the event of an emergency, the child can press the SOS button and connect directly to a control center. The addition of a microphone and camera to stream sound and pictures relating to the child's location is a very simple next step for a product like this.

The Business

How Will Anyone Make Money with Streaming Media?

The question addressed in this chapter is the burning issue for streaming media. Unless answers can be found, there will be no medium. At present, the streaming media business is caught in the chasm between early and mainstream adoption due to the absence of compelling applications and content. Bandwidth provision is both too patchy and too expensive. While the Internet has grown rapidly, television penetration grew faster and radio just as quickly. Bear in mind that streaming media penetration represents only a small fraction of current Internet usage, which is dominated by e-mail, instant messaging, and Web surfing. The penetration of streaming media is proving to be much slower.

However, the Web continues to grow, contributing to what may be the fastest-growing ad market ever, estimated to be worth $8.2 billion worldwide. AOL Time Warner and Yahoo account for about half this total. As a point of comparison, the total worldwide box-office revenue for the motion picture business exceeds $9 billion. Broadcast television and radio account for 20% and 8% share of total advertising expenditures respectively. By comparison, in mid 2000 the Net constituted only about 2.5% of total US ad expenditures. However ad spending for television, radio and cable all developed at a much slower rate than advertising on the Internet. What does this tell us for streaming media? It tells us that adoption has been slower than television and that it takes a while to make advertising-supported programming pay, particularly since the advertising inventory (places where an ad can be placed) on the Internet is practically bottomless. There is little competition for prime advertising inventory as yet, since there is so much of it and audience reaction is muted.

What is clear is that until the quality, reliability, price, and convenience of access of streaming media services reaches a level comparable to that of radio and television, consumers, content providers, and advertisers will have reasons to stay disinterested in the medium, while better alternatives already exist. Radio serves to illustrate the point. In the 1920s, you had to be almost a geek to operate a radio. They were complicated machines. Radio really only found its audience once receivers were simple to operate and well behaved. By 1930, 45% of American households were tuning in, compared to 10% in 1925.

It is estimated that Americans will spend 192 hours online annually, by 2003, assuming there is no significant streaming media content available. That's an average of about 3.5 hours a week. This compares with an average of 28 hours per week watching television. Americans rent 6 million videos a day. Total media consumption averages 11.8 hours a day, in the US. Streaming media has the potential to account for a significant proportion of that media consumption. Surveys show that Web use is already cutting into TV time, with Internet users spending 4.5 fewer hours a week watching television than non-users. Table 4.1 illustrates the average media consumption patterns in the US today.

TABLE 4.1

Average Annual Media Consumption per Person in the United States

	1998	1999	2000	2001	2002	2003
Total TV	1573	1579	1591	159	1605	1610
Radio	1050	1037	1024	1014	1003	992
Broadcast TV	884	840	805	773	751	729
Internet	74	97	122	146	168	192
Newspapers	156	154	152	151	150	149

Hours per annum.

This chapter will focus on potential ways to make money with streaming media. The return on investment report for enterprise streaming conducted by streamingmedia.com showed that businesses that had used streaming in a variety of contexts, including building revenue streams, increasing brand value, and generating more custom to retain audiences and to facilitate communications, had seen returns on their investments of between 38 and 72%. Enterprises who try streaming are, on the whole, happy, satisfied customers. The majority surveyed said they would continue with their streaming media efforts, given the success of their streaming media pilots.

Business models such as pay every day, pay per play, pay to own, prepay, pay what you think it's worth, pay if you liked it, or pay some other way are possible, but how you make that payment is a significant issue. The multitude of incompatible payment systems creates inconvenience. Who wants or needs 70 different e-money currencies in a virtual wallet (or, for that matter, 70 different e-wallets, as is the usual case)? Credit cards are not viable for making small payments because of the transaction charges involved, so they do not support many of the business models possible. What use is media commerce without a convenient way to pay? Microsoft's .NET claims to have an answer. Their "My Services" offering (also known as HailStorm) provides for convenient and transparent micropayments. Time will tell whether or not My Services proves to be a revolution in online commerce, or just another incompatible online payment scheme.

From the point of view of media commerce, the key questions to be solved are:

- How do consumers find media? (Napster proved to be a reasonably effective answer, but MPEG-7 enables many improved search techniques.)
- How do content owners protect their rights?
- How do consumers protect their rights?
- How do consumers make a payment for media, once it has been found?
- How do consumers obtain licenses to play media without trauma?

Was "Free to Air" Ever Really Free?

Despite appearances to the contrary, television never was free, nor was radio. One way or another, the consumer paid. Either the broadcast network was a business selling advertising space (advertising costs are merely added to the costs of all goods and services advertised) or else it was state funded, supported by taxes or compulsory subscriptions called *television licenses.*

What broadcast networks have going for them is that once the infrastructure and operating costs are covered, the incremental cost of adding another viewer is negligible, yet that viewer's attention translates into greater revenue for the broadcaster, who often sells advertising based on number of impressions. The cost of production and distribution is also amortized over a larger viewer base, so there are significant incentives for broadcasters to get as many people as they possibly can watching and listening. The limited choice of channels seems to hold viewers captive, but also acts to reduce the overall potential audience, since viewers disinterested in the limited number of program channels available ultimately switch off. Limited programming choice allows a broadcaster to hold a larger piece of a smaller pie. This push to maximize the number of viewers influences all program commissioning and scheduling decisions, with the goal of maximum audience share outweighing most other considerations.

Streaming media can adopt the broadcast model and try to compete on the same grounds as radio and television, but streaming currently has cost and quality disadvantages. However, Streaming Media can compete for niche audience satisfaction. While there has been talk of the resistance of consumers to paying for media access, the truth is that consumers have always paid for access to their media; it's just that the costs were often cleverly hidden. Also, many consumers were cross-subsidized by their wealthier audience peers. Commercial television and radio are, in fact, a form of disguised socialism, since the poor do not bear the same share of the cost of media access as do the wealthy. Affluent people, as a group, tend to buy more of what is advertised and also the more expensive items, which bear proportionately more of the manufacturer's advertising expenses.

Pay-per-View Streaming

Pay-per-view streaming is possible with e-commerce technology and digital rights management, but these media commerce systems are only just now becoming commercially available. The downside with pay-per-view and video-on-demand streaming is that the cost of each additional customer, in terms of bandwidth, scales with the number of viewers. The greater the number of viewers, the larger the content provider's bill for bandwidth. This is somewhat mitigated by the use of CDN edge networks, however. In this model, provision of each and every stream has to be profitable. With the quality and bandwidth limitations that currently apply to on-demand streaming, this model has not yet matured. However, in the future it may become one of the dominant models for streaming media access.

Streaming by Subscription

This is easier to do technically and administratively than pay per view. However, to get a consumer to sign up for a subscription is difficult, since customers require compelling reasons to trust in the quality of as-yet-unseen programming. Subscription services need to offer attractive media trailers and money-back guarantees to attract subscribers in the first place. Having signed up subscribers, they need to deliver on the customers' expectations, or else risk subscriber retention problems. With less stringent copy-protection systems, rather than full-blown digital rights management, this model can be realized quite simply. In common with all streaming media's current business models, however, this model is not mature. Bandwidth charges are still too high for content providers and end-user quality of experience is inadequate for most consumers.

Streaming Subsidized by Advertising

This is analogous to commercial television, where media access is free at the point of experience, but paid for by advertisers who interleave their ads with the content. Streaming advertising is already beginning to be trialed. There is still some resistance from advertisers who question the current viability of the medium, but streaming offers several new advertising methods, which are not as obtrusive as other types of inline advertising. Besides in-stream ads, streaming banners are feasible. The media player application also allows banner ads in the media player window and allows the use of the player border for advertising. Also, the response to streaming media advertising is directly measurable over a back channel, with appropriate software tools now becoming available. Ad personalization is also technically possible, allowing advertisers to target qualified niche audiences with finer granularity and less waste. Unfortunately, streaming players also make it relatively easy to skip over the ads. There may need to be an explicit agreement between the end-user and the media provider to watch the ads (this can be monitored by software in the player) in exchange for free access to content. The advertising models offered by streaming media technology are more flexible than the broadcast television model.

Streaming Subsidized by E-Commerce

Instead of making you watch ads for products, content providers may sell to you directly, making media access free, as long as enough people buy products alongside the media. This could work along the same lines as the cable QVC channel, except that the content need not be pure selling. For example, a band could make streaming video clips available, alongside an e-commerce site that offers CDs and T-shirts for sale. As long as the aggregate profits from merchandising covers the media distribution costs, this model is viable.

Trading Private Data for Streams

Often, the most valuable thing to a company is market research data. Rather than playing ads, the content provider could exchange access to streaming media for answers to market research questions. Consumers would be required to fill in answers and explicitly agree to allow the content provider to use that data for commercial purposes. Unfortunately, most people perceive the data they give up as more valuable than the media access they receive in return. They are also concerned that the data will make money for the content provider time and time again, whereas they only get access to the media once. In order to break this stalemate, consumers' rights to their private information and preferences need to be equally protected with digital rights management, so that the exchange can be made more symmetric and fairer. With right management of the answers to the market research questions, content providers could agree to certain terms of use covering the data. Consumer would have some confidence that their privacy requirements would be met, policed, and enforced by something stronger than the mere privacy assurances that Web sites offer today (i.e., digital rights management).

As with all rights-managed transactions, however, people of limited or no interest to market researchers, such as the poor, the elderly, or the unemployed, would be excluded from accessing media for free in exchange for data. This would lead to the anomalous situation where the poor pay more for media access than the rich.

Government Funded Streaming

Streaming media access could, in principle, be funded through taxation, either progressively or regressively. Progressive taxes would levy a charge based on income, whereas regressive taxes would resemble the UK's television license model, where consumers are compelled, by law, to pay what is, in effect, a subscription to media access. Perhaps the BBC will continue in this way, gradually migrating to the provision of streaming media (this is already happening—the BBC site streams many stories in Real Media format). The long-term viability of this funding model for streaming media looks doubtful to me.

Protecting Rights

When rights management becomes commonplace, there will be a role for "rights policemen," who monitor and detect rights abuses. They will also potentially be the royalty collection agencies. Because the explosion of choice in streaming media will lead to many more opportunities for rights abuse, companies that make money from protecting rights on behalf of content owners will make money. In fact, even in today's broadcast networks, where the number of available channels limits choice, copyright abuse is already fairly commonplace, according to industry insiders. Networks flagrantly copy and rebroadcast news and sports reports, with no acknowledgment. The broadcast industry does not even have a handle on the scale of the problem, since there are few ways to watch everyone else's program output to spot infringements of rights. With streaming media technology, agencies will be able to automate the monitoring of rights abuses and ensure compliance with the terms of the media licenses granted.

Better Bandwidth Utilization

Another way to make money with streaming media is to find better and more efficient ways to get more information through the network using less bandwidth. Better information coding means better video quality, faster response, lower bandwidth charges, and improved end-user quality of experience. Companies that create the technology to drive these improvements will undoubtedly make money from streaming media.

Multicasting

Multicasting in the public Internet is still relatively rare, since multicasting needs infrastructure in place to support multicast data transport. Unfortunately, much of the Internet's infrastructure is currently unable to support multicasting. Within a closed corporate network, this is less problematic. Multicast has the benefit of saving bandwidth charges, since it mimics broadcast, where the addition of incremental viewers adds negligible cost. However, there are technical challenges with multicasting inherent in its flow control. With multicasting, a single slow receiver can degrade the performance of the network for all receivers. Also, there are no handshaking protocols to cover the situation in which all listeners have ceased listening to a multicast group. The router must make the assumption that all listeners have disconnected, after some arbitrary timeout period. This leads to inefficient router and bandwidth utilization.

Stream Fountains

Digital Fountain, Inc., has taken an interesting approach to multicasting by figuring out a way of encoding a content package so that recreating the file at the end-user's player is just a matter of gathering the right amount of data. The data need not be streamed in order of playback. If packets are lost, it doesn't matter, since each packet's role is merely to contribute to the overall reconstruction of the content stream. Streaming is therefore connectionless. If you lose one packet, the next one you receive is just as useful. The only important criterion is that you have enough data packets to reconstruct the data. This seems to me to be an application of forward error-correction coding.

What a digital fountain does is make a piece of media available all the time on a particular multicast channel. Since there is no need for error checking, because packet loss is allowable, this form of multicast cannot be clogged by a slow client or broken network segment. End-users who want to tune in to the media content need not wait until the program starts again, as they would with regular multicast streaming; they just tune in and start gathering data packets. When they have enough packets, they may play the media back. Stream fountains are the devices that broadcast metacontent in this way. They can equally well be used to fill caches at the edge. This allows multicasting to the cache, then unicast of the content to individual clients.

The disadvantage with this scheme is that the player may have a large startup delay, while buffering, if the content provider has encoded a long piece of media for multicast. To get around this, the content may be segmented, by dividing the stream up into a short segment, followed by a longer one and so on. These segments can be multicast on a number of channels. This minimizes the time required to gather enough packets to play the first segment. As soon as the first segment begins playing, the player switches to the second stream segment's channel and begins reconstructing the second segment of the stream. If the player manages to reconstruct the second segment in time, just as the first segment finishes playing, the player merely continues across the boundary. The viewer sees a seamless playback of the stream with relatively low startup latency. If the original stream is segmented uniformly, with a relatively fine granularity, then locating to absolute positions within the stream becomes possible with low latency. The cost of the stream fountain technique is that the metacontent coding scheme takes more bandwidth than a raw stream, for a single unicast viewer. However, data delivery is perfect, as far as the viewer is concerned, even if packets are lost. Stream fountains work well in serving large audiences, since the fountain is always on—each additional client merely "drinks" data from the fountain representing a segment of a stream, whenever he or she happens to log on. The bandwidth used to transmit a fountain the whole time is much lower than the bandwidth required to serve each stream the traditional way. Hence, stream fountain infrastructure providers and content providers who use stream fountains can potentially earn good rewards, since the one enables the other to make a higher profit margin on content provision to large audiences.

Cheaper Bandwidth

Companies that find ways to provision the world with cheaper bandwidth will do well. We have outlined an application in this book that can potentially soak up a million times more bandwidth than is currently available. At the right price, the demand for bandwidth created by streaming media can be practically limitless. Companies that make bandwidth available cheaply will be in demand. This does not just mean the optical network equipment suppliers or the optical networked ISPs, but also the companies that lower the costs of provisioning and find cheaper ways to lay fiber in the ground to achieve widespread fiber to the home.

Mobile Connectivity

If mobile networks succeed in providing relatively cheap bandwidth to mobile users and if content is relatively cheap, mobile connectivity services will make money with streaming media. Consumer electronics companies will find many opportunities to create interesting and innovative appliances to take advantage of mobile streaming. Mobile streaming to the car or to an airline seat has not yet even started in earnest. The potential for consumer electronics manufacturers and mobile bandwidth providers has not even been adequately understood or quantified, but first-order estimates indicate that significant revenues could be generated.

Replacing Travel with Bandwidth

Streaming media is undoubtedly cheaper than other media distribution methods, since physical media need not be produced or transported. Hence, anywhere that streaming media can replace physical media delivery, the provider will be able to have higher profit margins. In addition, the value chain is truncated with streaming media delivery, since there are fewer agents in the chain trying to abstract profit. This gives the media provider greater opportunities to retain profits or else compete on the basis of end-user price.

Streaming media service providers who make it possible for business people to avoid having to make long and arduous international trips should also do well. The service only needs to save one face-to-face international meeting to begin paying for itself.

One of streaming media's most important attributes is that it increases reach. This could mean opening up new markets for products, providing a new distribution channel, or making a larger customer base aware of a company's offerings. Streaming media is a cost-effective marketing communications tool. Companies that use streaming media to increase their reach will make money with the technology, not from direct media commerce, but from the increased business for their other products and services brought about by marketing communications using streaming media.

Who Will Make Money?

While it is difficult to identify with any level of certainty those who will make money with streaming media, some indications and trends help to identify the likely winners. This section provides a necessarily speculative overview of the types of businesses that could benefit financially, if the medium achieves mainstream acceptance.

Content Owners and Creators

Irrespective of the breathtaking technical elegance and beauty of streaming media technology, the thing that will ultimately drive consumer adoption of the medium is compelling content. The higher the quality, the more innovative the interactivity, the more engaging the narrative, the better the medium will do. Content will not succeed if it merely repurposes "flat-earth" television programming. The most successful content will be that which makes the best use of the capabilities of streaming media technology. The content that matters is that which cannot be rendered adequately on any other medium or combination of media. To sell streaming media to a skeptical and jaded public, content will have to showcase streaming media technology in an exciting way. Content providers who grasp the unique characteristics of the medium will not only attract audiences to streaming media, but will also receive the best returns on their content creation investments. Online digital music sites showcasing unsigned artists are beginning to find this out.

Bandwidth Profligates

Streaming media is attractive, addictive, and powerful, with the capacity to deliver high-impact communications of unparalleled richness to a potentially global audience. Advertisers who burn as much bandwidth as possible to deliver their messages with streaming media will ultimately be inescapable. They will be noticed just because of the barrage of high-quality, high-bandwidth messages they put out using streaming media technology. This won't happen until bandwidth becomes more abundant, at much lower prices. When this occurs, as it undoubtedly will, those who burn the most bandwidth will reap the greatest amount of the public's attention and will consequently make the most money with the medium.

Optical Network Service Providers

Contrary to the prevailing view of the optical fiber communications industry, where companies are going bankrupt with alarming regularity and excess optical capacity is actually being dismantled when companies fail, we have demonstrated in this book that even though the telecommunications industry believes there is a bandwidth glut, the truth is that there is an appalling bandwidth shortage, when we consider the end-to-end path from streaming media server to individual consumers. The last-mile problem keeps many consumers from subscribing to all but the most meager connections. We have demonstrated that even with cable and DSL broadband, this amount of bandwidth to the home is at the bottom end of the range of usability, particularly when we consider the bandwidth requirements of home media gateways that must supply the entire household with streaming media on demand.

While it could be argued that the copper infrastructure is already in place, this infrastructure is also relatively old and crumbling. Achieving faster and faster data transfer over twisted copper pairs is possible, but the expense of the equipment required to achieve this technical feat is not insignificant. The neck of the technology can be wrung, but the fact that this infrastructure was designed for an entirely different purpose must be faced.

Optical fiber communications, by comparison, are potentially simpler and more reliable, and decreased costs of ownership for both the telecommunications supplier and the consumer should be possible. For this reason, I argue that optical network service providers hold the only viable technology for flawless streaming media delivery over wired networks, in the longer term.

Only when the phone companies get over their fixation with dedicated switched voice circuits and create an infrastructure purpose-built for the delivery of vast amounts of data with assured quality-of-service will streaming media hit its stride. The only technology that can really provide superabundance of bandwidth with quality of service guarantees is optical networking. For this reason, I believe that ISPs that embrace optical technology, bringing it as close to individual subscribers as possible, will make the most money from streaming media. These companies stand to do at least as well as the incumbent telephone companies did when they had a virtual monopoly on voice communications.

Optical Network Equipment Manufacturers

If we consider the scale of the network build-out needed to deliver high-definition streaming to everyone, it is clear that optical switching and routing elements will have significant cost advantages. In other words, as the massive optical network is constructed, it will replace and render obsolete most of the electrical switched-circuit technology currently in use, not to mention the existing electrical IP infrastructure. Both will have huge cost and capacity disadvantages, when compared to optically switched networks, which will make operation of those electrical networks no longer viable. Considering the amount of optical switching and routing equipment required, it is a reasonable bet that makers of that equipment will stand to benefit from the introduction of high-quality streaming media.

Consumer Electronics Industry

A multitude of appliances could effectively use streaming media. These devices are not only for entertainment in the home, but also for accessing streaming media in the car, while travelling, from your pocket, or even from your lawnmower. The potential for widespread use of wireless bandwidth to deliver streaming media to and from a wide range of consumer appliances is very large indeed. It would be incredible if the consumer electronics industry were not one of the main beneficiaries of the introduction of ubiquitous streaming media. Streaming media offers to consumer electronics vendors a virtual license to print money, since the imaginative uses of the technology have only just begun to be discussed. Each application requires streaming devices. There may be hundreds or thousands of devices for receiving streaming media in a number of scenarios in the long-term future.

Desktop Streaming Media Authoring Tool Vendors

Media authoring tool vendors are only just beginning to grapple with the requirements placed on authoring systems by streaming media. Flat-earth production suites will just not be adaptable to the requirements of full-blown MPEG-4 production and authoring. They don't do enough.

Another driver of the creation of innovative production tools is the production cost constraints that streaming media places on content authors. To address limited niche audiences, programs must be made cheaply and quickly, without sacrificing production values. These days, the most cost-effective media authoring tools are desktop tools that run on general-purpose PCs. The days of vast television production factories, equipped with multimillion-dollar production facilities are drawing to a close, as audience fragmentation makes the costs of production harder to justify. This would happen whether or not streaming media succeeded, since the number of available television channels has already fragmented audiences into smaller and smaller viewerships.

Desktop streaming media authoring tools that allow greater creative freedom in making content targeted for streaming media delivery, at lower cost, will be in demand and popular, since everyone can be a streaming media producer. Desktop streaming media authoring tool vendors should, therefore, make money from the widespread adoption of streaming media.

When Will Streaming Media Make Money?

This chapter is intended to act as a guide to business planning. While it would be wonderful to be able to give firm dates, so that businesses could plan for a cash flow until streaming media makes money, that simply is not possible. On many fronts, solutions are coming (glacially) slowly. This section focuses on the conditions which prevent streaming media from making money today. Highlighting the necessary and sufficient conditions for streaming media profitability should enable those with the power to make changes to focus their minds and activities on the things that currently hamper the industry as a whole. Whereas I would have liked to answer the questions posed in this chapter with definitive deadlines, the questions can be read as anguished, rhetorical cries for action.

Streaming media's profitability depends on bandwidth provision. So far, there is little evidence that providers have grasped the opportunity streaming media provides for their industry. In fact, they have a vested interest in not providing this bandwidth. In May 2001, *Wired* magazine published an article entitled "Telechasm," by Frank Rose, a damning indictment of wrong-headed executive (in)action, corporate complacency, and sclerosis amongst the ranks of the ILECs (Incumbent Local Exchange Carriers, also known as the local phone companies). Rose claimed that although the 1996 Telecommunications Act compelled ILECs to open up their local loops for access by competitive broadband providers, it was like asking McDonald's to open up half their freezer space, grill, and checkout counter to Burger King. The ILECs just decided not to cooperate. There were technical problems as well. The old Bell network was not designed to allow data access to outside companies and engineering retrofits were like wrestling with spaghetti. Then, provisioning requests from the CLECs (Competitive Local Exchange Carriers, also known as the broadband DSL startups) to the ILECs were not handled efficiently. DSL providers had to fax the orders in, and an employee of the ILEC would retype them, adding inevitable errors. Whereas the DSL providers wanted to process 10,000 orders a day, the ILECs could only manage 500. Some ILECs, such as Ameritech, had not even upgraded their network to the point where DSL installation was technically feasible. The entire broadband industry is at the mercy of the Bells because of the necessity for facilities upgrades. Upgrades cost money and the Bells spend their profits paying stock dividends to their shareholders. If they stopped doing so, investing their profits into network upgrades instead, their stock value would crash. Bells only increase their capital spending by 10 to 15% a year. The Bells, which control 88% of the 185 million local lines in the US, also offered broadband, but were more interested in selling T1 lines to corporate clients at $1200 a month than offering DSL to consumers at $50 a month, according to Rose. The article further suggests that the first commandment to any executive in the Bells is to avoid risk, since risk taking could lead to mistakes and mistakes could thwart a climb up the corporate ladder to "Carpetland" (the head office in the rolling hills of New Jersey's hunt country was so deeply carpeted, that it muffled any discordant sound. Rose's article said that you could take a nap in there and nobody would ever disturb you). In such a well-cushioned and protected environment, is it any wonder that the build-out of the broadband network is not proceeding with anything approaching enthusiasm or vigor? The hide-bound culture of the ILECs is the primary obstacle to the success of streaming media.

Similar horror stories can be told regarding the rollout of broadband in the UK. The incumbent network, British Telecom, now a privatized utility, was also compelled to open up its local exchanges to competitive service providers. However, so few providers have succeeded in installing broadband equipment into the decrepit old exchanges that serve the local loops that the government has been forced to take note. It has been reported anecdotally that when the network was surveyed for DSL upgrade, British Telecom rediscovered long-forgotten exchanges and found that many of these were unsuitable for DSL upgrade simply because the fabric of the exchange building was no longer sound and would require rebuilding to allow outside companies to place their equipment in those exchanges with safety. In other words, the allegation is that the cost of the building work prohibited the opening up of the local loop to competition in several areas. I live in the county of the UK that has the highest concentration of millionaires per capita anywhere in the country, yet sitting here in leafy stockbroker-belt Surrey, I am unable to get DSL connection. My local exchange, Clandon, is "not suitable for upgrade," according to British Telecom. They can offer me no date for when this might not be so. Indeed, even my existing and expensive ISDN connection is unreliable and problematic. Cable television is nonexistent, since our residential housing estate is deemed "too rural" to be viable. I am in a broadband blackout zone.

We have noted elsewhere in this book that the "whole product" of streaming media, including literally everything you need to receive streaming media, is not complete. It appeals only to the early adopters who are prepared to engage in the integration work necessary to get a media stream to play at home. It does not appeal to the mass majority of consumers. Until the consumers can connect simply, streaming media will struggle to make money. Figure 4.1 shows the adoption of types of broadband access. It is clear that while there are many ways to receive a broadband connection, adoption is at an early stage.

Figure 4.1
Adoption of types of broadband access.

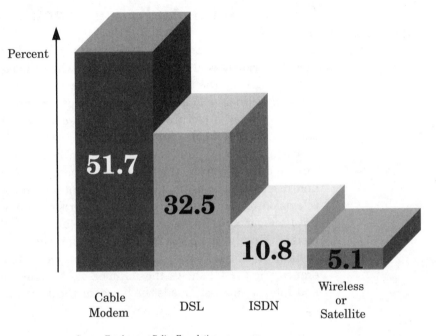

Percent

51.7

32.5

10.8

5.1

| Cable Modem | DSL | ISDN | Wireless or Satellite |

Source: Employment Policy Foundation

In business LAN/intranet environments, bandwidth provision is not as much of a factor. There is an audience for streaming media, the bandwidth is usually available, and connectivity has been provided by virtue of the fact that the IT department has connected each PC to the LAN. Legal issues surrounding the ownership of the content are not as significant. So why don't corporations routinely stream? The answers are that the addition of streaming places added demands on the IT staff, most of whom struggle to keep the e-mail system running. Also, content production is difficult and expensive. Content that is made very cheaply is not compelling, even to a captive audience of employees required to watch it. Also, the quality of streaming playback is not guaranteed, even on corporate LANs, without the addition of quality-of-service assurance techniques. In short, corporations still wind up with significant integration issues to make streaming media work within their corporate IT environments. You can't just wheel it in and turn it on. For this reason, it is difficult to make money with streaming media in corporate environments, though this is changing rapidly.

The following sections examine key obstacles that prevent streaming media from making money.

When Will the Audience Reach Critical Mass?

The short answer to this question is that it will happen when the technology works as well as television does, when the content becomes more compelling than the television experience and when the costs are comparable to those of television. To make the experience comparable to television, technically, pictures must look as good at full-screen resolution, they must play back without glitches, continuity must be preserved between program items, the players must be simple to install and maintain, and the receivers must be easy to operate, turn on and off instantly, and disguise their computer origins. Content cannot just be as good as television (or be repurposed television programming), it must be better than television. If it is not, why is there any reason to adopt streaming media? Finally, the costs cannot be out of line with consumer expectations of what it costs them to access digital media today. As a point of reference, Figure 4.2 shows the number of US households with computers and Internet access. This is the basis for the streaming media audience.

When Will Practical, Inexpensive Receivers Arrive?

The consumer electronics industry has much to learn about streaming media and about interactive design in general. Every year, it seems, home audio and video gadgets provide ever more functions that nobody can use without a manual, since all the function keys are overloaded, sometimes several times over, and displays are often limited to a few lines of glowing characters. Until grandma can use a streaming media receiver, streaming will remain an unprofitable medium appealing to gadget and fashion lovers only.

Careful receiver design can be achieved. Techniques for product design can ensure that the product is usable by the majority of consumers. Consumer electronics companies are focused on cost reduction, but they ought to shift their attention to the usability of their products. Tech-

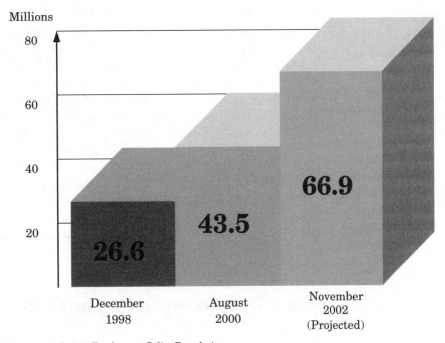

Figure 4.2
US households with
computers and
Internet access.

Source: Employment Policy Foundation

niques like quality function deployment have been successfully applied to
the design of both high-ticket consumer durable goods and other industrial products, to produce elegant products that work simply. There is an
opportunity here for a consumer electronics company to start making
media receivers that combine media types. At present, the average home
consists of a veritable forest of equipment for receiving television, satellite, cable, playing DVDs, VHS videos, CDs, and so on. I can count no less
than nine boxes in my living room, with six remote control handsets.
There must be at least 24 separate cables connecting them all.

Unfortunately, I can see no sign that the established consumer electronics industry will devise practical and inexpensive streaming media
receivers anytime soon, even though these are absolutely required to
bring streaming media into the mainstream.

When Will Bandwidth Be Cheap?

Technical progress is bringing the cost of bandwidth down at a rapid
rate. However, the price of bandwidth is not determined solely by the

characteristics of the underlying technology. Bandwidth price decreases depend on policy decisions and on the ability to deliver higher productivity from every employee of the bandwidth provision companies. Dramatic productivity improvements require capital investment. Bandwidth will only become cheap when the bandwidth providers make the investments necessary to make it so. Before they come to that point, they will need to be convinced of the riches awaiting them should they enable bandwidth-hungry applications such as streaming media. The stock market, in turn, needs to be convinced of the returns on their investments to be made.

There is hope, however. Fiber optical communications systems can require far less maintenance than copper switched-circuit technology. Once the capital equipment costs have been amortized, it should be possible to provide flat-rate, unmetered access to consumers, at prices between $25 and $40 a month.

When Will Connectivity Be Easy?

The customer requires a single phone call to summon a single person from a single organization, who will arrive on site and "get the streaming media receiver working." Anyone who has tried to connect to DSL will know that it sometimes involves several organizations and multiple home visits by technicians. Figure 4.3 shows the percentage of European households connected via broadband. The adoption rate has been slowed by the difficulties faced by consumers who want to get a connection. This is a considerable impediment to the profitability of the streaming media industry.

Wireless broadband provision has advantages over wireline in terms of simplicity of connection. However, the wireless third-generation network operators are focused on the wireless mobile applications of their broadband networks. One day, it may be as simple to connect to streaming media services as buying a home media gateway box, bringing it home, and punching in the prepaid access codes, just as one does with prepaid cellular telephones.

When Will Compelling Content Be Produced?

Fortunately for the streaming media industry, compelling content is already being produced. The popularity of Macromedia's Shockwave for-

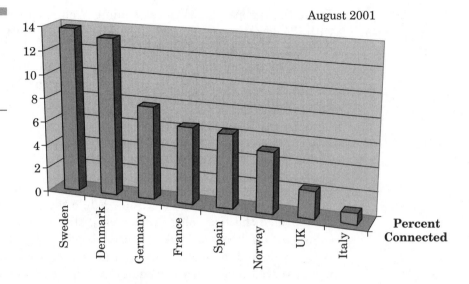

Figure 4.3
Percentage of
European
households
connecting via
broadband.

mat testifies to the fact that an industry already exists to produce attractive interactive digital content. Streaming media production may attract traditional audio and video producers, but designers from the computer games industry and interactive 3D Web designers may also cross over to streaming media content production. In addition, desktop production tools will continue to improve, as will the authors' understanding of the creative possibilities offered by the medium. The final impediment to the provision of compelling content seems to be related to rights management, where solutions are being developed rapidly.

When Will the Quality of Service Be Acceptable?

Raw bandwidth is not enough to guarantee good quality of service to the end-user. Streaming media is also sensitive to data delivery jitter, network delays and packet loss. There are a great many techniques in development to create acceptable streaming media experiences, but these have yet to be deployed widely. There is pressure for ISPs to introduce quality-of-service techniques into their IP networks. The manufacturers of streaming media players and servers have also understood and responded to this requirement.

What consumers require is a service that plays without glitches and is available on demand. Meltdowns, such as occurred during the World

Trade Center attack, where streaming Web sites were unable to cope with the demand, are unacceptable. The network must scale and remain resilient to load and fault conditions.

Once again, all the necessary techniques are known and ready for deployment. The question is whether or not the investments will be made to provide assured quality of service.

When Will Standards Prevail?

Consumers require a single standard to emerge. If one does not, the majority of consumers will elect one. While there are vigorous efforts to standardize the technology underlying streaming media, proprietary solutions largely prevail today. This is a good thing when the technology is emerging and innovation is required, but a poor thing when the technology is seeking to attract the early majority of consumers.

MPEG-4 seems to be gaining momentum as the emergent streaming media standard, but there are still issues surrounding Web transport standards. In particular, edge caching is still in flux. Thereafter, the biggest obstacles to standardization concern IP routers, whose vendors have not widely adopted either multicasting capabilities or quality-of-service protocols. Because of the sheer size of the installed base, it may take many years to phase out or upgrade legacy routing equipment.

When Will Cost-Effective Production Techniques Arrive?

Content producers are still in love with analog production methods and quality requirements. We have seen that these are not well suited to digital content production for niche audiences and are overkill for production and distribution systems that do not progressively degrade the content via generation loss and induced system noise. In other words, audiovisual production techniques are too slow, unwieldy, and needlessly expensive for streaming media applications. Streaming media audiences are much too fragmented to support expensive production methods, though the on-demand nature of streaming media means that the content can go on earning revenue long after its first release.

Production tools need to be highly automated and desktop tools are required for cost effectiveness. For production operations that require a great deal of computing power, such as video effects processing or ray tracing and rendering, it is likely to be more cost effective to rent time on a distributed application provided by a specialist application service provider and running on remote server farms, rather than for each production house to buy the necessary equipment and have it sit idle for much of the time.

There has been too much investment in industrial-age television production facilities to make the switch to low-cost, high-value streaming media production overnight. The change in production techniques will, more than likely, be led by guerrilla outfits: small production companies willing to experiment with new technologies to produce their product. There are already entertainment news crews making streaming media reports with nothing more than DV cameras and laptop computers. With their particular production method, they can cut the story in the back of the Jeep on the way back to the office. This not only gives them cost advantages, but their reports are also on air (i.e., on the Web) faster than those of many other electronic news gathering operations. For streaming media to make money, production techniques like these will be required.

When Will The Legal Issues Be Solved?

Laws, especially those surrounding digital copyrights, change slowly and global harmonization takes even longer. Unfortunately, the lack of globally enforceable digital rights protection threatens the viability of streaming media as a profitable industry. It is not clear how these issues will be solved, since some nation states do not even respect patents, let alone digital copyrights. The solution would require intellectual property treaties to be signed, but there is little evidence that this will happen soon. Digital copyrights could be the issue that prevents streaming media from ever achieving its commercial potential. The only reason to suppose that the copyrights issue might not be a severe threat is the experience software companies have had with digital piracy, the experience of the music industry with illegally produced CDs, and Hollywood's losses due to DVD and video counterfeiting. All these industries currently deal with illegal copying as a cost of doing business. In fact, honest media consumers effectively subsidize the dishonest. Whether the presence of streaming media technology will worsen this situation remains to be seen. However, there is a perception that streaming media will encourage disastrous levels of illegal digital copying.

Content owners are working with legislators and law enforcement agencies to clamp down on the pirates. Indeed the US government recently announced that the lack of effective digital rights management solutions is delaying the uptake of broadband Internet connections, and has threatened to impose a solution on the industry, if the industry does not work to agree to one. This is an indication of how seriously the government of the US, for example, takes the issue of digital copyrights and how strategically vital it deems broadband connectivity.

Why Will Streaming Media Make Money?

For any system to be a compelling consumer proposition, it needs to have unique selling points. This section examines streaming media from the point of view of what characteristics of the medium make it so compelling. In demonstrating the medium's irresistibility, we thus demonstrate why it will inevitably make money (provided that all the other conditions for its success, discussed elsewhere in this book, are met). Many of the characteristics cited here will remind us of our discussions of why audiences will watch and why streaming media is better than television. They are aggregated here as a ready guide for business planning.

What You Want, When You Want It, Wherever You Are

This is the characteristic of streaming media that makes it a killer application. It provides consumers with access to whatever they want, whenever they want it, wherever they happen to be. Provided that receivers are made convenient and unobtrusive and that some kind of intelligent roaming and synchronization software is developed, programming favorites can literally follow the view from device to device, via wireless mobile networks.

There are some dangers, however. We are already painfully aware that mobile phones are a source of great annoyance, if used in inappropriate locations or times. The same will apply to streaming media receivers. There will be people who seek to ban their use in particular locations and social situations. However, streaming media offers consumers freedom from the whims of television program schedulers. For the first time, a highly portable, interactive, live video medium, whose content can be up to the minute, with an unprecedented range of instantly accessible choice, is available. VHS, DVD, cable TV, satellite TV, and terrestrial TV cannot compete on the same terms.

Your Personal Data Shadow

Personal data shadows do not exist commercially yet. However, we already have the underlying technologies necessary to realize them. A data shadow, as well as referring to data kept about you on various computers as you engage in transactions with companies or the government, also refers to the ability to access any of the digital data you deem important, from wherever you are, on whatever machine you happen to be using, regardless of which machine the original is on. MPEG-4 technology allows receivers to filter the data from the source according to the display and interaction facilities offered by the receiver, rendering only those data streams that it can. Thus, if my Word document is on my desktop PC and I am using a mobile PDA, I should be able to stream a view of the document onto my PDA directly from my desktop PC.

This is an oblique application of streaming media technology, to be sure, but in the future, when Web sites and "documents" actually become rich multimedia pieces, it will represent a much-needed application. If I am showing a client a PowerPoint slide show, I don't need to have the file with me, I just need to direct the viewer to a stream of the slide show I produced. The key to enabling this application is keeping some sort of online database of my assets, somewhere in my distributed multimedia world, so that I can always locate the original of the asset and either stream or copy it to my current receiver. In this way, every streaming media device I own becomes a piece of my own personal, miniature edge-caching network. Another way of looking at this is that the edge of the public, streaming-media Internet is in my hand or on my desk.

Feel the Quality

As compression technology continues to improve, you get better quality pictures and sound for the same bandwidth (or the same quality, for less bandwidth). Bandwidth becomes more abundant and cheaper over time as well. With improvements in both compression and bandwidth technology, you eventually get better pictures and sounds for less cost than any other digital communications system can offer. Evidence of this is that there are already moves to incorporate MPEG-4 decompression in DVD players. Audio MP3 players are also widely available today. This means that systems using older compression (or no compression) techniques will, over time, become increasingly uncompetitive, since they will offer poorer end-user experiences at a much higher cost than could be achieved with streaming media technology.

It is possible, using today's streaming media technology, to encode HDTV-like video at a bandwidth of 3 megabits per second. Compare this with the raw 270 megabits per second for native HDTV pictures and the savings begin to look significant. Indeed, if you are able to allocate perhaps 30 or even 300 megabits per second to a stream, you could theoretically get super high-resolution video at very high frame rates. It may even be possible to transmit 360-degree panoramas in resolution so high that the user feels he or she is actually standing in the space. Telepresence is made possible by streaming media technology.

The other aspect of streaming media technology that is peerless is the flexible way bandwidth can be used and allocated. Low-cost streams, for niche audiences, can be offered using low bandwidths, whereas premium content, commanding a higher price, can use more bandwidth to create an ultimately more compelling end-user experience. Television, and especially digital television, are not as flexible, since their airwave bandwidth allocations saddle them with fixed bandwidth limits. The receivers and the entire television transmission infrastructure have also more or less frozen MPEG-2 compression into the stone of the system. Streaming media technology allows for compression upgrades. The typical digital television system does not.

Honest, Doc

The highly interactive and participative nature of streaming media makes it possible for users to check sources online. You can verify the truth of a story with a quick search and some light research. You can also know the identity of storytellers or reporters, checking interests and affiliations, which may taint their viewpoint or lead them to tell their stories with a particular distortion or slant. Stories can also have peer reviews appended to them (though these are often misused by unscrupulous marketers to provide a faux-objective testimonial to the quality of a work, when in fact it is almost a paid advertisement).

The other aspect of streaming media technology that tends to keep people honest is that it lends itself to low-cost, direct surveillance applications. Also, the sheer amount of editorial choice available with streaming media (the million-news channel universe?) makes it harder for anyone to slant and control the media's overall agenda. Diversity protects dissent.

Trust Me, I'm Streaming Media

As with the vanilla World Wide Web, trust networks of people who share similar views and values can form almost spontaneously, using streaming media video conferencing technology. However, unlike the Web, with streaming media you can directly see and hear other members of your trust network. It is much harder to masquerade as something you are not, when the camera and microphone are on you.

Today, powerful interests and political despots can influence the entire television output of their spheres of influence, hiding inconvenient stories and making it impossible for dissenters to find a voice. Streaming media content is produced by a wide range of program makers, so the likelihood of the medium's falling under the control of special interests, whether corporations or governments, is reduced. Propaganda, spin doctoring, and other officially sanctioned fictions are harder to maintain when digital media spans borders.

Television brought the benefit of being able to see people, and judge their characters by their body language and the movement of their eyes. Politicians once able to hide behind the veil of radio were suddenly revealed in full-motion pictures. Streaming media adds the ability to freeze or slow down the action, review the video, and even to examine images in detail, zooming in on things of interest. This added level of control and the ability to examine video in fine detail helps viewers either establish trust in those they are watching, or gives them additional tools with which to scrutinize those they do not trust. The hand may be quicker than the eye, but streaming media allows viewers to slow down the hand.

Everyone's a Media Mogul

When television was the only available audiovisual mass medium, if you had a program you wanted to make, it was difficult to get it made and aired. The cost of production meant you either had to have deep pockets, or a generous benefactor, or else you made the program as a commissioned piece for one of the major broadcast networks. Unless the program was self-funded, the injection of outside finance necessarily circumscribed limits on the range of editorial opinion that could be expressed. Some points of view were inexpressible. A program maker could go only so far, but no further. Thereafter, to get the program scheduled for airing meant another round of negotiations and crossed fingers. Unpopular points of view or taboo subject matter meant that the program aired only in the graveyard hours if at all. I recall once seeing a wonderful exposé of the activities of certain multinational companies that, while compelling and authoritative, was relegated to debut in a 3AM time slot. Hence, while freedom of speech is enshrined in the US constitution, the commercial nature of the television medium imposes practical limits to the range of topics that can be aired. Freedom of speech is not protected when such filters and censorship, however benevolently applied, exist institutionally. The insidious aspect of television is that it leads viewers to believe that allowable dissent (those viewpoints considered radical, yet still acceptable for broadcast) represents the real limits of the range of viewpoints possible on a given subject. In reality, the range is often greater, but some viewpoints are considered too extreme or unpopular to air, so they never receive the oxygen of publicity. It's as if those opinions don't exist at all, because they cannot be rendered to a television audience. This is a distortion of the public debate and an erosion of democratic freedoms, whether or not people like to acknowledge the fact.

With streaming media technology, the scarcity of airtime is a thing of the past. Everyone who has enough money for the equipment and bandwidth can have a channel (this is still a democracy-distorting limit, but not as severe as the limits imposed by television). Making the program is also much cheaper and within an individual's means, since desktop editing tools and high-quality consumer-grade digital video cameras are now available. Just as desktop publishing expanded the range of publications, so desktop video production will expand the range of programs.

Although large media conglomerates can still exercise considerable influence and control over streaming media, because they have the financial resources to control access to the major content-delivery networks, no media mogul can own all the content ever created, nor can that mogul create an impervious gateway barring access to the streaming media network. The edge of the network is actually very porous. It was designed to be so, in the event of nuclear war, back when the Internet was first posited by DARPA. Hence, while the incumbents in the media industry can seek to control access to the network, as they have sought to control magazine distribution, for example, they can never achieve it in practice, or even approach the level of control they still have over access to the broadcast television networks. From the point of view of democracy and freedom of speech, streaming media is less bad than television.

Involving, Immersive, and Interactive

With streaming media, you don't just sit and watch, taking dictation, as it were. Instead, there are opportunities to get involved, immersed, and emotionally engaged. Streaming media technology also makes it possible to interact with other people in a number of ways, as you navigate through the streaming media program. Of course, streaming media can be watched in a passive way, just like television, if that is the viewer's preference. The point is that the medium does not limit viewers to one option.

Overturning the Old Order

Streaming media has some revolutionary appeal. There is a sense in which people come to resent those who exercise excessive power or whose wealth accumulation exceeds what any normal human being can usefully spend in a lifetime. The power of Hollywood over the range of movies that get made is not universally popular. Many do not like extremely wealthy content owners' using their influence to pay so little tax. People do not like to be at the apparent mercy of such modern barons.

Egalitarianism is something many people aspire to and consider just. It is for this reason that Bill Gates and Rupert Murdoch must spend so much time and effort on public relations. It also partially explains the popularity of Linux and Napster. These technologies are, to many users, subversive; using them an act of brave, righteous, and noble rebellion. It explains, in part, why users of these technologies are so zealous and fervent about them. Streaming media's ability to overturn the established order within the media industry has the potential to exercise a similar appeal.

Rights Guarded

With appropriate rights-management technologies, the average consumer and content owner alike can achieve better protection over rights, copyrights, sensitive personal information, and privacy than can be achieved with other popular, but totally unprotected, digital data-delivery systems, such as the telephone system or the raw public Internet. Using phones and faxes (digital audio and text) is a very insecure way to deliver sensitive data. There are no rights-management facilities to speak of.

Streaming media, on the other hand, will soon have very robust rights-management technologies and laws, which ought to be available to everyone. Thus, the streaming media infrastructure will provide a secure communications facility to every user. Content owners can deliver content to consumers in safety, with very little risk of piracy or violation of the consumer's privacy. Compared to the piracy that occurs with videos and CDs, streaming media will be more difficult and expensive to compromise. Even home taping will require specific permissions from the content owners. It might not be in the content owners' best interests to deny all home copying, but at least they will be able to charge for it appropriately or else limit its scope. This is not possible with other copy protection schemes, which deny all copying. With streaming media, copying can be allowed, but only on the rights owners' terms.

Free Samples

Expanding on the possible rights-management scenarios that streaming media technology makes possible, content owners can use "heroin economics," where samples of some digital media asset are given away to build an audience for other similar streaming media assets. Because digital rights management allows content owners to limit the scope of the copying of the free sample, it is possible both to build an audience and to protect the value of the asset. Beyond a certain point of wealth accumulation, or some time after initial release, content owners can even opt to donate the proceeds of their properties to good causes, or else allow the media to be effectively free (or very much cheaper to access). Rights management supports many different business models, some of which are entirely charitable.

Uncensored

The lack of any censorship in streaming media is a double-edged sword. On the one hand, there is access to literally anything, which can be a very liberating force, but on the other hand some material that most people would regard as unsavory would also be available. A lack of censorship is, nevertheless, a compelling driver for uptake of the medium by consumers. The allure of forbidden fruit is a strong one. Streaming media almost uniquely allows the global movement of information, without censorship.

Technical solutions to filter out potentially unsavory material exist, but are unreliable, filtering out innocent material inadvertently, and they are ultimately easily bypassed. Privacy preferences can be set with modern browsers, placing the onus on the individual consumer to reveal tolerance to adult-only material. There is a growing realization that censorship is only part of the solution to maintaining moral standards in society. Education and societal attitudes have a role to play as well. Reliance on technical solutions to safeguard the moral fiber of society is not only naïve but also somewhat dangerous. The relevance to streaming media is that censorship, or the lack of it, will tend to make streaming media profitable. Evidence for this is the current earnings of the online adult entertainment industry, some of the earliest adopters of streaming media technology. Online adult entertainment is unarguably successful, in pure financial terms.

An Enriched World Wide Web

Another way that streaming media can be regarded (and marketed) is as an enhanced World Wide Web, but more fun and with better and more interesting content. Streaming media has the potential to be more entertaining and informative than all the content already on the Web. Besides allowing the creation of entirely new content, streaming media also unlocks a vast archive of audiovisual material currently only rarely accessed. Old film stock, once seen as expensive to store, now becomes an opportunity for revenue generation. The streaming media experience can be thought of as the next generation of the World Wide Web.

Why Digital Television Can't Compete

Digital television, as we know it, is expensive. Multimillion-dollar production facilities are the norm. Quality standards in production are far beyond what the consumer actually receives or requires. More importantly, digital television is extremely wasteful of scarce radio frequency resources. When it is possible to deliver high-definition streaming media pictures over fiber, without impacting any other communications service, how can the use of hundreds of megahertz of the airwaves per channel be justified? Surely there are other more worthy uses of that spectrum. Terrestrial transmission of uncompressed high definition video is no longer a technical necessity. There are other ways. The real question is how the broadcast industry can get away with such waste and profligacy.

Digital television is largely didactic. You watch and listen while "they" tell you what they want to tell you. There are few ways to shape the information delivered, to interact with the program makers as the program comes together, or to delve into the inside and background stories surrounding any given television report. It echoes the attitudes of a long-gone age, where deference to authority was the norm and the common man did not question what he was told. Modern media consumers are not as accepting. Digital television struggles to adjust to the new expectations of a questioning, inquiring, sometimes cynical audience. Its technical features make this adjustment difficult.

Digital television is undoubtedly under the control of relatively few organizations. This limited diversity poses several potential problems, some of which society is not even aware of. There are well-documented instances of media owners' using their media properties to influence the masses, so as to pervert the course of ordinary political decisions. Media owners desirous of particular policy decisions have coerced politicians into making laws more favorable to themselves, through the simple expedient of threatening public exposure of some less-than-honest politicians. Some call this kind of activity "blackmail." These same media owners own and control significant digital television properties globally. For this reason and because viewers can never know or find out why the television media holds and broadcasts certain opinions, television is no longer a trustworthy source of information.

Television shows are not the "events" they once were. To many television viewers today, the medium has become boring, primarily because of its passive nature. The term "couch potato" sums up the backlash.

Digital television, though a relatively recent technical system, is actually based on a much older technology. Interlaced pictures are a workaround for the problems with modulating television pictures on analog transmission systems, dating back to the 1930s. Interlacing, which has some severe visual artifacts associated with it, was meant to mask the effect of limitations of the analog transmission system. Yet, today's digital television systems still slavishly broadcast interlaced content, even though there is no longer any technical reason to do so. This is because of the desire to maintain backward compatibility with older television receivers. The American NTSC television broadcast system was deployed about 1954 and today's television sets remain compatible with their ancestors of that era. Picture quality of VHS recorders is hardly different from that in the first VHS machines of 1981. The image quality that consumers receive is hardly better than that of yesteryear, because of the need to conform to the now elderly standard. Streaming media, on the other hand, can redefine quality and cost tradeoffs, allowing the consumer and content provider to make different compromises.

Digital television may respond to the emergence of streaming media technology by grafting some of the more exciting capabilities onto the existing infrastructure, but such retrofits are rarely elegant or successful. Anyone who has experienced digital interactive television or the text-based services delivered with the video signal (such as Teletext) cannot help but be aware of the technical limitations of adding data transmissions to what is a purpose-designed video delivery system. The digital television delivery networks cannot be used as flexibly as the IP delivery networks employed with streaming media.

Indeed, as MPEG-4 begins to gain popularity, receivers will hide the individual delivery systems. MPEG-4 allows digital video to be delivered using the traditional digital television delivery system, while the interactive graphical and text elements are delivered over an IP network. It is entirely possible that tomorrow's digital television sets will be MPEG-4 media players under the hood. Once this happens, where does digital television end and streaming media begin? The distinction becomes blurred. MPEG-4 allows streaming techniques to subsume television delivery methods. Ultimately, streaming media will eat television.

Upsides—
Downsides

How Significant Is Streaming Media?

It is always a good idea to place any technology or system into some sort of context, so that industry practitioners can keep one eye on the significance of what they are doing. Such perspective informs good judgment. At the time of writing, streaming media is not all that significant, in the grand scheme of things. There are many other industries and existing media distribution systems that command greater viewership and a larger share of global gross domestic product. However, as we have seen throughout this book, we have so far only experienced the tip of the iceberg. Streaming media has the potential to be a significant social force and a large industry, if not a huge one. It could be the medium of choice for the new century. If the technologies required are developed, as it seems they will be, and if the commercial and other problems are solved, as might happen, then the medium could, indeed, replace television, over time. Streaming media could also be the driver of consumer demand for third-generation wireless mobile services and for ever-faster optical networks. Streaming media technologies could underpin many of the next waves of "killer applications."

While streaming media may not directly provide a cure for cancer or solve world hunger, it will perhaps make significant contributions to progress in solving these problems. Its societal impact, like that of television and film, is likely to far exceed the monetary value of the streaming media industry. This section briefly examines the ways in which the media will be significant.

Sizing the Potential Market

This section contains some extrapolations of samples of current market size, to guesstimate future market size. The data are, therefore, subject to all the usual limitations of crystal balls. However, data like these are often good for performing "back of the envelope" calculations, in order to gain an order-of-magnitude appreciation of the scale of the industry. Make of these data points what you will.

In November 2001, the number of broadband users in the US reached 21 million, representing 20% of the 106 million Internet users in the US. It is estimated that the number of World Wide Web users will reach 320 million in 2002, accessing the Internet using some 515 million devices.

By 2004, there are projected to be 90 million broadband homes globally, rising to perhaps 200 million by 2006. Figure 5.1 illustrates the number of household Internet connections, worldwide.

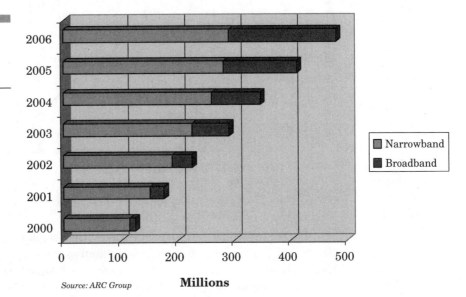

Figure 5.1
Worldwide household Internet connections.

Source: ARC Group   **Millions**

According to some surveys, consumers will spend an average of $40 a month each on broadband access and content, though AOL projects $159 per user per month in the future, including $24 for online services, $20 for music, $20 for Web access via mobile phones and other devices, and $15 for games and entertainment.

Home media receivers and gateways may range in price from $100 to perhaps $400, and there are likely to be at least two or three devices per broadband household. Content producers will be spending on the order of $50 per hour for production facilities, with each finished minute of content requiring an hour of production time. Advertising on the net was worth $4.6 million in 2001, equal to 2.5% of advertising expenditure in the US.

The 25- to 34-year-old age group tends to embrace technology fastest. They are also the most significant demographic group, because they command high disposable incomes, have fewer financial responsibilities than older age groups, and are more numerous than other age groups. Figure 5.2 shows the percentage of European mobile phone owners, by age group. Streaming media adoption is likely to mimic this pattern.

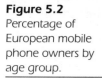

Figure 5.2
Percentage of
European mobile
phone owners by
age group.

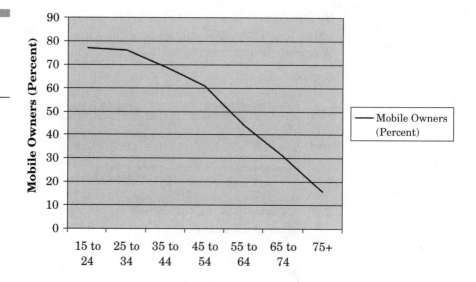

Fifty percent of Internet users have tried streaming media. In fact, the size of the monthly Internet radio audience nearly equals the terrestrial radio audience. Thirty-four percent of all Americans over the age of 12 have watched or listened to streaming media (78 million people). Internet users are projected to spend 192 hours online annually by 2003.

It is useful to examine the data for video consumption using television technology, since it can provide upper bounds for streaming media consumption. Americans rent 6 million videos a day. One-year-old children watch 50% more television than the national average, at 6 hours per day. The 2- to 17-year-old age group watches a staggering 19 hours 40 minutes a week, or 1023 hours a year. This compares with the 900 hours a year they spend in school.

Spending on online digital music is $1 billion today and is projected to be $6.2 billion by 2006. Cahners In-Stat predicts that 2004 annual revenues for consumer-oriented multimedia broadband services will be about $15.7 billion, with the market for streaming media content growing to $2.5 billion by the same time.

We know from the analysis of streaming media technology in this book that streaming media will only just begin to attract a mainstream audience over the next two to five years. The real gains will come over a ten-year time period. Unfortunately, few studies are willing to predict this far into the future. The only model that can inform the potential size of the market is the S-curve analysis typically applied to the life cycle of innovations. Over the next two to five years, streaming media will only begin moving up its S-curve. Significant market size is not likely to be realized until the technology reaches 10 to 15 years in maturity.

Sizing the Potential Audience

Approximately 12.7 million people consumed streaming media in 2001. According to Neilsen/NetRatings, the streaming population is 40.7 million, growing at 18% per year. Compare this with the 750 million people worldwide who watched John F. Kennedy's funeral some 10 years after the NTSC television system was first deployed, or the 600 million that watched man set foot on the moon. Today there is one television set produced for every child born. A good estimate of the size of the audience that might one day be reached with streaming media is over a billion souls. Television ownership represents over 95% of households in the US; it will be quite some time before streaming media penetration is so comprehensive.

During the attack and its aftermath at New York's World Trade Center on September 11, 2001, CNN's streaming media site registered 9 million page views per hour compared the usual 11 million they get on a normal day. By some estimates, 200 million stream views were generated across all the streaming news sites available. This did not fulfill demand.

Hence, the potential audience that might be reached by streaming media is in the hundreds of millions. This will be limited mostly by the penetration of streaming media receiver technology.

Streaming Media and Democracy

Streaming media, as a communications technology, is potentially very democratizing, because it allows greater access and participation than existing global mass media does. This will tend to allow orthodoxies to be challenged. However, the cost of streaming media access technology, for both consumers and content producers, is still far beyond the means of most people in the Third World, the majority of whom, as I am fond of reminding the reader, have never even made a phone call. This cost will still act as an almost impenetrably high barrier to the majority of the earth's citizens. On the other hand, in regions of high illiteracy, audiovisual information is perhaps the most effective means of education and this is where the cost saving associated with compressed streaming media might help spread democratic thought, rather than stifle it.

Streaming Media and Ignorance

One of streaming media's strengths is in distance-learning applications. There is some hope that with mobile and satellite Internet coverage, streaming distance learning applications could deliver enlightenment to even the remotest reaches of the planet, cross-fertilizing ideas and cultural influences internationally. Education is one of the most effective tools in eradicating ignorance, prejudice, bigotry, and parochialism. Streaming media provides an effective tool for education.

Richard Stiennon, research director for Internet security with Gartner, cites lack of education and wholesale illiteracy as reasons why terrorist groups are readily able to find legions of disaffected people willing to sacrifice themselves in terrorist actions. He also says better education and higher literacy have the potential to lead to a more peaceful age and that easy access to information has already proved to be a liberating influence. To this I would add that the ability of streaming media to provide telepresence experiences helps to dispel notions of how different from one's own the lives of one's perceived enemies are. When the Berlin Wall came down, people on each side were astonished to learn that those on the other side were more similar to themselves than they had been brainwashed to believe. With understanding comes empathy. With empathy comes compassion. With compassion comes the cessation of inhuman acts toward other human beings.

Streaming Media and Knowledge Capital

For centuries we have stored our collective learning and knowledge in books. With a book, an author can speak to a reader, mind to mind, across the centuries. Streaming media technology affords us an improvement on this knowledge recording technique. What better way to document the intellectual artifacts of our civilization than using streaming media? It is a digital representation, so long-lived methods of storage, relatively immune to degradation, can be sought. Streaming media will record not only the words of an author, but also the modulations of his or her voice and even the nuances of body language. Not only is streaming media a potentially robust and rich way to store and archive our knowledge capital for posterity, it is also a very good way to disseminate knowledge today. Companies often "wish they knew what they already know," meaning that it is difficult to teach a company's collective experience to new employees. Streaming media offers a way.

Streaming Media and the Speed of Business

Streaming media undoubtedly has the potential to speed up the pace of business. In the first place, audiovisual material often delivers messages more succinctly and with greater clarity than the written word does. Secondly, avoiding travel and using the instant global reach and high richness of streaming media, cross-site decisions can be made in minutes rather than days. The ability to collaborate globally means progress comes more rapidly. Streaming media, as a business tool, acts as an accelerant.

Streaming Media and Privacy

Streaming media allows two-way surveillance, but sometimes one of the parties doesn't make the other party aware he or she is being watched. In fact, a third party can also tune in without either party's consent, unless digital rights management and encryption are used. Therefore, streaming media has the potential to seriously degrade individual privacy, especially because streaming media technology also lowers the cost of surveillance. On the other hand, it is possible to trade personal data for media access or other goods and use cloaking proxies like www.anonymiser.com, which allows streaming media consumers to hide their identities.

With all transactions that record data about consumers in databases, there is the risk of stealing people's digital identities, data corruption and inaccuracies, unauthorized data access, and malicious use of data. This issue is not peculiar to streaming media. Data abuse and the overly simplistic interpretation of data about people is something that the Internet, in general, will have to solve, as will the offline users of such data in everyday life. Where streaming media technology could help is in providing rights protection for individual consumers of their private data. Symmetric digital rights management is considered important by both the MPEG-21 standards body and the IETF. This is a healthy sign, though only time will tell if rights management becomes truly symmetric, or if the powerful parties prevail at the expense of individual consumers.

Streaming Media and Community

Because streaming media can cater to special interest groups that span the globe, the potential for the creation of strong virtual communities is high. Streaming media's interactivity features aid in interactions between real people, so communities of purely online acquaintances can form quickly. The return to notions of community is healthy, in times where physical communities are made impossible due to commuting and the pressure of modern lifestyles.

People already turn to cell phones, e-mail, instant messaging, and logging on to news Web sites (in addition to watching broadcast television reports) in times of crisis. Streaming media will provide yet another way for people in crisis to find missing persons, organize relief efforts, coordinate resources, check on relatives and friends, and give comfort and support to one another.

Streaming Media and Advertising

Because streaming media makes it easy to avoid ads designed to steal your attention in some way, the only advertising that has a hope of working, in the long term, is that which seeks a consumer's permission to be shown. Opt-in advertising actually benefits both consumers and advertisers. Consumers suffer from less information overload, can exercise choice over how their attention is spent, and never need feel that they are being imposed upon, since they are only shown advertisements of genuine interest to them. The advertiser wastes less money pitching to disinterested consumers, can individualize the messages, can directly address smaller demographic groups, can measure instantaneously the effect of advertising expenditure on sales, and avoids annoying and alienating potential customers. Streaming media technology has the potential to revolutionize advertising, changing it into something unrecognizable to the advertising and marketing executives of the television era.

What Could Go Wrong?

When a technology is poised on the edge of becoming a mainstream product, which appeals to the majority of consumers, as streaming media currently is, there are no guarantees that the offering will make it. Many products and technologies have failed to make the leap into widespread acceptability. Recalling CB radio, the Sega Dreamcast games console, and the Betamax video format serves to illustrate that point. These technologies all worked well enough, but other factors relegated them to history.

This section will discuss some of the factors that could stall the uptake of streaming media technology. This list could never be considered complete, but it represents some of the higher risk factors for the evolution of streaming media technology into a widely used communications medium.

Streaming media's success depends almost entirely on the success of broadband. However, broadband is in trouble at present. Excite@Home, one of the US's more high-profile broadband providers went broke in spectacular fashion. Other broadband providers in the US such as Covad, Northpoint, and Rhythms are all dead or dying, according to pundits like Robert X. Cringely,* killed by phone companies with no interest in helping to create broadband competition. The incumbent local phone companies aren't doing well with their DSL efforts either. Their motivation for getting into broadband in the first place was to comply with the 1996 Telecommunications Act and to keep competitive local exchange carriers from stealing their business. With competition severely weakened and the 1996 law merely permission for them to spend lots of money getting into the long distance business, where they make no money at all, they have little incentive to put more money into broadband infrastructure. DSL equipment makers are seeing flat sales and declining profits. Venture capital has cut back investment in companies that might have entered the market to take over from the ones that are dead or dying. There also isn't a single company providing high bandwidth content to mass consumers that is making any money on it. Third-generation mobile systems are being scaled back and canceled the world over. The simple fact seems to be that we really can't afford all that bandwidth at the price it is today. The effect of broadband's slow takeup was felt by one of streaming media's leading companies, Real-Networks, which cut 15% of its workforce in July 2001.

What is really going on with streaming media is a lack of industry coordination. Broadband is useless without applications. Applications require content. Content producers require an audience. Audiences need reasons to buy into broadband. Thus the stalemate perpetuates. While the streaming media industry cycles from a situation where one sector is forging ahead while the other supporting sectors are "waiting and seeing," to the opposite extreme of doom, gloom, and severe cutbacks, the pioneers will continue to take arrows in their backs.

---

*Robert X. Cringely, the author of the best-selling exposé of silicon valley's conspicuous success stories, *Accidental Empires*, has been a long-time commentator on high technology, presenting his sometimes irreverent and contrary views on PBS television and his website "I, Cringely," His views are reportedly based on information he gleans from unnamed industry insiders.

If the entire streaming media system, including delivery technology, player technology, and content were introduced in a coordinated manner, with growth managed so that nobody's cash flow went too far off scale, the streaming media communications system would not only grow, it would absolutely flourish. But what are the chances of such coordination?

At the time of writing, the biggest overall threat to the success of streaming media is the state of the worldwide economy and of the technology and entertainment sectors in particular. The risk is that the companies creating this exciting new medium will falter and fail. Thereafter, nobody will ever go back to the technology, once its credibility has been tainted, because the "tried it and it didn't work" syndrome will prevail.

It is also possible that something better than streaming media will be invented before the technology takes root. It may be that digital television technology can be adapted to create the sort of streaming media experience envisaged in this book. It is very hard to imagine what could replace streaming media technology, but then that's what the promoters of Betamax and CB radio said.

Quality Never Improves

Let's face it. Most of the quality-of-service issues with the broadband Internet come down to the ancient copper wiring used in the local loop. The quality of the streaming media experience is most severely downgraded by the copper infrastructure connecting us to the bandwidth pool that is the Internet.

People who have adopted DSL technology have experienced more than their fair share of teething problems. Service providers have not given a bandwidth guarantee to their consumers, so it is not uncommon for a DSL link's speed to drop down to an agonizing slowness, comparable to dial-up, as more and more consumers connect to the local loop. Many service providers also fail to keep their e-mail and HTTP servers running with anything like acceptable availability. Services simply shut down for days at a time. Cable modems from cable providers have fared a little better, but cable availability is not as widespread as the telephone infrastructure. Dealing with the cable company for technical support is often not much more fun than dealing with the multitude of companies involved in establishing DSL links. Whether or not the broadband connection is cable or DSL, consumers have found themselves in the sticky position of suddenly being forced to become system administrators for their homes. The level of knowledge and expertise needed to effectively administer a network of present-day computers is higher than can be reasonably expected of individual householders. Yet, the computer industry and the telecommunications industry have done precious little to make the adoption and administration of broadband connections simple, foolproof, and painless.

Computers and network routers, essential elements in a broadband connection today, are designed with the assumption that there will be a highly technology-literate operator available on site. This might be so in the business world (though it often isn't), but it won't wash with the general public. Yet, designing systems to be administered by grandma is barely on the radar of most computer software vendors. In November 2001, Microsoft announced the creation of a new business unit called "eHome," charged with developing software and working with partners on Web-enabled products aimed at consumers living in an Internet-connected home. Yet, at the heart of the offering will be the PC with the Windows operating system running on it. Administration of the forthcoming .NET or XP server (code-named Whistler) is not a simple matter. If this is the platform upon which the eHome vision is built, grandma will be excluded from receiving her information, news, and entertainment via streaming media.

What the streaming media industry needs, in the long run, is widespread FTTH. However, either through lack of investment, lack of vision, or lack of policy to make it happen, the local loops may stay with antique copper connections, making it impossible for consumers to go cheaply enough to higher and higher bandwidths.

If the quality of bandwidth provision to the home doesn't improve radically, then the quality of streaming media services will not be acceptable to most consumers, who are already familiar with the quality offered by digital television services. If the quality of the overall streaming media service, including bandwidth provision and the quality of the streaming media experience, remains as low as it currently is, nobody except curious technophiles will watch. It's as simple as that.

Abuse of Privacy

For streaming media to succeed, people will need to trust the system to guard their privacy. In every transaction involving streaming media, there will be opportunities for private and personal information to be recorded and accessed. Consumers need to feel confident that unauthorized access to that information will never take place.

They must also be confident that simple-minded interpretation of the data gathered will not lead to nightmare consequences. Already today there are people whose lives have been ruined by "identity theft," in which a person gains access to another person's "data shadow," simply by knowing his or her social security number, address, and mother's maiden name. Thereafter, the criminal passes himself off as that person, running up debts and never paying them. If the data held on file for the person whose identity had been stolen were interpreted in a simple-minded way, as it often is today, the victim could spend the rest of his or her life trying to prove to people that they are of good character and credit rating and that the violations occurred because of identity theft. There is no easy way to erase the misdeeds of felons done in your name (and with your digital identity).

Another example of the abuse of privacy because of simplistic interpretation of data gathered about a person involves credit references. In the UK today, consumers can be refused credit, even though they have the ability to pay and an immaculate credit history, simply because more than the "right" number of credit references has been recently requested. This is a good protection against consumers or fraudsters applying for excessive credit, but it is common for people buying a house to run many credit checks while shopping around for the best mortgage. Yet, by overly simplistic interpretation of the number of credit reference requests for an individual consumer, they can find themselves in a store, in public, being refused credit as if they were a criminal or unable to pay. This is an abuse of privacy that cannot be condoned, yet is the prevailing situation today in the UK.

What most fail to realize, when they interpret data about a person and infer some form of suspicion or guilt, is that people are innocent until proved guilty. Data in a database provide no such proof of guilt, unless an unbroken and untainted chain of evidence, sufficient to stand trial in court, can be established, linking the data held to the person's actual activities. Otherwise, the data merely serve to form a not necessarily well-founded allegation. The number of credit checks registered on your file says almost nothing about you, if they can equally accumulate through a legitimate search for the best mortgage as well as through the activities of a fraudster, for example. Companies or agencies that treat individuals with prejudice, on the basis of assumptions made according to data held about them, or those that deliberately contaminate data held about a person, ought to be prosecuted with the full force of the law. However, I know of no such case ever being successfully prosecuted, nor if there are even sufficient legal powers in existence to prevent this kind of abuse. In any case, proof of simplistic data interpretation leading to prejudice or of deliberate data contamination is exceedingly difficult for the individual to obtain. How can you know you are on a secret black list if you can't see the list? How can you discover who put you on the list and their reasons for doing in the first place? How can you challenge and correct an unfair listing, or limit the extent to which the faulty list is propagated and perpetuated? The answer is that data about an individual is content and should be subject to digital rights management, with every access authorized and granted according to each individual's terms of use of that data. Why should media producers get all the protection?

Imagine if the character of a content producer were judged according to the storylines of fictional productions. Who would ever invite Stephen King to be a houseguest, if that situation prevailed? Yet, individual consumers have their character assessed according to digital artifacts about their lives stored in vast, anonymous databases, every day.

The Kafkaesque burden of guarding privacy is too onerous to fall on the shoulders of the individual. Rather, the state must provide protections, so that companies, individuals, or government agencies that abuse an individual's privacy are subject to prosecution under criminal law, with the guiding minds individually accountable. Digital rights management is not only about protecting commercial interests from theft of their content, it is also about managing the individual's rights to private data, which is just another form of digital content.

Streaming media has the potential to provide much richer data about a person, which can be gathered by companies, individuals, or other agencies. With the richness comes some protection against simplistic interpretation, but also other opportunities for that data to be used against the person by persons or agencies unknown. For streaming media not to fail, consumers need watertight assurances that their data will never be abused. A privacy statement on a Web site does not suffice. Digital rights management systems being developed for streaming media applications must also include methods to protect consumers' rights to control access to and use of their private data.

Laws Lag Behind Technology

Laws relating to obscenity, libel, data protection, consumer protection, fraud, wire fraud, state secrets, and treason have failed miserably to keep up with the global nature of the Internet. State or national jurisdiction is nonsense, when information can be distributed globally in under eight seconds, via a series of anonymizers that make it impossible to locate the source. Streaming media merely introduces richer data to the mix.

The Internet is already hopelessly lawless, since there are no appropriate bodies exercising legal authority over all online activities globally (AOL is *not* an appropriate body for the administration of international law). Unfortunately, most places that are lawless, even virtual places like the Internet, eventually fail because of the lack of legal protections for individual citizens and corporations trying to operate there. One only has to observe the chaos in provinces of the former Soviet Union for evidence of that truism.

When legal cases have been brought involving unacceptable activity on the Internet, the operation has either required careful coordination between authorities in different jurisdictions, as was the case when several online pedophiles were arrested in raids coordinated globally, or else the authorities shoot the messenger, as was the case in France when portals found themselves having to remove from their services anti-Semitic and pro-Nazi material posted by their subscribers.

If a racist organization were to create a streaming media supremacist race-hate site, or insert backward satanic messages into all their audio content, the authorities would not easily know how to cause the site to be removed legally, or how to identify the content's authors, or even which country the site was in. Their only resort would be to mount expensive and complex operations, such as those that net child pornographers. This could cause the Internet and with it, streaming media, to become a technology that the majority of people shun.

Bandwidth Revolution Stalls

We have shown in this book that streaming media could potentially absorb staggering amounts of bandwidth, if higher and higher-quality experiences are to be made available. However, even though the growth of bandwidth provision has tended to outpace Moore's law, we are at an impasse at present. There is surplus bandwidth, with no immediate prospect of applications or consumers to use it, yet insufficient bandwidth to fuel the growth of streaming media. We are in the gulf between sufficient bandwidth to meet current demand and the bandwidth necessary to firmly establish a new application, streaming media, which will *create* demand for staggering amounts of bandwidth. In this stalemate, telecommunications companies feel no need to invest in more bandwidth, when what they already have is underutilized, yet there is less bandwidth available than what we really need to allow streaming media to appeal to the majority of consumers.

Perhaps an all-optical Internet 2 is a step too far, in the current climate. The silicon that goes into the production of fiber optic cables is far cheaper, ton for ton, than the copper from which our existing networks are built. However, those copper wires are already in place. Optical fibers that remain dark are also in place. The telecommunications companies had the good sense to reduce future costs by burying more optical capacity in the ground than they currently need, yet investors have tended to punish them for what is, in fact, sound foresight. Perhaps optical network provision can replace existing legacy copper networks, but a different investor and pricing mindset will need to prevail. Charging for optical network bandwidth on the basis of equivalent bundles of 64-kilobit-per-second voice circuits doesn't make any sense for streaming media applications. The United States will require a communications capacity of around 35 terabits per second by the end of 2002, just to serve the Internet. This is an order of magnitude more than is required by all the voice traffic in the world today.

For optical networks to deliver vast amounts of bandwidth to consumers at prices they can afford, the signals must remain photonic during switching and routing operations. Conversion to and from electronic signaling adds cost and slows the overall network, since optical switches and routers are much faster than electronic ones. Some companies are already developing all-optical network routers and switches. However, we are a long way from seeing those photonic network devices deployed in our local neighborhoods.

Internet traffic currently doubles every nine months, but Internet service provider operating costs are the key to lower and lower prices for bandwidth. It is these operating costs that limit further broadband build-out. Equipment only represents some 20% of a telecommunication company's cash flow. Eighty percent of the cash flow goes into covering provisioning and maintenance. These are primarily people costs. For these costs to drop, as will be necessary if the bandwidth revolution is to continue, network technology must require less maintenance. Provisioning is the work that takes place from the time a customer places an order for connectivity to the time the circuit is turned up and tested. The process involves order validation, design activities to provide the physical connectivity to satisfy the customer's order, building the designed network segment, design verification, test and turn-up,* and order finalization. The provisioning process must become cheaper, maximizing resource utilization and minimizing expenditures, while also becoming faster, improving on-time performance and cycle time, while minimizing rework. Companies like Adventis are providing computer-assisted provisioning systems that can bring about cost savings in provisioning.

About 40% of the money you pay for your phone service goes toward itemizing your bill, according to some estimates. With streaming media, does it really make sense to spend this much money to tell you when and for how long you took streaming media packets from your service provider's network?

For streaming media to succeed, the cost of bandwidth and connectivity must become cheaper. Keeping your service running also has to be less costly. Advances in optical fiber capacity are insufficient to guarantee the medium's future success.

Business Models Never Mature

Today, the industry is in flux. Business models that work have been the exception. Nobody has made streaming media pay yet, because the quality of experience is much lower than for cable television. Once the content served is good enough, or at least as good as television, business models might begin to make sense.

People are not used to paying for streaming media content, either by subscription or on demand, but this could change quite rapidly. When the cable television infrastructure was first rolled out, many of the same concerns over whether or not the consumer would pay for cable content were voiced. Blockbuster video was also initially questioned about the viability of charging for video rentals. In time, both these distribution channels showed that there is a market for content. Online digital music services such as Napster also demonstrated a market for streaming digital music.

Today, even if a profitable payment model for streaming media content could be established, bandwidth charges are still too great a percentage of the outlay. This situation will need to change before business models can mature. Many individual businesses up and down the streaming media value chain will need to transform the economics of their operations before the industry as a whole can succeed.

---

*The term "turn-up" is telecommunications industry jargon for the process of bringing a new connection into operation. In the old days of analog telephony, there was a potentiometer in an equipment rack in the local exchange that literally had to be "turned up" to the correct level in order to bring the new analog connection into service.

Receivers Never Materialize

The consumer electronics industry has been slow to embrace streaming media. They have not been flooding the market with designs for non-PC streaming media appliances. Perhaps the prevailing notion is that the market is too immature to make the investment in streaming media products worthwhile. Since the bandwidth needed to create good streaming experiences is not yet available, perhaps the consumer electronics industry is right to maintain a watching brief. Imagination in the solutions so far prototyped is evident, but the industry does not seem poised to release products as soon as there is sufficient demand. Unfortunately, if there are no players other than PCs to produce acceptable streaming experiences, the full potential of streaming media as a communications system cannot be realized. Demand growth will depend on the availability of good players. Hence, the consumer electronics industry that waits for demand before designing and releasing new types of streaming media receivers may wait indefinitely.

Another risk with streaming media receivers that could kill the medium as a whole is the possibility that the devices, appliances, and connections may never become simple enough for grandma to use. To some degree, this is true of the current crop of video recorders, digital cameras, DVD players, and CD players. The consumer electronics industry does not have a great track record in producing devices that interact elegantly with their users. As we mentioned earlier, a critical factor in the success of the medium is in the complexity of the software that provides the end-user with the streaming media experience. If consumers are required to understand the intricacies of network administration and firewall port configuration, the majority of consumers will never be able to get their systems to work. Beloved senior relatives in my family routinely accidentally erase vital system files from their computers. If streaming media receivers require high levels of technical skill, training, understanding, and expertise just to get them to work, consumers will never be able to adopt streaming media, whether or not they wish to.

Content Owners Don't Trust the Channel

If content owners are never satisfied that the medium is a safe place to entrust their media assets, there will hardly be much available via streaming media that is worth watching. Good content costs money to make. If it is open to immediate theft and redistribution by the first person to access it, there will be a strong incentive for content owners to avoid the medium.

There were once, allegedly, clauses in the licenses issued by streaming media encoder, server, and player software vendors that made content owners flinch. In order to transcode media assets into proprietary streaming formats, the content owner was effectively required to sign away partial rights to his media property. With Hollywood properties costing hundreds of millions of dollars to produce, these content owners were naturally hesitant to use any technology that required them to sign away any rights whatsoever, merely to access another distribution channel.

We can only speculate whether or not the streaming media software vendors intended to scare content owners away like this. Indeed, in the absence of professional media law opinion and access to the contracts and licenses in question, it is difficult even to verify whether or not such restrictive clauses were ever suggested. However, it would be a great pity if the streaming media industry failed to gain critical mass because of overzealous lawyers trying to overprotect the interests of their employers, the streaming media software vendors. If the clauses in question were inserted out of pure greed to acquire streaming rights to media properties by stealth, this was an unbelievably counterproductive position for these vendors to take.

The point of this discussion is that until content owners can trust the medium and the digital rights management systems that will be necessary to protect their interests, the medium will struggle to achieve the sort of audience penetration that television and radio already enjoy. As we have seen elsewhere in this book, digital rights management solutions are far from mature, though much of the work being done under the auspices of the MPEG, for MPEG-21, holds promise. Consumer acceptance of rights management also remains a concern.

The Audience Is Busy Doing Other Things

Ironically, a significant risk to the success of streaming media is the backlash against television and the music industry. Television audiences and advertising revenues are falling and audiences are more fragmented than ever. Fewer people buy the number-one single in any given week because people have many more choices for how to spend their leisure time. They can play computer games or mess around with the computer, rather than watch streaming media. Alternatively, they can engage in real-life pursuits, rather than the more passive entertainment pursuits represented by broadcast television and radio.

Because many people are more affluent, they have more toys and hobby options. These activities compete for consumer attention. Because many people have more to distract themselves with, it will be harder to get them to engage with streaming media than it was to create the initial television audiences. Teenagers, in particular, have begun to turn the Internet off. What was once a fascinating pastime has become just another thing to play with. The Internet wasn't what they wanted. The waiting was just too much.

For streaming media to succeed, providers of the service will need a full understanding of the competition for attention that exists today so that they can market their services appropriately. There is little evidence that this is happening.

The Audience Rejects Rights Management/E-commerce Security

As we mentioned, if digital rights management and protection are seen as too obnoxious and obtrusive, consumers will simply reject streaming media and stick with other more familiar forms of entertainment. Consumers are rightfully concerned that rights management seems to be a one-way street, with protection for the content owners, but little to protect the consumers. They are also concerned that rights management will merely oblige them to pay again for music they have already bought on CDs. They fear that they will have to pay again and again to access music or videos they love. In other words, they fear that the content production industry is using the opportunity to compel them to pay more for their digital media.

Any major record label or Hollywood studio that sees rights management as a way of extracting greater profits from media assets, through excessive access charges, could be in for a rude shock. There is already evidence that the movie industry has used DVDs to charge more for media content than they could with VHS tapes, even though the costs of production and distribution are somewhat lower for DVDs. What record labels and movie studios need to understand is that they no longer enjoy a de facto monopoly or control over the production, distribution, and marketing of digital media. Streaming media makes it possible for people with modest production budgets to produce quality entertainment. Distribution is much simpler with streaming media than mastering and pressing a CD or DVD ever was. There is no longer a need to truck product to distribution centers nor to compete for shelf space in music and video retail outlets. No intermediaries need to make a profit, since streaming media allows content producers to sell directly to end-users worldwide. As for marketing media properties, major labels once used their stranglehold on radio to promote only their product. Studios used the limited number of film exhibitors to keep alternative media product out of circulation. Streaming media allows marketing directly to the consumer, bypassing these marketing monopolies. Building an audience for a media asset is more difficult for the incumbents, due to audience fragmentation, but easier for independent producers, since all the same marketing avenues and many new ones are now available to them. The playing field has been leveled as streaming media technology has reduced the cost of entry into the media industry.

If through greed or misjudgment, the incumbent media companies maintain too large a gap between the cost of producing a media property and the prices they charge consumers for access to it, other media publishers, producing products and experiences equivalent in quality, but at lower prices, will enter the market and steal market share. Any attempt to use digital rights management to generate excessive profits will be countered by other media producers happy to satisfy demand for digital media while making relatively less profit. Streaming media will therefore make media products commodities. However, until other media producers enter the market to fill the need, any attempt to gouge consumers may harm the streaming media industry irreparably.

Allied to these concerns are the same concerns that consumers already have over e-commerce, where credit card fraud and other abuses of consumer rights have proliferated. If all their entertainment is subject to e-commerce transactions, they might get cheated much more frequently, with no right of redress or body to enforce fairness. Until consumers find the rights management and e-commerce security measures proposed by industry attractive enough to accept, the streaming media industry will struggle to attract significant participation.

Lowest Common Denominator Programming Prevails

The unfortunate consequence of fragmented audiences is that it puts great pressure on content producers to lower their costs of production. This can result in the proliferation of "reality TV" style productions, where cameras are positioned in a house, videoing how real people live together, with the result edited into a television program. If content producers always opt for the cheapest form of populist production, streaming media will be saturated with choice, but the choice will be limited to reruns, game shows, and *Wayne's World* wannabes. That uniformity may, in the end, drive consumers away from the medium, once the initial novelty has worn off.

Paradoxically though, the low cost of entry for content producers may actually stimulate adoption of the medium as more and more producers search for an elusive hit. For a model of how this might go, surveying what is happening with independent music production online is instructive. Bands and artists already have the means to record and distribute product of a quality similar to major-label CDs. This has tended to make many more interesting bands offer their art directly to audiences rather than to spawn a million Britney Spears clones. Perhaps streaming media will breed diversity in programming rather than uniformity.

It's Outlawed

If streaming media merely becomes a conduit for "unacceptable," yet uncensored content, because that is the sector that seems to be the most profitable soonest, lobby groups may move to outlaw the medium, or else severely restrict and regulate it. Knee-jerk reactions are possible when there is a policy vacuum.

Governments and regimes around the world that don't like their citizens to be informed and empowered may also seek to legislate to turn streaming media off within their spheres of control. Satellite television is already banned in some less-liberal Middle Eastern states.

If enough opposition is brought to bear, resulting in legislation to ban or severely restrict streaming media, this may all but kill the medium at birth. It is arguable that supersonic passenger transport never found favor because of lobby groups that petitioned against the noise and pollution of the Concorde. Would supersonic air travel be as rare and expensive as it is today, if it had not been for the activities of those lobby groups and the politicians who acted to satisfy them? We'll never know, but we do know that public opinion can outlaw a technology. Streaming media is at risk of being banned if all it succeeds in providing profitably is pornography.

Conclusion

Streaming media is a technology and a medium that could be tremendously important, generating growth and providing end-users with much better access to information and entertainment than they ever had before. Streaming media could be a much bigger business and have a greater societal impact than television. However, while the potential is great, delivery of the system, to date, has been less than what is required to cross the chasm between early and mainstream consumer adoption.

It is an open question whether or not people will spend more time looking at screens (in other words, consuming more media) as a result of streaming media technology. I believe there are reasons to expect that they will, but perhaps in smaller increments of time: more sessions of shorter duration. Will streaming media cannibalize radio, television, and computer-game audiences? I think the answer is that it inevitably will. Will streaming media enrich society? From the point of view of increased richness and reach of communications, the indications are positive. Will streaming media make money? Again, it appears that the potential exists, but that the limit will be related to the increase in the average number of hours we spend consuming media. There is no bottomless pit of attention to attract. All media consumers are mortal. Attention is finite and increasingly precious.

Why hasn't the potential of streaming media been realized to date? Besides the sheer degree of the technical challenge and the scale of the logistics required to build out the system, I believe the main impediment to progress has been the lack of coordination between sectors of the industry. As one sector has developed delivery technology or built excess optical capacity, other sectors such as the consumer electronics industry or the content providers, have waited on the sidelines. Every company acted rationally, within its own legitimate concerns, but the result of this piecewise progress was that those forging ahead, unsupported by other sectors, lost money. The net effect on streaming media technology has been slow progress from promising technology to everyday use.

When the broadcast systems were introduced, a small number of companies supplied the first content, production facilities, transmission infrastructure, receivers, and customer support. Look at how RCA in the US and the BBC in the UK went about introducing television. Companies were more vertically integrated in those days (or aspired to be).

Even in recent times, the UK's Sky Digital is a case in point. Connecting to Sky requires a man from Sky to install a Sky-supplied set-top box in your home. He doesn't leave until it works. Thereafter, Sky delivers the programming from its network operations center, including content it produces with its own Sky branding. For the consumer, converting to digital television reception is a one-stop shopping experience. Contrast this with the plethora of companies a consumer must deal with in order to adopt streaming media. Even with all the requisite contracts, equipment and arrangements in place, the consumer is still left to figure out the integration details for himself, with no idea whom to call when something goes wrong and no way to prove it, assuming the guilty party can be found. Would the adoption of digital television in the UK have been as rapid if consumers had faced the same obstacles as those interested in streaming media reception face? It is doubtful.

Perhaps streaming media needs a single organizing, vertically integrated force, supplying the full end-to-end solution, to help it take root. If one organization supplied the content, production, and authoring facilities, broadband distribution networks, receivers and nice, friendly people to help consumers over the phone when they had difficulties, the system may achieve faster, more widespread, mainstream adoption. The company might subcontract the individual pieces to specialist companies and suppliers. The point is that consumers would see a single guiding intelligence, with which they could maintain a simple and direct relationship, assured that their particular problems and concerns would be addressed and solved by the monolithic body. Perhaps vertical integration still has some merits as a business model.

Technically speaking, the streaming media industry needs to deploy cache management for edge streaming systems; forward error-corrected transmission; intelligent load balancing; re-engineered media players that deliver better quality of playback experience and presentation continuity; MPEG-4-compliant infrastructure from end to end; synchronization mechanisms for multiply sourced streams; simulcast capability; synchronized multimedia; multicast technology, assured quality-of-service content delivery networks; home media gateways to bring the edge of the network into the home; imaginative receiving devices; bulk optical; wireless and mobile bandwidth; symmetrical rights management that protects all parties to a media commerce transaction; laws to support global copyrights; laws against abuse of consumers' rights and; finally, licensing technology that does not annoy or intrude. Content providers must also resist the opportunities for rapaciousness that digital rights management technology presents. That is a long "to do" list, but I believe it is realistic.

For streaming media to fulfill its promise, the industry will need a great deal of will, vision, imagination, and courage; qualities that can be lacking during periods of economic uncertainty.

To summarize, then, streaming media can be characterized quite simply and succinctly. Streaming media is almost great.

APPENDIX A

MPEG-4 PROFILES

MPEG-4 was designed to allow efficient audio-visual coding in a wide variety of applications, using a multitude of transport technologies and systems. For this reason, the MPEG-4 standard provides a deep and rich set of tools for coding audio and video objects. This means that just about any practical system and application constraints can be met, but also that compliance to the standard is almost always going to be partial. Typically, no single encoder or decoder will be capable of producing MPEG-4 streams using any of the available tools. The complexity involved in creating a complete, ideal, universal coder would make it uneconomic for individual applications. So, the usual situation is going to be that the MPEG-4-compliant coder will implement a subset of the standard, cherry-picking various coding tools and techniques suitable for the application in question.

MPEG-4 has defined what are called *profiles*. These are subsets of the MPEG-4 tools, which have been identified by the Motion Picture Experts Group (MPEG) that can be used for specific applications. Profiles limit the number of tools a particular decoder needs to implement, while providing a template for interoperability between individual vendors' implementations. In principle, any vendor's coder than complies with a given profile will encode or decode audiovisual material that can be used with another vendor's coder. Remember that the details of how encoding is done is outside the scope of MPEG-4, which primarily defines how decoding is done. The point of profiles is that they provide target technical specifications that, if met, should ensure standards compliance.

Within each profile defined in the MPEG-4 standard, there are also one or more "levels." Whereas profiles limit the selection of tools, levels limit computational complexity. This is roughly tantamount to specifying maximum throughput bit rates. From an application point of view, levels are typically matched to the transport system's available bandwidth.

Hence, a combination of profile and level should allow a codec constructor to implement only the tools needed for his or her target application, while maintaining internetworking with other MPEG-4 devices built to the same specification of profile and level. It also allows for conformance testing, to check whether or not devices comply with the standard. End users, therefore, should always look for the profile and level specifications before accepting a vendor's word that their product is MPEG-4 compliant!

There are profiles for audio material, visual material, graphics content, and scene descriptions. The MPEG does not prescribe or even advise on combinations of these profiles, though some care has been taken to ensure that good matches exist between the four different profile areas. Considering that there are some 30 available video profiles, 8 audio profiles, 4 graphics profiles, and 5 scene graph profiles (not to mention the additional profiles for portable, Java-based devices—so called MPEG-J), the number of possible combinations and permutations is large! In practice, codecs will often support a number of different profiles and levels, so that the problem of identifying which vendor's codec works with another becomes somewhat complex.

Profiles and levels are not new to MPEG-4. MPEG-2 defined a number of profiles and levels. Like any successful idea, the use of profiles and levels has proliferated in the MPEG-4 standard. Only time will tell whether or not the explosion in choices will limit or enhance the usefulness of profile and level compliance descriptions in the real world.

Visual Profiles

In visual profiles, it must be remembered that these were added to the standard in three stages. Version 1 of the MPEG-4 standard defined the initial visual profiles, Version 2 added more, and additional profiles have been added subsequently. Compliance is, therefore, also a matter of specifying which version of the MPEG-4 standard is used, though later versions are supersets of earlier ones.

Visual profiles exist for natural visual objects (typically things captured with a camera), synthetic visual objects (computer generated video and sprites), and hybrids of natural and synthetic visual content.

Visual Profiles for Natural Video Content

MPEG-4 Version 1 defined five visual profiles for natural video content. These are:

- **Simple**—The simple visual profile provides for efficient and error resilient coding of regular rectangular video objects, for applications on mobile networks.
- **Simple Scalable**—This profile adds temporal and spatial scalability to the simple visual profile. Frames can be discarded or added and the number of pixels in the image can be varied. This profile is useful on the Internet and in applications where the bit rate available can vary. It provides for continued operation, through fallback to lower frame rates and image sizes, if the transport network encounters congestion or the decoder runs out of resources.
- **Core**—The core visual profile provides support for coding of arbitrary-shaped objects and temporal scalability as additions to the simple profile. Simple interactive applications are the target of this profile.
- **Main**—Main profile adds support for interlaced, semi-transparent (alpha-blended), and sprite objects to the core profile, targeted at interactive and entertainment-quality applications, such as DVD and interactive digital television.
- **N-Bit**—This profile adds the ability to quantize the image anywhere from 4 bits per pixel to 12 bits per pixel to the core visual profile. The ability to reduce the number of bits per pixel is particularly useful in surveillance applications, where storage space or transmission bandwidth is limited.

In Version 2 of the MPEG-4 standard, three additional profiles for natural visual content were added. These are:

- **Advanced Real-Time Simple (ARTS)**—This profile provides error resilient coding of rectangular video objects, using a back channel to provide feedback to correct errors. It also has improved temporal resolution stability, with low buffering delay. In other words, it can maintain its frame rate better, in the face of network congestion and packet loss, without the decoder having to store a vast buffer of data to provide for error resilience. Low buffering delay means that users can flip from stream to stream without incurring a long delay. This profile is useful for applications that must maintain real-time performance, such as videophone, teleconferencing, and remote telepresence.

- **Core Scalable**—This does for the core profile what the simple scalable profile did for the simple profile, in that it adds support for coding of temporally and spatially scalable objects of arbitrary shape to the core profile. The core scalable profile finds application in broadcast, mobile, and Internet video transmission applications, where the flexibility of adjusting the signal to noise ratio, or the temporal or spatial resolution of individual video objects or regions of interest in a scene is desirable.
- **Advanced Coding Efficiency (ACE)**—The advanced coding efficiency profile improves the coding efficiency of both rectangular and arbitrary-shaped objects. This profile is targeted at mobile broadcast reception applications and camcorders, where high coding efficiency is required, but where the device's footprint need not be small.

Visual Profiles for Synthetic and Synthetic/Natural Hybrid Visual Content

MPEG-4 Version 1 defined the following profiles for synthetic visual objects and combinations of natural and synthetic visual objects:

- **Simple Facial Animation**—MPEG-4 includes a simple facial model, which may be animated under this profile, to simulate a talking person. Reasons to do this include creating audio and video presentations for the hearing impaired. Other applications include animated avatars, designed to speak words that may have been synthetically generated, for example.
- **Scalable Texture**—This profile provides scalable spatial encoding of still images (called textures, in MPEG-4 parlance). Applications that need to scale textures, such as those that map the texture onto wire frame models and for high-resolution digital still cameras, are targeted by this profile.
- **Basic Animated 2D Texture**—A visual profile that provides spatial and signal-to-noise ratio scalability for textures, along with mesh-based model animation. Hence, a still image can be animated by mapping it onto an animated mesh. The example always given is that of a flag waving in the breeze, comprising a texture representing the flag and a mesh, animating its movement in the breeze. This profile can also be used to wrap a texture around an animated facial model, thereby allowing somebody's actual face to be superimposed on a facial model and then animated. "Synthespians" (actors that look real, but who are animated) can be created this way.

- **Hybrid**—This profile is essentially a combination of the core visual profile for natural visual objects (including arbitrary-shaped natural objects) with the basic animated 2D texture visual profile for synthetic content. This profile allows the composition of content-rich multimedia presentations, where photo-realistic animated synthetic characters share the scene with natural video objects, for example.

Version 2 of the standard added the following profiles for synthetic and hybrid visual content:

- **Advanced Scalable Texture**—A visual profile that supports decoding or arbitrary-shaped textures and still images, with scalable shape coding, wavelet tiling, and improved error resilience. This allows fast random access of still images, so it supports browsing of still image libraries well. Other applications include multimedia enabled PDAs and Internet-ready high-resolution digital still image cameras.
- **Advanced Core**—This profile is actually a combination of the core visual profile for natural video objects with the advanced scalable texture profile. It combines arbitrary-shaped, scalable natural video objects with arbitrary-shaped, scalable still images. Interactive streaming multimedia over the Internet is the main application for this profile.
- **Simple Face and Body Animation**—Adds body animation to the simple facial animation profile, using a body model that can be animated parametrically.

Additional Visual Profiles

Subsequent to Version 2 of the standard, the following visual profiles were added:

- **Advanced Simple**—A tweak to the standard simple visual profile, which makes rectangular video coding more efficient. It adds B-frames (bi-directional prediction frames), quarter pel (picture element) motion compensation, and global motion compensation.
- **Fine Grain Scalability**—This allows for the progressive improvement in signal-to-noise resolution, as bandwidth and resources allow. The idea here is to start with either the simple or advanced simple profiles as base layers and to add up to eight enhancement layers, which consist of additional data to improve the resolution of the

displayed image. This allows the delivery quality to easily adapt to transmission and decoding circumstances.

- **Simple Studio**—Why should the multimedia application vendors have all the fun, right? This profile is specifically designed for very high quality usage, such as in studio editing application, or anywhere that MPEG-2 currently finds professional applications, such as in broadcast television production. It is a fully-fledged MPEG-4 profile, not just an MPEG-4-compliant version of the MPEG-2 professional profile, high-level combinations. The simple studio profile eschews predication frames, based solely on I-frames (intraframes—where the entire frame can be decompressed and rendered from data contained in that frame, without reference to other frames in the sequence or group of pictures). It also supports arbitrary shapes and multiple alpha channels (for flexibility in layering, image compositing, and use of transparency). Bit rates of up to 2 gigabits per second are supported, with image sizes up to 4000 by 2000 pixels (the typical size of a scanned film frame) and pixel depths of up to 12 bits per component (RGB or YUV descriptions). Additionally, the simple studio profile allows for lossless transcoding from the 4:2:2 profile for MPEG-2. The simple studio profile provides for both 4:2:2 and 4:4:4 coding (these describe the sampling resolution and periodicity of the luminance and two color difference signals).

- **Core Studio**—Adds P-frames (prediction frames) to the studio profile, plus sprites. This makes it more efficient to handle high quality material, but at the cost of greater complexity in implementations. There are no backward prediction frames (B-frames) in studio profiles.

Audio Profiles

MPEG-4 Version 1 identified four audio profiles. These are as follows:

- **Speech**—Provides a very low bit rate parametric speech coder called HVXC (Harmonic Vector Excitation Coding), a CELP (Code Excited Linear Prediction) narrowband and wideband speech coder, and a text-to-speech interface. It can code speech at rates down to 2 kilobits per second, with the primary applications being voice mail, voice chat, high-capacity voice archives, and memo-style voice recording.

- **Synthesis**—The synthesis profile provides score-driven synthesis of music and speech, using the Structured Audio Orchestra Language

(SAOL), as well as wavetables and a text-to-speech interface. Structured audio techniques in MPEG-4 allow the transmission of synthetic music and sound effects at bit-rates from 0.01 to 10 kilobits per second, and the concise description of parametric sound post-production for mixing multiple streams and adding effects processing to audio scenes. The text-to-speech interface allows speech to be produced at exceptionally low bit rates.

■ **Scalable**—A superset of the speech profile, it allows for scalable coding of speech and music for networks like the Internet and the Narrowband Audio Digital Broadcasting (NADIB) network. Bit rates between 6 kilobits per second and 24 kilobits per second are available, with bandwidths in the audio pass band of between 3.5 kHz and 9 kHz.

■ **Main**—This is a rich superset of all the other audio profiles, containing tools for natural and synthetic audio.

Version 2 of the MPEG-4 standard added the following audio profiles:

■ **High Quality Audio**—This profile contains the CELP speech coder and the Low Complexity Advanced Audio Coder (AAC), including Long Term Prediction. Scalable coding can be performed using the AAC scalable object type. This profile also provides optional error-resilient bit stream syntax. It provides the highest fidelity audio coding available in MPEG-4.

■ **Low Delay Audio**—Intended for voice interactive applications, like voice over IP, this profile contains the HVCX and CELP speech coders (with optional error recovery syntax), the low-delay version of the AAC, and the text-to-speech interface.

■ **Natural Audio**—Contains all of the MPEG-4 coding tools for natural audio, but none of the synthesis tools. Intended for applications where the computational complexity or additional hardware for sound synthesis is prohibitive or unnecessary.

■ **Mobile Audio Internetworking**—This profile contains the low-delay and scalable AAC object types, including Twin VQ and BSAC. BSAC stands for *bit sliced arithmetic coding* that provides one of the forms of scalability in MPEG-4 audio. In order to make the bit stream scalable, BSAC uses an alternative to AAC noiseless coding module, although the other coding modules are identical to AAC. A bit stream encoded by AAC can be transcoded to a BSAC bit stream noiselessly. BSAC is capable of generating a bit stream with a precise bit rate control in the range of 16 kilobits per second to

64 kilobits per second per channel. This bit rate enables the decoder to stop anywhere between 16 kilobits per second and the encoded bit rate with a 1 kilobit per second step size. Through use of this scalability, the user can experience nearly transparent sound quality at 64 kilobits per second and graceful degradation at lower bit rates. BSAC is best performed in the range of 40 kilobits per second to 64 kilobits per second, even though its operating range extends down to 16 kilobits per second. Twin VQ refers to Vector Quantization, wherein Twin VQ has different VQ tables for different audio sampling rates and bit rates. In vector quantization, the speech is coded as matches to pre-defined wave shapes (or vectors). The mobile audio internetworking profile is intended to extend communication applications using non-MPEG speech coding algorithms with high quality audio coding capabilities.

Graphics Profiles

These profiles define which graphical and textual elements can be used in an MPEG-4 scene.

- **Simple 2D**—This profile provides for only those graphics elements that are necessary for placing one or more visual objects in a scene.
- **Complete 2D**—This profile provides two-dimensional graphics functionalities and supports features such as arbitrary two-dimensional graphics and text, whether or not they are used in conjunction with visual objects.
- **Complete**—This profile provides advanced graphical elements, such as elevation grids and extrusions. It also allows the creation of content with sophisticated lighting. The target application is the creation of rich and complex virtual worlds that exhibit a high degree of realism.
- **3D Audio**—This strangely named profile is all about providing a geometric description of an audio environment, to define the acoustical properties of the scene. It includes information about the virtual acoustic environment's geometry, acoustic absorption, diffusion, transparency of materials, etc. It is used for applications that do environmental spatializations of audio signals, such as surround sound.

Scene Graph Profiles

Also known as Scene Description Profiles, these profiles allow audiovisual scenes comprising audio only, two-dimensional scenes, three-dimensional scenes, and mixed 2D/3D content.

- **Audio**—If the application is audio-only, such as broadcast Internet radio, for example, this is the profile to use. This can also be used for advanced interactive music presentations.
- **Simple 2D**—This profile allows the inclusion of audiovisual objects in a presentation, without any interactivity. Thus, the objects can be placed, but the user cannot modify them or his point of view of them. This profile is used to create a broadcast television experience, for example.
- **Complete 2D**—This profile provides all of the two-dimensional scene description elements of the BIFS (Binary Format for Scenes) tool. It includes two-dimensional transformations (e.g., translation, scaling, rotation, skew) and alpha blending (for object transparency). It is used for two-dimensional applications that require extensive and customized interactivity.
- **Complete**—Three-dimensional elements are added to the complete 2D profile in this profile. This profile includes all of the tools available in the BIFS. It includes perspective manipulations and viewpoint relocation, as well as three-dimensional audio interactions. This profile is used for applications that create dynamic virtual three-dimensional worlds and for interactive three-dimensional games.
- **3D Audio**—This profile is useful for creating immersive virtual audio environments, providing tools for three-dimensional sound source location and positioning, either in relation to the scene's acoustic parameters or its perceptual attributes. The user can interact with the scene by changing the position of the sound source, changing the characteristics of the room, or by moving the listening point. This profile, clearly, is for audio-only applications.

APPENDIX B

MPEG-2 AND MPEG-4 CODING COMPARED

	MPEG-2	MPEG-4
Image shape	Rectangular	Rectangular or arbitrary-shape
Image elements per frame	Single	Limited by decoder capability
Image spatial dimensions	Constrained in standard	Variable
Text overlay	No	Yes
Lowest bit rate	1.5 Megabits per second	5 kilobits per second
Highest bit rate	300 Megabits per second	2 Gigabits per second
Number of profiles	6	38
Basic image unit	Group of Pictures (GOP)	Group of Video planes (GOV)
File format	Not defined in standard	MP4 format based on QuickTime
Rights management	Hooks provided, but not defined	Hooks provided, but not defined
Interactivity	None	2D or 3D
Editable	I-frame profiles only	I-VOP profiles only
Lip sync	Audio and video sync hardwired	Resynchronized at decoder
Wire frame models	None	Face and body
Mesh warping	None	2D
Text to speech capability	None	Available
Music synthesis capability	None	Available

continued on next page

	MPEG-2	MPEG-4
3D audio spatializations	None	Yes
Error resilience	Some	Strong feature of some profiles
Alpha blending and transparency	No facility	Available
Background sprite support	None	Progressively downloadable
Transport interfaces	One transport stream at a time	Multiple simultaneous interfaces
Audio	Mono or stereo or 5.1 surround	Multichannel AAC

APPENDIX C

AUDIO AND VIDEO SWEETENING TECHNIQUES

There are a number of signal processing techniques, borrowed from post-production, which can be used to enhance the audio and video passed to a streaming media encoder. Enhanced signals sound better or look better, subjectively speaking. The colloquial term for this process is "sweetening." If the enhanced signal is then heavily compressed, in order for it to be transmitted through a limited bandwidth channel, the resultant media stream has a better chance of looking and sounding better than an untreated one. The objective of sweetening prior to compressing is to reduce noise, which would otherwise consume precious bits to represent, or else to emphasize, aspects or the sounds and pictures that are considered important at the expense of de-emphasizing residual details. In some senses, signal cleaning is all about signal simplification.

Noisy signals are problematic for data compression for two reasons. First, why spend computation cycles and payload bits to encode image and sound components that shouldn't be there in the first place? Second, because noise is generally superimposed, it breaks the statistical consistency of a natural video image. Because it "breaks the rules" about what the compressor can assume about the data, it actually takes a disproportionately large number of bits to represent noise.

One caveat with recommending signal conditioning prior to encoding is that the range of tools available, in the hands of the inexperienced operator, has the potential to make the image and sound far worse and much harder to compress. Signal cleaning is more art than science. What looks and sounds great to one person can be utterly objectionable to another person. Taste and discretion must be applied, as with all essentially artistic endeavors.

Audio Sweetening

The conventional wisdom in media production is that ears are more sensitive to defects in sound quality than eyes are to defects in image quality. Therefore, although the audio only represents a small proportion of the total information, in terms of bits, it can have a marked effect on the quality of experience of an audiovisual stream.

One of the most common faults with an audio signal is incorrect audio level. This results in degradation of the signal-to-noise ratio and in extreme cases, clipping and distortion. If the clipping occurs when the signal is analog, unwanted higher order harmonics are added to the signal, which are harder to compress and encode. Intelligibility of the signal is also adversely affected. If level clipping occurs after the audio has been digitized, the audible artifacts are extreme and universally objectionable. Taking good care with audio level settings is the best answer, but in cases where the encoder is presented with an already degraded audio signal, there are some corrective tools that can be applied. In the first case, gain can be normalized. The entire audio track, if it exists as a file, can be analyzed and gain makeup or attenuation applied to ensure that the average level of the overall track falls within acceptable maxima and minima. Gain normalization also has another beneficial effect. End users will typically play one stream after another, continuously (as with Internet radio, for example). For users to obtain a high quality of experience, there cannot be sudden jumps or drops in audio level, from stream to stream, especially where these are butt-edited together. Gain normalization minimizes the sudden changes in audio level from clip to clip (wouldn't it be nice if television ads were normalized to the program material?).

In localized regions of the audio track there are further processing options. Audio level compression (not data compression, but gain compression), limiting, expansion, and noise gating all form a continuum of local audio level modification techniques that can have beneficial effects. These techniques are now so commonly applied in tandem that devices have been designed to implement all of these functions with a single gain modification element, thus reducing the overall noise added by the signal processing itself.

In audio level compression, the average root mean square power of the signal is continuously analyzed. As the power of the signal increases, the gain is decreased (in actual fact, the attenuation is increased). Thus, low-power audio is boosted in level, relative to the high-power portions of the track. Attack and release time constants can be adjusted so that the change in attenuation has some lag and recovery time, in response to changes in audio signal power. What this process sounds like, to the ear, is as if the audio track is apparently louder and clearer, even though the average signal level still falls within normal limits. Limiting is a more severe form of audio level compression in that, beyond a selected signal power, the audio level is constrained at any cost, even if it introduces signal distortion. Limiting is most commonly used to squash transients without affecting anything else, so that the signal does not distort in response to sudden loud noises, yet does not require that the gain is set so low to leave sufficient headroom for the spikes. The tradeoff here is that transients no longer have as much energy, in relation to the rest of the audio content, but the benefit is that the remainder of the audio content can make better use of the available audio headroom, thus improving the signal-to-noise ratio and making the audio track sound clearer and louder.

Expansion is the obverse of compression. As the root mean square audio signal power gets smaller and smaller, the assumption is made that the signal consists of ambient and background noise, which the listener isn't all that interested in. An expander works to increase the attenuation for low-level audio, thus rendering it more inaudible. It works to discriminate between the wanted audio and the background noise, de-emphasizing the extraneous noise in order to make the featured audio more intelligible. In live interview situations, where there may be low-level background traffic noise, for example, expansion can work to mask this intrusive noise source. Noise gating, as limiting is to compression, is a more severe form of expansion, where the audio signal is totally silenced below a lower audio power threshold.

Hence, it is possible to set no fewer than four distinct threshold points—one above which the signal is hard-limited, one above which the audio signal is progressively attenuated, one below which the signal is progressively reduced, and a final point below which the audio is completely silenced. The slope of the gain attenuation versus audio power can be set for the compression and expansion thresholds and the attack and release constants can be set for each of the four thresholds. This complex nonlinear gain modification structure works to try to make semi-intelligent choices on how to set the audio gain, in response to real time changes in root mean square audio power. The effect sought is to remove audio components that aren't worth encoding and transmitting while maximizing the quality of the audio signal presented to the encoder, at all times. The trick with setting the thresholds, gain reduction slopes, and time constants is to make the operation of the gain modification element unobtrusive to the listener. This is dependent on program material and not as easy as it sounds. Skilled audio engineers spend years finding out how to get these settings right for given program types.

Digitization of audio presents the operator with several opportunities to severely degrade the audio signal inadvertently. In the first case, the operator must select an appropriate sample rate, so that sufficient bandwidth is allowed to capture all of the features of the audio required. Secondly, the quantization level must be chosen to represent the audio with minimal quantization noise. If the audio is digitized at a low sample rate, using only eight bits per sample, for example, the signal to noise ratio will be limited by the small number of bits used in the representation and the higher frequency components will be entirely absent due to the low sampling rate. If this audio track is subsequently encoded using one of the higher-quality encoding profiles (such as MP3, for example), the audio will still sound bad to the end user, simply because so much quality was thrown away at the digitization stage. In general, it is always better to digitize source material in the highest quality possible.

Another nasty problem that can be introduced when audio is digitized is that if the sampling rate (the rate at which snapshots of the instantaneous audio waveform height are taken) is less than twice the highest frequency component present, *aliasing* noise can be introduced. This is because the digitization process is uncertain about how to represent frequencies higher than the Nyquist limit. In other words, given the limited number of data points output from the digitizer per unit time, a number of possible waveforms can be superimposed to satisfy those data points. The ear actually hears these other possible waveforms. Aliasing noise sounds like an in-band whining sound, which varies according to the highest frequency component present (its frequency is actually the difference between the sampling frequency and the high frequency component). For clean digitization at a certain sampling rate, it is necessary to first entirely remove frequency components above half the sampling rate. This is done with an anti-aliasing filter. In the past, these filters had to be of high-order, for good out-of-band rejection, but the higher the order, the more likely the filter was to overshoot and ring, introducing its own objectionable audio noises. These days, audio is typically oversampled at some multiple of the required sampling frequency, so that the analog filter used to remove out-of-band components can be designed using lower-order filters that introduce fewer artifacts of their own. Once digitized at several times the required sample rate, the data is reduced (decimated) using digital filtering techniques to eliminate the frequency components outside the required audio band. Hence, part of the anti-aliasing filtering takes place in the analog domain, prior to digitization and the remainder of the audio anti-aliasing takes place in the digital domain, after digitization. What this means to the streaming media compressionist is that they need to select digitization hardware that has been designed to eliminate quantization and aliasing noise. For cost reasons, some commonly available sound cards do a relatively poor job at reducing these digitization artifacts than costlier equipment designed for professional applications. Not all audio digitization devices are the same. Choose a good one.

Unless you have been living in a cave for the last twenty years, you would have encountered *graphic equalizers*. These devices allow parts of the audio signal to be enhanced, while other things are de-emphasized. These are frequency-selective devices that can be used to make the audio signal sound subjectively better. Other types of equalizer are available for professional use, the most common of which is the four band parametric equalizer. Skilled compressionists can use equalization techniques to produce more acceptable-sounding audio streams, at least partly compensating for the losses introduced by the data compression process itself.

Audio frequency equalization can also be used to reduce audible artifacts in low bit rate audio encoding. For example, there is a well-known warbling effect that happens on highly compressed, low bit rate audio streams, which has the effect of making the audio sound like it is being transmitted from the moon. The remedy is to roll off frequency components above 9 kHz. This roll-off could have been achieved during digitization, but a single audio digitization may need to be encoded at multiple bit rates, so for the lower bit rate encoding, further reductions in audio bandwidth must be made.

Another frequency-selective correction that can be made is *notch filtering*. If the audio signal also includes a fixed frequency noise component, such as AC mains hum, for example, a notch filter can be used to reduce the apparent level of that frequency component.

Today, much audio equalization and filtering is done after digitization, using digital filter implementations. These digital filters must be designed to use sufficient bit widths in their internal arithmetic to avoid introducing rounding errors, which sound subjectively like quantization noise. Digital filters can be a source of noise as well as a remedy for unwanted audio artifacts.

An interesting variant of frequency-selective audio enhancement is the dynamic noise reduction process. This process makes the assumption that the audio of interest typically has lower-frequency spectral components than extraneous background noise. So, as the root mean square audio power reduces, the audio pass-band is progressively reduced. This tends to filter out background noise when the main program audio is quiet, without reducing the crispness and intelligibility of the desirable audio components. Once again, as with non-linear audio gain correction, the time constants governing the response of the pass-band limiting element must be selected carefully, according to the program material being processed, for the noise reduction to sound unobtrusive.

Another device that resembles an audio level compressor, but which works frequency-selectively is the *de-esser*. Whereas the audio level compressor determines audio power in a broadband way, using information from the entire audio pass-band, a de-esser is especially sensitive to frequency components in the 5kHz region. It was noticed by audio engineers that when a speaker says an "ess," the audio level goes up appreciably. Because these sibilant sounds are comprised of relatively high frequency components and have high energy, they take a lot of bits to encode as audio stream data. The de-esser detects the presence of sibilant utterances and rapidly turns down the audio gain, releasing just as rapidly. The effect is to de-emphasize the sibilant sounds, without affecting the overall apparent audio signal level. If overused, the de-esser tends to make "ess" sounds turn into "effs." However, if used with taste and discretion, the signal level of the sibilant components can be reduced without affecting intelligibility, with the added benefit of making it possible to encode the audio with fewer bits.

Taking the audio level compression and frequency-selective techniques to a logical conclusion, the audio band can be split into say four bands (low frequencies, low-mid frequencies, high mid frequencies and high frequencies) and each individual band subjected to all the non-linear gain corrections that can be applied individually. After gain correction, the four bands are then recombined, creating an audio signal that has enhanced intelligibility and much less noise than would have been obtained with broadband nonlinear gain corrections applied uniformly across the entire audio pass-band. Nearly every commercial radio station on air uses signal-processing devices that work this way to enhance their station's output. The strident, punchy sound is familiar to any fan of American FM radio programming. These signal processors are now available in digital implementations, suitable for use prior to audio stream encoding and also in silicon, for design into streaming media players. The company at the forefront of development of these devices is OCTiV, with their product OCTiMax (see www.octive.com).

There are some sweetening techniques that are mainly used artistically, simply because they make audio program material sound subjectively "better." They are effects, more than signal conditionings. These include aural exciters, suboctave generators, reverberation devices, and echo-cancellation processes. Aural excitation and suboctave generators work to extend the apparent audio spectrum. The aural exciter works to synthesize higher order harmonics related to the audio program material. It makes the audio program sound brighter. The voice-over artist who does all of the Hollywood trailers has a particularly impressive rasping quality to his voice, which audio exciters respond well to. Many radio personalities use audio exciters to make their voices more intelligible and to cut through traffic noise. The suboctave generator synthesizes low frequency components below the actual program content, making the audio signal sound much more powerful and authoritative. It is like turning up the bass on a graphic equalizer, except that it actually creates bass components to emphasize.

The addition of artificial reverberation can serve to create the effect of a more pleasing audio space, thereby enhancing the audio program material. Because it adds a diffuse sound field, it can actually help the encoder compress the signal more efficiently, since it makes the audio signal somewhat more statistically consistent, with adjacent audio waveform segments more closely resembling each other (though in extreme cases, added reverberation can make the signal harder to compress, because of the added apparent randomness of the reverberant signal). The reverse process is echo removal, wherein the signal is processed to remove unwanted delayed reflections of the desired audio program. If the original recording has separable quadrature components (such as with crossed microphone pair stereo recording or middle/side recording techniques), removing the echoes is much easier, algorithmically, than if cross correlation is needed to identify the echoes for removal. When the original audio signal is awash with echoes and reflections, it is sometimes desirable to remove those to enhance intelligibility and clarity.

Occasionally, audio digitized at one sample rate must be resampled at another. Unwary operators can introduce a great deal of noise and distortion in this process. Firstly, the audio sample clock needs to be synchronized to the new sampling time base, so that there is no drift in the relative sample rates. The simplest sample rate converter simply converts the original digital signal back to the analog domain and then re-digitizes the audio at the new sample rate. This process can suffer from errors introduced by the conversion process and the new anti-alias filtering required for the new sample rate. However, sample rate conversion often takes place wholly in the digital domain. Unfortunately, not all sample rate conversion algorithms are the same. Some introduce more noise than others. High quality sample rate conversion is actually computationally expensive, requiring eleventh order filters at a minimum. A reduction in sample rate conversion filter order or oversampling rate serves to directly degrade the signal-to-noise ratio. The advice here is to examine the specifications of any sample rate conversion processor very carefully before trusting it to resample a digitized audio signal faithfully, without adding additional noise.

In streaming media encoding, much of the audio material will come from archives. Recordings may have been stored as shellac or vinyl records, reel-to-reel tape, or magnetic film stock. These recording media had particular limitations and shortcomings and the media age in characteristic ways that serve to degrade the quality of the audio and to introduce noise. Sometimes the noise added all but overwhelms the original audio program. It goes without saying that the compressionist should always try to find the highest quality archive source available, when there is a choice of archived originals. However, the usual case will be that only a single source exists and that the compressionist will have to work with it, regardless of how degraded it is.

There are a number of companies that have produced signal-processing devices to aid in audio restoration. Techniques to remove clicks and pops, such as those that result from scratches on vinyl records, for example, are commonly used. Dropouts in the audio program, which can happen when magnetic stock breaks or sheds its magnetic material, can be corrected using waveform interpolation and reconstruction techniques. A number of signal restoration processors can correct clipping errors, and others are useful for spatially enhancing the signal, creating a pseudo-stereo effect from monaural source material. All of these restoration techniques can greatly improve the chances of acceptably encoding the audio material as an audio stream.

The ultimate audio restoration technique is to either dub new sounds over a copy of the existing archived material, refurbishing the audio track by the addition of new sound components, or else to re-master the audio track entirely, using the archived source merely as a guide. Sometimes it is possible to access the original sound elements and simply remix them, but other times the sound design will need to be reconstructed from scratch, using new sonic elements. For dialog, automatic dialog replacement techniques can be used to re-voice the original narrative, for example. When the audio material is being repurposed for streaming media presentations that will include 5:1 surround sound (such as when high quality, DVD-like streams of old feature films are made), then remixing or reconstruction of the sound track can often yield pleasing results, albeit at a high cost. When the streaming media presentation will make a lot of money, the cost may well be worthwhile.

Video Sweetening

Just as with audio sweetening, there is a vast array of techniques available to make video look better or compress more efficiently (or both). There are also video restoration techniques that can help produce acceptable streams from archive material. Video comes from a variety of sources, including analog and digital video cameras, analog and digital videotape, and film stock. Signal conditioning techniques can be analog, digital, or a combination of the two. Appropriate video sweetening can radically improve the efficiency of the streaming media encoding processor and enhance the end-user's quality of experience for a given media stream bandwidth budget.

If analog signals are involved, the first thing to take care of, prior to digitization and streaming media encoding, is to make sure the analog video signals are well conditioned. There should be no impedance matching problems, video level problems, or unwanted signal components. Black and white levels should be corrected as well. There are modular signal conditioners, such as those made by the Grass Valley Group, which can be used to correct these problems. In many cases, it is necessary to use a time base corrector, in order to time the video signal correctly. If digital signals are input, the main thing is to ensure that the digital word clocks are synchronized correctly.

Streaming media encoders require stable images in order to compress optimally, so video signals must be deinterlaced prior to encoding. Inverse telecine processing should also be applied to any telecine video material, to remove repeated images. Deinterlacing is a complicated process, if done correctly, involving specialist equipment such as that made by Faroudja Laboratories, Inc. (see www.faroudja.com). There are simpler deinterlacing techniques that give less acceptable results than the patented motion compensated techniques applied in the Faroudja products. Once again, the compressionist needs to be aware that not all deinterlacing processes are the same and that there are quality sacrifices involved in choosing cheaper or simpler deinterlacing processors.

As with digitizing analog audio, digitization of video requires the appropriate choices for pixel depth and quantization, correct choice of sample rate, and care with anti-aliasing. Oversampling is more difficult for video than it is for audio, because of the much higher sampling rates involved. However, there are oversampling solutions available. *Brick wall filters* are often applied to the video pass-band, in order to remove all unwanted higher frequency components, but because these are higher order filters, ringing and overshoot is a potential problem. High order filters can introduce artifacts in the video signal that make it much harder to encode for streaming media. As with audio, pass-band limiting can also be performed in the digital domain, after digitization.

Video signals often suffer from chroma phase errors and ringing artifacts, due to the number of signal handling stages the video may have passed through from its initial creation to the digitization process. Sometimes it is possible to correct these phase errors and overshoots in the digital domain, after the signal has been digitized.

As with audio, sometimes video sampled at one frame rate and sampling frequency must be resampled to another. The general form of this conversion is called *sample lattice conversion* and it is roughly analogous to audio sample rate conversion, except that it can convert digitized images to images with different frame rates, pixels per line, and lines per frame. Sample lattice conversion finds application in standards conversion (converting PAL to NTSC, for example), image cropping, image resizing and aspect ratio correction. Standards conversion, in particular, often benefits from motion-compensated and motion adaptive processing, in order to preserve the quality of the image when high motion content is present. It is important to note that the quality of the sample lattice conversion process depends on the complexity and order of the algorithm used to perform the task. Hence, not all sample lattice conversions are the same and the compressionist needs to understand the specifications of the one in use or else suffer image degradation.

In typical digital video production environments, video signals can be compressed and decompressed a number of times before they are repurposed for streaming media. By the time the streaming media encoder sees the video signal, it may have been through several compress/decompress processes (as is the case when a television production plant, wired for MPEG-2 digital video production, wishes to transcode the video into MPEG-4 for streaming). To avoid image degradation, it is important that the macro blocks of every compression stage are aligned. Any shift in macro block alignment only serves to degrade image quality over successive compression/decompression cycles. Unfortunately, to this author's knowledge, there are no streaming media encoders available today that are responsive to macro block alignment parameters. Many MPEG-2 converters, however, allow for macro block alignment and these should be used throughout the video production chain if MPEG-2 is used.

Once the video signal has been digitized, the video can be further cleaned and enhanced using two-dimensional filtering techniques, such as low pass filtering and median filtering. These intraframe filters are useful for removing various noise sources, such as speckle noise (a form of pink noise) that can be introduced by satellite transmission paths, for example. These filters have the effect of softening the image, but in the case of low pass filtering, can introduce image smearing. Hence, these filters must be applied judiciously.

Another class of two-dimensional video filter uses information from other frames of the video sequence to clean up the video image. These techniques are sometimes called *interframe filters*. Many image-enhancement filtering techniques were originally developed to clean up satellite photographs and medical images. Some of these techniques yield very good results when applied to video sequences, prior to streaming media encoding. In particular, there are some contrast and resolution enhancement techniques that make use of multiple samples of particular features of interest in the video sequence, correlating the image data from adjacent video frames. The company 2D3 (www.2d3.com) publicly demonstrated a software application to do this at the most recent International Broadcasting Convention in Amsterdam, September 2001. Some two-dimensional filters are motion adaptive, meaning that they filter moving features in the images differently to static elements. Motion adaptive filters can soften backgrounds without smearing high motion features.

When Hollywood feature films are prepared for release, one of the important post-production stages, which can give a film its distinctive appearance, is colorizing. In this process, a skilled colorist makes primary and secondary color corrections, using specialized color correction equipment, to enhance the overall appearance of the images. Contrast changes can be made and color corrections applied selectively to regions of interest in particular image sequences. These enhancements often greatly benefit the streaming media encoding process, since the images start out looking better than real life. Losses introduced by the compression process do not subjectively seem as bad, as a consequence. However, as with many other sweetening processes, color corrections need to be applied subtly to avoid producing hideous colors and images.

Gamma corrections are also often made. It is a fact that not all display devices display images the same way. What looks right on a television may look too dark on a computer monitor. Gamma is a property of the phosphors used in the display technology, but also of the overall video delivery system and even the color of the surround of the display device. Gamma correction applies a look-up table to the digitized pixel values in order to correct them for display on different target display devices. Gamma correction is particularly important in streaming media applications, since most video material is originated using television production equipment, yet most users stream to a computer monitor today.

Another video signal processing technique commonly applied to video signals prior to streaming media encoding is *blurring* and *unsharp masking*. These techniques are sensitive to edges in the video image and work to reduce the suddenness of the transition. This makes pixels on either side of the edge more like each other. Compression algorithms tend to work best when neighboring pixels are similar. Other image detail reduction techniques can often yield good results prior to compression. Pixels may be quantized differently in a process called *posterization*. This reduces the range of colors in the image and so provides less challenge to the streaming media encoder, at the expense of introducing visible contours in the image. Colors in the image can also be corrected to conform to the nearest web-safe colors, which can be a consideration for some streaming media players launched from browsers. By choosing the nearest web-safe color from the palette, the ultimate appearance of the streamed images can at least be judged at the encoding stage.

Compression techniques also work best when there is no extraneous motion to encode. For this reason, motion jitter introduced by a telecine process (film weave, for example) or by camera shake can cause the video compression algorithm to spend unnecessary payload bits and encoder computation cycles representing motion that is of no interest to the viewer. Image stabilization algorithms can be used to resynthesize video frames, in order to reduce this unwanted motion jitter. A number of manufacturers like discreet (www.discreet.com) and Quantel (www.quantel.com) make commercial systems that can take motion jitter out of a video sequence, either fully automatically, or else with some human intervention to nominate anchor points. They are typically very expensive devices.

When film from archives must be encoded for streaming, a number of mechanical and digital film restoration techniques can help produce better encoding efficiency. Scratches on the film can be automatically recognized and removed. Color variations over successive frames due to age can be automatically corrected. Film grain can be filtered out. Missing frames can be resynthesized from available adjacent frames. The weave of the film as it passes through the film scanner can be compensated. There are specialist companies, like Rank Cinetel and Digital Vision, which provide high quality digital film restoration solutions to perform these enhancements. Once again, for streaming media applications, the cost is only justified if the streaming media asset will attract a large audience or generate significant revenues. Sometimes the restoration process serves to renew the copyright on expired works, allowing the copyright owner to extend their monopoly franchise over what they feel to be a valuable media property. This is because extensively remastered works can count as new works. The process of sweetening is not only to make things look better or encode more efficiently, but it can be a way to extend copyright protection.

The final resort, when the material is important but damaged beyond the capabilities of image restoration software, and it positively must be encoded for streaming media regardless of cost, is to retouch or repaint video frames by hand, one-by-one. This is painstaking and labor-intensive work, but sometimes necessary. Packages like Adobe After Effects can be used as a tool for frame retouching and repainting. If original sources are still available, but the edited archived piece is badly degraded (as when film stock is used to create an edited program that was conformed to video tape, for example), it may be possible to autoconform the program from original sources and the original edit decisions.

Hence, there is a large and ever-growing collection of audio and video enhancement techniques that can be applied to audiovisual material in order to produce better media streams, which give end-users superior viewing and listening quality. The ultimate in audio and video quality enhancement does not come cheaply, however, since the complex and often patented processing devices, as well as the skilled operators, command premium prices. Enhancement can also be a laborious and slow process. In most cases, then, streaming media encoding will need to work with signals that have been suboptimally cleaned, conditioned, and enhanced.

APPENDIX D

REAL WORLD STREAMING MEDIA ENCODING

The main body of this book concentrated on the general characteristics of streaming media, but people must use specific products to stream media in the real world. Consequently, this appendix has been included in order to serve as a brief industry guide. These products are not necessarily recommended, since every application will have its own specific requirements and different products will more satisfactorily match these requirements better than others, for each given application. The list is also not exhaustive. It would also be inappropriate for me to make specific recommendations, as I was closely involved in the design of the Grass Valley Group's Aqua streaming media encoder, so I must declare an interest.

This appendix presents some current streaming media encoding products. The streaming media industry is a fast moving one and has seen many companies introduce and withdraw streaming media products. For this reason, it is important to note that while the products listed here represent some of the available choices at the time of writing, new products will constantly be introduced and older products retired. The reader is advised to consult the manufacturers or else an industry overview resource such as www.streamingmedia.com before making any equipment decisions.

Although there are a number of vendors providing streaming media encoding tools, most (if not all) include formatting and compression software written by Microsoft, RealNetworks, Apple, and Sorenson, which provide software development kits and re-distributable software modules for streaming media encoding into their particular formats. Exceptions are products that encode into MPEG-4 format, where there are a larger number of encoder vendors selling compression and formatting software modules to encoder original equipment manufacturers. The main differences between these encoding tools, then, is the quality of the interfaces and the hardware they run on.

Encoding Machines

- **Anystream**—(www.anystream.com)—Agility Workgroup and Agility Enterprise Encoders—These are industrial strength streaming media encoding solutions designed to serve the needs of medium to large streaming media production workgroups respectively. The systems allow batch digitization and batch streaming media encoding to the popular streaming media formats. Throughput is limited only by the power of the server upon which the software runs. They have demonstrated the system running with state-of-the-art multiprocessor servers and have announced support for Intel's Itanium 64-bit processor.
- **Avid**—(www.avid.com)—Unity Pro Encode—Avid has been at the forefront of digital media post-production tool design for over a decade. The Unity Pro Encode is a machine capable of supporting large workgroups and also includes batch mode encoding. When used with the Unity workgroup, it can support up to 10 encoding stations, interfacing with the Avid nonlinear editing interface to allow background encoding without operator intervention. Each encoding station is, in fact, a dual 800MHz Intel Pentium server, which launches discreet's Cleaner 5 software (see below) to perform the actual media encoding. The Pro Encode software, which Avid provides, acts to orchestrate the encoding operations for a workgroup comprised of Avid editing machines.
- **Chyron**—(www.chyron.com)—Clari.net DualStreamer and Scheduler—The DualStreamer is a turnkey two-rack unit box that contains video capture hardware and the streaming media encoding processor. The Scheduler allows multiple DualStreamers to be ganged together to increase the number of simultaneous streams that can be encoded and ads a batch encoding capability. The design of this unit is to make the machine look to the operator like any other signal processing unit, which the company advertises as "plug and stream." Simplified operator controls are featured.
- **GlobalStreams**—(www.globalstreams.com)—GlobeCaster—A complete television production system and Internet encoder in a box. It includes a multicamera switcher, video effects, on-air graphics and text generator, and encoder, to allow high-production-value broadcasts live to the Internet. The beauty of this solution is that it can be operated with a minimum of operators. Whereas most other encoders take their input from the back end of a regular television production environment, this solution provides the television production facilities.

- **Grass Valley Group**—(www.grassvalleygroup.com)—Aqua—A multiformat, multibit-rate real time encoder that can produce multiple streams (up to 12) simultaneously. The solution differs from other encoding machines in that the system runs on telecommunications grade Compact PCI servers, which are very small and require less power and cooling than regular servers. The solution also allows for signal cleaning, using hardware, prior to encoding and allows owners to simply add processing power when they have a requirement for more simultaneous streams. The software allows batch encoding as well as live encoding.
- **Pinnacle Systems**—(www.pinnaclesys.com)—StreamFactory and StreamGenie—The StreamFactory is a dedicated rack-mounted unit that can encode two streams simultaneously, similar to the Chyron DualStreamer. Units can be ganged together for greater real time throughput. The StreamGenie is a portable streaming encoder suitable for live, on-the-road applications. It provides production facilities similar to the GlobeCaster.
- **Philips**—(www.mpeg-4.philips.com)—WebCine Encoder—One of the very first MPEG-4 encoding stations available commercially. The WebCine encoder produces only MPEG-4 streams, at up to 1.5 Megabits per second. The system is sold as a turnkey system, running on a dual processor, rack-mounted server platform.
- **Sonic Foundry**—(www.sonicfoundry.com)—MediaSite Live—A turnkey self-contained webcasting solution designed for videoconferencing and other live communication applications. It encodes using Windows Media Format only.
- **Telestream**—(www.telestream.net)—FlipFactory—A server-based software application that can encode into streaming media formats from most professional video standards. It scales by adding servers and load balancing. There are solutions in the FlipFactory family that allow media to be passed and transcoded in both directions between professional video servers, such as those made by Omneon Networks, and streaming media servers.

Software-Based Encoding Tools

- **Discreet**—(www.discreet.com)—Cinestream and Cleaner family—Cleaner was originally developed by Terran, sold to Media 100, and now owned by discreet, a company in the Autodesk stable. Cleaner was one of the first streaming media encoding packages and produces output in all the popular streaming formats. The Cleaner family includes live and workgroup versions and the software technology is used in other products, such as Avid's Unity Pro Encode (see above). Indeed, the Cleaner technology is leveraged in the Cinestream product, which adds editing capabilities and interactive authoring to encoding. Discreet also make high-end graphics workstations that are suitable for film restoration.

- **Microsoft**—(www.microsoft.com/windows/windowsmedia)—Windows Media Encoder—This is Microsoft's free, downloadable encoding tool, which produces live and on-demand streaming media in Windows Media Format, on the desktop. The latest versions allow encoder ganging and control over videotape machines, to allow encoder farms to be integrated. It also provides deinterlacing and inverse telecine processing and can produce media files up to 30GB in size (the previous limit was 2GB—not enough for a feature length film).

- **RealNetworks**—(www.realnetworks.com)—Real Producer Plus—This $200 desktop software package is also available as a stripped-down, free version. It is used to create streaming media presentations in RealNetworks' format, including synchronized multimedia elements. As with the Microsoft tool, the Real Producer Plus package can perform deinterlacing, inverse telecine, and can work with files larger than 2GB.

- **Sorenson Media**—(www.sorenson.com)—Video 3—This package is a popular choice for creating QuickTime streams. Although other encoding machines and software tools like Cleaner can produce streams in QuickTime format, Sorenson has the advantage of having developed the most popular QuickTime Streaming media video codec, so that control over the encoding parameters is comprehensive.

Encoding Labs

- **EncodeThis!**—(www.encodethis.com)—Los Angeles-based encoding lab. Recently changed their name to Digital Media Broadcast (www.digitalmediabroadcast.com). They can encode media into any of the popular streaming media formats, using their proprietary encoding chain, assembled from many of the best-of-breed signal conditioners and encoding tools. Because this lab serves the Hollywood feature film industry, encoding film trailers to promote blockbusters, for example, they take particular care with signal cleaning, using skilled compressionists.

- **Loudeye**—(www.loudeye.com)—Seattle-based encoding lab. One of the longest established streaming media encoding labs conveniently located close to the Windows Media Technologies and RealNetworks development teams. They have assembled their own bulk encoding plant, which can handle high volume encoding tasks. The company also preconditions signals, prior to encoding. Loudeye offers live and on-demand webcasting services as well. Through their subsidiary, VidiPax, they can provide audio and video restoration services.
- **MediaWave**—(www.mediawave.co.uk)—In the UK, one of the more prolific encoding labs. MediaWave is actually a full service corporate intracasting services provider, from production to distribution. Encoding is performed using their encoding rack, integrated from popular encoding tools and high-spec servers.

Precompression Signal Conditioners

- **Cedar**—(www.cedar-audio.com)—Cedar for Windows—The company provides a comprehensive range of audio restoration products, which can dramatically improve the sound of degraded archive material. Cedar for Windows is their flagship product, equipped with a powerful digital signal processing rack to speed the processing to real time and beyond.
- **Colorfront**—(www.colorfront.com)—star*dust—A suite of software tools for film restoration. The company also makes color correction software, and their tools can be used to greatly improve the appearance of video and film-based material.
- **Digital Vision**—(www.digitalvision.se)—DVNR—High-end digital signal conditioner and film restoration processor, with a vast array of available processing options. This machine has plenty of processing power and is used by professional compressionists to produce the most pleasing media streams.
- **Snell and Wilcox**—(www.snellwilcox.com)—Prefix—This standalone signal conditioning box is specifically designed for precompression signal cleaning. It offers a suite of audio and video filters, noise reduction techniques, and gain controls to allow the trained operator to condition media in a variety of ways.

- **TC Electronic**—(www.tcelectronic.com)—Finalizer—The Finalizer is an audio processor intended for CD mastering. It provides a number of signal processors to improve the overall sound of pre-mixed audio content. The company also provides a full range of production signal processors, which can be used during sound design and dubbing.
- **Teranex**—(www.teranex.com)—Video Computer—This processing platform has outrageous throughput capabilities, with processor cycles to burn, it seems. The company offers a range of signal cleaning tools (as software packages) for this proprietary signal-processing platform. The Teranex is expensive, but powerful.

Video Capture Cards

- **Viewcast**—(www.viewcast.com)—Osprey Series—Osprey capture cards are popular among the streaming media encoding fraternity, mainly because of the full range of features that they offer, their easy interface to most encoding software applications, and because they are cost-effectively priced. Cards range from cheap and simple to more expensive and full-featured. The company also recently released their turnkey streaming media encoder, Niagara.
- **Winnov**—(www.winnov.com)—Videum Series—Another popular range of video capture cards popular with streaming media compressionists. They have recently introduced their Videum StreamEngine and Xstream Multicaster encoding appliances.

APPENDIX E

VIDEO AND AUDIO COMPRESSION

Jerry C. Whitaker, Editor
Adapted by Michael Topic

Introduction

Applications that use audio and video to communicate deal with an enormous amount of data. For this reason, compression is an integral part of most of those applications, whether the application is dealing with high definition digital television pictures [1] or streaming media. Streaming media, as we noted in the main body of this book, would not be possible at all without compression techniques.

Modern video compression schemes employ a combination of processing techniques. Something like an MPEG-4 encoder actually uses a veritable arsenal of basic compression techniques, in concert, to maximize data reduction. Those compression systems that have been widely adopted enjoy economies of scale and reduced market confusion, becoming the obvious choice for new applications. The process of standardization is not without risk. If the timing of the standards effort is poor, the standard may not become widely used. For example, if a standard is defined well ahead of market demand, other more cost-effective or higher performance approaches may emerge before the market takes off. On the other hand, if a standard is ratified long after alternative schemes have become well established in the marketplace, the exercise might as well have been academic.

What can be stated with certainty is that any applications that use digital audio and video will be highly dependent on compression technology and improvements in compression techniques for the foreseeable future.

This appendix presents a number of fundamental compression techniques and demonstrates how these are combined to create compression standards such as MPEG-2. We will also show how a compression standard such as MPEG-2 evolves into successors, namely MPEG-4.

Transform Coding

In the technical literature, countless versions of different coding techniques can be found [2]. Despite the large number of techniques available, one that comes up regularly (in a variety of flavors) is *transform coding* (TC).

Transform coding is a universal bit rate reduction method that is well suited for both large and small bit rates. Furthermore, because of several possibilities that TC offers for exploiting the visual inadequacies of human sight (psychovisual limitations), the subjective impression given by the resulting picture is frequently better than with other methods. If the intended bit rate turns out to be insufficient, the effect is seen as a lack of sharpness, which is less disturbing (subjectively) than coding errors that result in frayed edges or noise with some repetitive structure. Only at very low bit rates does TC produce a particularly noticeable artifact: the *blocking effect*. Streaming media coders often have to work with extremely limited bandwidth, so a number of very low bit rate techniques have been invented and included in the MPEG-4 standard. So, although streaming media encoding schemes make heavy use of transform coding, they cannot rely solely on TC to produce acceptable image quality at very low bit rates.

Because pictures have differing statistical characteristics, in terms of frequency and energy distribution, the optimum transform is not constant, but depends on the momentary picture content that has to be coded. It is possible, for example, to recalculate the optimum transform matrix for every new frame of video to be transmitted, as is performed in the *Karhunen-Loeve transform* (KLT). Although the KLT is efficient in terms of ultimate performance, it is not typically used in practice because analyzing each new picture to find the best transform matrix is usually too computationally expensive when a "good-enough" compromise will suffice. Furthermore, the matrix must be indicated to the receiver for each frame, because it must be used in decoding the relevant inverse transform. In a streaming application, the matrix selection data could add a significant bandwidth overhead to the streaming transmission, thereby somewhat defeating the original reason for compressing the video in the first place. A practical compromise is the *discrete cosine transform* (DCT). This transform matrix is constant and is suitable for a variety of images; it is sometimes referred to as "quick KLT." Other transforms that can be used in data compression include the popular discrete wavelet transform (DWT) and the less-commonly used discrete Hartley transform (DHT), the slant transform, the Haar transform, and the Hadamard transform [3].

The DCT is a close relative of the *discrete Fourier transform* (DFT), which is widely used in quadrature signal analysis. Similar to DFT techniques, DCT offers a reliable algorithm for the quick execution of matrix multiplication.

The main advantage of DCT is that it *decorrelates* the pixels efficiently; put another way, it efficiently converts statistically dependent pixel values into independent coefficients. In so doing, DCT packs the signal energy of the image block onto a small number of coefficients. Another significant advantage of DCT is that it makes available a number of fast implementations. A block diagram of the DCT-based coder is shown in Figure E.1.

Figure E.1
Block diagram of a sequential DCT codec: (a) encoder, (b) decoder. (From [2]. Used with permission.)

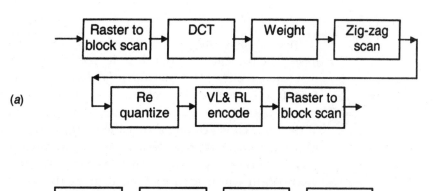

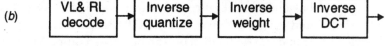

Planar Transform

The similarities between neighboring pixels in a video image are not only line or column oriented, but also area oriented [2]. Said another way, pixels in a video image typically have similar pixels all around them. To make use of these *neighborhood relationships*, it is desirable to transform not only in lines and columns, but also in areas. This can be achieved by a *planar transform*. In practice, *separable transforms* are used almost exclusively. A separable planar transform is nothing more than the repeated application of a simple transform. It is almost always applied to square picture segments of size $N \times N$, and it progresses in two steps, as illustrated in Figure E.2. First, all lines of the picture segments are transformed in succession, then all the rows of the segments calculated in the first step are transformed.

Figure E.2
A simplified search of a best-matched block. (From [2]. Used with permission.)

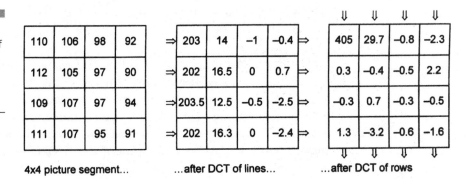

4x4 picture segment...   ...after DCT of lines...   ...after DCT of rows

In textbooks, the planar transform is frequently called a *2D transform*. The transform is, in principle, possible for any segment forms, not just square ones [4]. Consequently, for a segment of size $N \times N$, $2N$ transforms are used. The coefficients now are no longer arranged as vectors (i.e., a list of values), but as a matrix (i.e., a table). The coefficients of the i lines and j columns are called $c_{ij}(i,j = 1...N)$. Each of these coefficients no longer represents a basic vector, but a *basic picture*. In this way, each $N \times N$ picture segment is composed of $N \times N$ different basic pictures, in which each coefficient gives the weighting of a particular basic picture. Figure E.3 shows the basic pictures of the coefficients c_{11} and c_{23} for a planar 4×4 DCT. Because c_{11} represents the dc part, it is called the *dc coefficient*; the others are appropriately called the *ac coefficients*.

The planar transform of television pictures in the interlaced format is somewhat problematic. In moving regions of the picture, depending on the speed of motion, the similarities of vertically neighboring pixels of a frame are lost because changes have occurred between samplings of the two halves of the picture. Consequently, interlaced scanning may cause

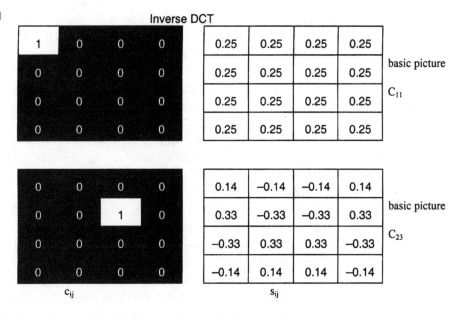

Figure E.3
The mechanics of motion-compensated prediction. Shown are the pictures for a planar 4×4 DCT. Element C_{11} is located at row 1, column 1; element C_{23} is located at row 2, column 3. Note that picture C_{11} values are constant, referred to as dc coefficients. The changing values shown in picture C_{23} are known as ac coefficients. (From [2]. Used with permission.)

the performance of the compression system (or *output concentration*) to be greatly weakened, compared with progressive scanning. It is for this reason that most commercial streaming media encoders include a de-interlacing option, prior to compression. MPEG-2 and MPEG-4 also require deinterlaced images. Well-tuned compression algorithms try to detect stronger movements and switch to a transform in picture one-half (i.e., field) for those picture regions [5]. However, the coding in one-half of the picture is less efficient because the correlation of vertically neighboring pixels is weaker than in the full picture of a static scene. Simply stated, if the picture sequences are interlaced, the picture quality may be influenced by the motion content of the scene to be coded.

Interframe Transform Coding

With common algorithms, compression ratios of approximately 8:1 can be achieved while maintaining good picture quality [2]. To achieve higher compression ratios, the similarities between successive frames must be exploited. Not only does every pixel in a frame of video have similar pixels all around it, it also has similar pixels in the same relative screen positions as itself, in previous and subsequent video frames. The nearest approach to this goal is the extension of the DCT in the time dimension. A drawback of such *cubic* transforms is the increase in calculation effort, but the greatest disadvantage is the higher memory requirements: for an $8 \times 8 \times 8$ DCT, at least seven frame memories would be needed. Much simpler is the *hybrid DCT*, which also efficiently codes pictures with moving objects. This method comprises, almost exclusively, a motion-compensated *difference pulse-code-modulation* (DPCM) technique; instead of each picture being transferred individually, the motion-compensated difference of two successive frames is coded.

DPCM is, in essence, the predictive coding of sample differences. DPCM can be applied for both *interframe coding*, which exploits the temporal redundancy of the input image, and *intraframe coding*, which exploits the spatial redundancy of the image. In the intraframe mode, the difference is calculated using the values of two neighboring pixels of the same frame. In the interframe mode, the difference is calculated using the value of the same pixel on two consecutive frames. In either mode of operation, the value of the target pixel is predicted using the reconstructed values of the previously coded neighboring pixels. This value is then subtracted from the original value to form the differential image value. The differential image is then quantized and encoded. Figure E.4 illustrates an end-to-end DPCM system.

The JPEG Standard

Before compression schemes for video were being standardized, the Joint Picture Experts Group (JPEG) was formed to standardize the compression of static images. Video compression standards, such as MPEG-1, MPEG-2, and MPEG-4, can trace their origins to this standards effort. The JPEG standard is enjoying widespread commercial use today in a bewildering variety of applications. Because JPEG is the product of a committee, it isn't surprising that it includes more than one fixed encoding/decoding scheme. It can be thought of as a family, or as related compression techniques, from which designers can choose, based upon suitability for their particular application. The four primary JPEG family members are [2]:

- Sequential DCT-based
- Progressive DCT-based

Figure E.4
Overall block
diagram of a DPCM
system: (a) encoder,
(b) decoder. (From
[2]. Used with
permission.)

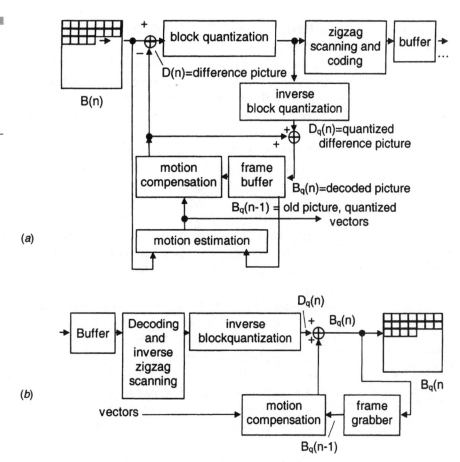

- Sequential lossless
- Hierarchical

As JPEG has been adapted to other environments, additional JPEG schemes have come into practice. JPEG is designed for still images and offers reduction ratios of 10:1 to 50:1. The algorithm is symmetrical, meaning that the time required for encoding and decoding is essentially the same. There is no need for motion compensation and there are no provisions for audio in the basic standard.

The JPEG specification, like MPEG-1 and MPEG-2, is often described as a "tool kit" of compression techniques. Before looking at specifics, it will be useful to examine some of the basics. Remember that many of these underpin current streaming media compression techniques.

Compression Techniques

As discussed briefly in previous sections, a compression system reduces the volume of data by exploiting spatial and temporal redundancies and by eliminating the data that cannot be displayed suitably by the associated display or imaging device. The main objective of compression is to retain as little data as possible, but just sufficient to reproduce the original images without causing unacceptable distortion of images [1]. To paraphrase Albert Einstein, the aim is to make things as simple as possible, but no simpler. A compression system consists of the following components:

- **Digitization, sampling, and segmentation**—Steps that convert analog signals on a specified grid of picture elements into digital representations and then divide the video input—first into frames, then into blocks.
- **Redundancy reduction**—The decorrelation of data into fewer useful data bits using certain invertible transformation techniques.
- **Entropy reduction**—The representation of digital data using fewer bits by dropping less significant information. This component causes distortion; it is the main contributor to *lossy* compression.
- **Entropy coding**—The assignment of code words (bit strings) of shorter length to more likely image symbols. This minimizes the average number of bits needed to code an image.

Key terms important to understanding video compression include the following:

- **Motion compensation**—The coding of video segments with consideration to their displacements in successive frames (in other words, coding segments of the picture according to how they move over a series of frames).
- **Spatial correlation**—The correlation of elements within a still image or a video frame for the purpose of bit rate reduction (in other words, looking for segments of the picture that are the same or similar, so that you don't need to fully describe each and every similar segment, you can just instruct the decoder to duplicate a single fully described segment).
- **Spectral correlation**—The correlation of different color components of image elements for the purpose of bit rate reduction (you guessed it—looking for segments of the picture than have the same or similar

color, so that you don't need to fully describe the color for each instance).

■ **Temporal correlation**—The correlation between successive frames of a video file for the purpose of bit rate reduction (or to paraphrase, looking for parts of successive video frames that are the same or similar, so that you don't need to fully describe those features in every frame—you can just instruct the decoder to obtain the data from a single frame to decode successive frames).

■ **Quantization compression**—The dropping of the less significant bits of image values to achieve higher compression (this is tantamount to picking colors that are near enough to the actual colors, but which can be described with fewer bits).

■ **Intraframe coding**—The encoding of a video frame by exploiting spatial redundancy within the frame. Once you have found spatial correlations, you code so that you describe duplicated picture regions only once.

▸ **Interframe coding**—The encoding of a frame by predicting its elements from elements of the previous frame. The idea is to describe a visual element just once and then pass information about how it moves (translates) in two dimensions.

The removal of spatial and temporal redundancies that exist in natural video imagery is essentially a lossless process. Given the correct techniques, an exact replica of the image can be reproduced at the viewing end of the system. Such lossless techniques are important for medical imaging applications and other demanding uses. These methods, however, may realize only low compression efficiency (on the order of approximately 2:1). For video, a much higher compression ratio is required. Exploiting the inherent limitations of the *human visual system* (HVS) can result in compression ratios of 50:1 or higher [6]. These limitations include the following:

■ Limited luminance response and very limited color response. There are more colors than we can perceive and there is a greater range of light than we can see. If we could see a greater range of luminance, we would be better able to discriminate objects in the dark and to differentiate objects in very bright glare, for example.

■ Reduced sensitivity to noise in high frequencies, such as at the edges of objects.

■ Reduced sensitivity to noise in brighter areas of the image

The goal of compression, then, is to discard all information in the image that is not absolutely necessary from the standpoint of what the HVS is capable of resolving. Such a system can be described as *psychovisually lossless*.

DCT and JPEG

DCT is one of the building blocks of the JPEG standard. All JPEG DCT-based coders start by portioning the input image into nonoverlapping blocks of 8 × 8 picture elements. The 8-bit samples are then level-shifted so that the values range from -128 to $+127$. A fast Fourier transform is then applied to shift the elements into the frequency domain. Huffman coding is mandatory in a baseline system; other arithmetic techniques can be used for entropy coding in other JPEG modes. The JPEG specification is independent of color space or gray scale. A color image typically is encoded and decoded in the *YUV* color space with four pixels of *Y* for each *U, V* pair.

In the *sequential DCT*-based mode, processing components are transmitted or stored as they are calculated in one single pass. Figure E.5 provides a simplified block diagram of the coding system.

Figure E.5
Block diagram of a DCT-based image-compression system. Note how the 8 × 8 source image is processed through a forward-DCT (FDCT) encoder and related systems to the inverse-DCT (IDCT) decoder and reconstructed into an 8 × 8 image. (From [1]. Used with permission.)

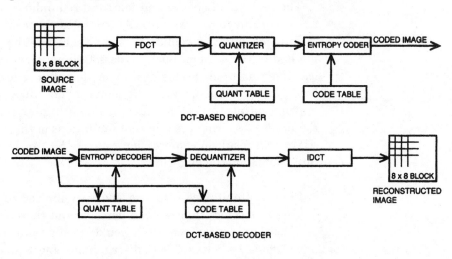

The *progressive DCT*-based mode can be convenient when it takes a perceptibly long time to send and decode the image. With progressive DCT-based coding, the picture first will appear blocky and the details will subsequently appear. A viewer may linger on an interesting picture and watch the details come into view or move onto something else, making this scheme well suited, for example, to the Internet.

In the *lossless* mode, the decoder reproduces an exact copy of the digitized input image. The compression ratio, naturally, varies with picture content. The varying compression ratio is not a problem for sending still photos, but presents significant challenges for sequential images that must be viewed in real time.

The efficiency of JPEG coding for still images led to the development of *motion* JPEG (M-JPEG) for video applications, primarily studio use (early nonlinear editing systems, such as Avid and Lightworks, were based on M-JPEG). Motion JPEG uses intraframe compression, where each frame is treated as an individual signal; a series of frames is basically a continous stream of JPEG-compressed images. The benefit of this construction is easy editing, since video sequences can be "cut" at frame boundaries, without any further computation or compression of the images. Also, any individual frame is self-supporting and can be accessed as a stand-alone image. The intraframe system is based, again, on DCT. Because a picture with high-frequency detail will generate more data than a picture with low detail, the data stream will vary, in terms of bits per second. This is problematic for most real-time systems, which would prefer to see a constant data rate at the expense of varying levels of quality. In practice, many systems that played back motion JPEG video sequences, such as nonlinear editing machines, employed sophisticated multilayer buffering schemes and limited the amount of data allocated for the encoding of each image, so that while the data rate varied according to image detail, the variation was not completely unbounded. This allowed these systems to maintain crisp response to play, stop, fast-forward, and rewind commands, but also limited the data rate (and hence quality) of highly detailed images, while compressing images with low detail with high quality. Because nonlinear editing systems need to record as well as play, the symmetry of the complexity of encoders and decoders was also an important advantage of motion JPEG.

The major disadvantage of motion JPEG is bandwidth and storage requirements, which are a direct consequence of the limited range of compression ratios. Because stand-alone frames are coded, there is no opportunity to code only the differences between frames (to remove redundancies). Large compression ratios, which exploit this interframe correlation are not available to motion JPEG systems.

M-JPEG, in its basic form, addresses only the video—not the audio—component. Many of the early problems experienced by users concerning portability of M-JPEG streams stemmed from the methods used to include audio in the data stream. Because the location of the audio may vary from one unit to the next, some decoder problems were experienced [7]. In actual fact, there was virtually no interoperability between non-linear editing systems from different manufacturers at a video and audio data level, because each manufacturer had to independently solve the problem of synchronizing audio to the video and they all did it in different ways. The best that could be done is that an edit decision list could be exported from one system and imported into another (theoretically) and the audio and video redigitized into the destination machine from the original source video tapes, whether or not M-JPEG data files representing the video were in existence on the first system.

The MPEG Standard

The Moving Picture Experts Group (MPEG) was founded in 1988 with the objective of specifying an audio/visual decompression system, composed of three basic elements, which the sponsoring organization (the *International Standards Organization*, or ISO) calls "parts." They are as follows:

- **Part 1—Systems**—Describes the audio/video synchronization, multiplexing, and other system-related elements
- **Part 2—Video**—Contains the coded representation of video data and the decoding process
- **Part 3—Audio**—Contains the coded representation of audio data and the decoding process

The basic MPEG system, finalized in 1992, was designated MPEG-1. Shortly thereafter, work began on MPEG-2. The first three stages (systems, video, and audio) of the MPEG-2 standard were agreed to in November 1992. Table E-1 lists the companies and organizations participating in the early MPEG work. Because of their combined efforts, the MPEG standards have achieved broad market acceptance.

As might be expected, the techniques of MPEG-1 and MPEG-2 are similar and their syntax is extensible.

TABLE E.1

Participants in Early
MPEG Proceedings
(After [2])

Computer Manufacturers	IC Manufacturers
Apple	Brooktree
DEC	C-Cube
Hewlett-Packard	Cypress
IBM	Inmos
NEC	Intel
Olivetti	IIT
Sun	LSI Logic
	Motorola
	National Semiconductor
Software Suppliers	Rockwell
	SGS-Thomson
Microsoft	Texas Instruments
Fluent Machines	Zoran
Prism	
Audio/Visual Equipment Manufacturers	**Universities/Research**
	Columbia University
Dolby	Massachusetts Institute of Technology
JVC	DLR
Matsushita	University of Berlin
Philips	Fraunhofer Gesellschaft
Sony	University of Hanover
Thomson Consumer Electronics	

Basic Provisions

When trying to settle on a specification, it is always important to have a target application in mind [1]. The definition of MPEG-1 (also known as ISO/IEC 11172) was driven by the desire to encode audio and video onto a compact disk. A CD is defined to have a constant bit rate of 1.5 Megabits per second. With this constrained bandwidth, the target video specifications were:

- Horizontal resolution of 360 pixels
- Vertical resolution of 240 pixels for NTSC, and 288 for PAL and SECAM
- Frame rate of 30 Hz for NTSC, 25 for PAL and SECAM, and 24 for film

A detailed block diagram of an MPEG-1 codec (coder-decoder) is shown in Figure E.6.

MPEG uses the JPEG standard for intraframe coding by first dividing each frame of the image into 8 × 8 blocks, then compressing each

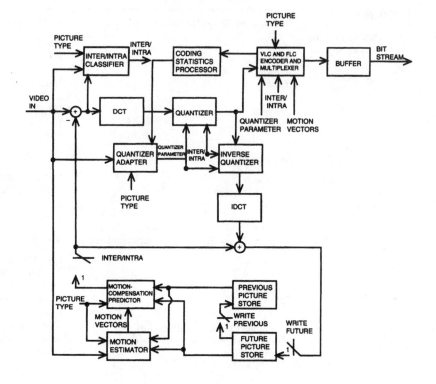

(a)

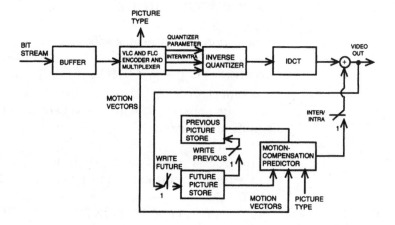

(b)

block independently using DCT-based techniques. Interframe coding is based on *motion compensation* (MC) prediction that allows bi-directional temporal prediction. What this means in practice is that if a visual element is partially obscured in the current video frame, but it appears unobstructed in a previous or future frame, the data from either frame can be used to fill in the detail in the current frame. The direction of time in the video sequence is not important when searching for redundancies. A block-matching algorithm is used to find the best-matched block, which may belong to either the past frame (*forward prediction*) or the future frame (*backward prediction*). The best-matched block may, in fact, be the average of two blocks, one from the previous and the other from the next frame of the target frame (*interpolation*). In any case, the placement of the best-matched block(s) is used to determine the motion vector(s); blocks predicted on the basis of interpolation have two motion vectors. Frames that are bi-directionally predicted are never used themselves as reference frames.

Motion Compensation

At this point, it is appropriate to take a closer look at MC prediction [1]. For motion-compensated interframe coding, the target frame is divided into nonoverlapping fixed-size blocks and each block is compared with blocks of the same size in some reference frame to find the best match. To limit the search, a small *neighborhood* is selected in the reference frame and the search is performed by *stepwise translation* of the target block.

To reduce mathematical complexity, a simple block-matching criterion, such as the mean of the absolute difference of pixel values, is used to find a best-matched block. The position of the best-matched block determines the displacement of the target block and its location is denoted by a (motion) vector—a pair of relative X,Y coordinates representing the translation of the block's position (note: no scaling, skewing, or rotation information is used).

Block matching is computationally expensive; therefore, a number of variations on the basic theme have been developed. A simple method, known as *OTS* (one-at-a-time search), is shown in Figure E.7. First, the target block is moved along in one direction and the best match found, then it is moved along perpendicularly to find the best match in that direction. Figure E.7 portrays the target frame in terms of the best-matched blocks in the reference frame.

Figure E.7
A simplified search of a best-matched block. (From [1]. Used with permission.)

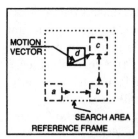

a: INITIAL POSITION OF TARGET BLOCK DURING THE SEARCH

b: BEST MATCH ALONG *x*-AXIS

c: BEST MATCH ALONG *y*-AXIS GIVEN *b*

d: LOCATION OF TARGET BLOCK IN THE TARGET FRAME

OTS SEARCH FOR THE BEST MATCHED BLOCK

Putting It All Together

MPEG is a standard built upon many elements [1]. Figure E.8 shows a *group of pictures* (GOP) of 14 frames with two different orderings. Pictures marked *I* are intraframe-coded. A *P*-picture is predicted using the most recently encoded *P*- or *I*-picture in the sequence. A macroblock in a *P*-picture can be coded using either intraframe or the forward-predicted method. A *B*-picture macroblock can be predicted using either or both of the previous or the next *I*- and/or *P*-pictures. To meet this requirement, the transmission order and display order of frames are different. The two orders are also shown in Figure E.8.

Figure E.8
Illustration of I-frames, P-frames, and B-frames. (From [1]. Used with permission.)

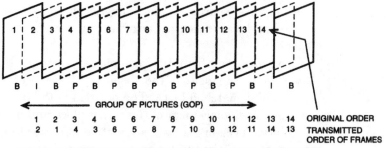

TRANSMISSION ORDER AND DISPLAY ORDER OF I- P- AND B- FRAME IN MPEG

The MPEG-coded bit stream is divided into several layers, listed in Table E-2. The three primary layers are:

■ **Video sequence**—The outermost layer, which contains basic global information such as the size of frames, bit rate, and frame rate.

TABLE E.2

Layers of the
MPEG-2 Video
Bitstream Syntax
(After [2])

Syntax layer	Functionality
Video sequence layer	Context unit
Group of pictures (GOP) layer	Random access unit: video coding
Picture layer	Primary coding unit
Slice layer	Resynchronization unit
Macroblock layer	Motion-compensation unit
Block layer	DCT unit

- **GOP layer**—Contains information on fast search and random access of the video data. The length of a GOP is arbitrary.
- **Picture layer**—Contains a coded frame. Its header defines the type (I,P,B) and the position of the frame in the GOP.

Several of the major differences between MPEG and other compression schemes (such as JPEG) include the following:

- MPEG focuses on video. The basic format uses a single color space (Y, C_r, C_b), a limited range of resolutions and compression ratios, and has built-in mechanisms for handling audio.
- MPEG takes advantage of the high degree of commonality between pictures in a video stream and the typically predictable nature of movement (*inter-picture encoding*).
- MPEG provides for a constant bit rate through adjustable variables, making the format predictable with regard to bandwidth requirements.

MPEG specifies the syntax for storing and transmitting compressed data and defines the decoding process. The standard does not, however, specify how encoding should be performed. Such implementation considerations are left to the manufacturers of encoding systems. Still, all conforming encoders must produce valid MPEG bit streams that can be decompressed by any MPEG decoder. This approach is, in fact, one of the strengths of the MPEG standard; because encoders are allowed to use proprietary but compliant algorithms, a variety of implementations is possible and, indeed, encouraged.

As mentioned previously, MPEG is actually a collection of standards, each suited to a particular application or group of applications, including:

- MPEG-1, the original implementation, targeted at multimedia uses. The MPEG-1 algorithm is intended basically for compact disk bit rates of approximately 1.5–2.0 Megabits per second. MPEG-1 supports 525- and 625-type signal structures in progressive form (i.e., non-interlaced) with 204/288 lines per frame, sequential-scan frame rates of 29.97 and 25 frames per second and 352 pixels per line. The coding of high-motion signals does not produce particularly good results, however. As might be expected, as the bit rate is reduced (compression ratio increased), the output video quality gradually declines. The overall bit rate reduction ratios achievable are about 6:1 with a bit rate of 6 Megabits per second and 200:1 at 1.5 Megabits per second. The MPEG-1 system is not symmetrical; the compression side is more complex and expensive than the decompression process, making the system ideal for broadcast-type applications in which there are far more decoders than encoders.
- MPEG-2, which offers full ITU-R Rec. 601 resolution for professional broadcast uses and is the chosen standard for the American ATSC DTV system and the European DVB suite of applications.
- MPEG-3, originally targeted at high-definition imaging applications. Subsequent to development of the standard, however, key specifications of MPEG-3 were absorbed into MPEG-2. MPEG-3 was stillborn and is no longer in use.
- MPEG-4, a standard that uses very low bit rates for teleconferencing, streaming, and related applications requiring high bit efficiency and user interactivity. Like MPEG-2, MPEG-4 is a collection of tools that can be grouped into profiles and levels for different applications. The MPEG-4 video coding structure ranges from a *very low bit rate video* (VLBV) level, which includes algorithms and tools for data rates between 5 kilobits per second and 64 kilobits per second, to ITU-R. Rec. 601 quality video at 2 Megabits per second and beyond to 2 Gigabits per second streams with 4k × 2k image size. MPEG-4 does not concern itself directly with the error protection required for specific channels, such as cellular radio, but it has made improvements in the way payload bits are arranged so that recovery is more robust (error resilience).
- MPEG-7, is not a compression scheme at all. It is a metadata description scheme for media assets. MPEG-7 is an attempt to provide a standard means of describing multimedia content.

- MPEG-21, is not a compression scheme either. It is an attempt to standardize media transaction information, such as when the owner of a piece of streaming media content sells a viewer the right to view it, for example.

Profiles and Levels

For historical completeness, MPEG-2's *profiles* and *levels* are discussed below. The MPEG-4 profiles, including those most relevant to streaming media, are described in Appendix A. MPEG-2's profiles are found in television production plants and much of this content is repurposed for streaming media application, so it is important to understand these upstream compression profiles. Streaming media encoders that simply transcode between MPEG-2 and MPEG-4 without decompressing are still rare. In cases where MPEG-2 is used in production, the video is typically decompressed and rendered to baseband prior to re-encoding into MPEG-4. This is wasteful, since much of the required math to compress the video into MPEG-4 format has already been done to get the video into MPEG-2 format. Unfortunately, this data cannot yet be directly reused by most commercially available streaming media encoders.

Six profiles and four levels describe the organization of the basic MPEG-2 standard. A *profile* is a subset of the MPEG-2 bit stream syntax with restrictions on the parts of the MPEG algorithm used. Profiles are analogous to features, describing the available characteristics. A *level* constrains general parameters such as image size, data rate, and decoder buffer size. Levels describe, in essence, the upper bounds for a given feature and are analogous to performance specifications.

By far the most popular element of the MPEG-2 standard for professional video applications is the *Main Profile* in conjunction with the *Main Level* (described in the jargon of MPEG as Main Profile/Main Level), which gives an image size of 720 × 576, a data rate of 15 megabits per second, and a frame rate of 30 frames per second. All higher profiles are capable of decoding Main Profile/Main Level streams.

Table E.3 lists the basic MPEG-2 classifications. With regard to the table, the following generalizations can be made:

- The three key flavors of MPEG-2 are:
  - Main Profile/Low Level (source input format, or SIF)
  - Main Profile/Main Level (Main)
  - Studio Profile/Main Level (Studio)

TABLE E.3 Common MPE Profiles and Levels in Simplified Form (After [2])

Profile	General Specifications	Parameter	Level			
			Low	Main (ITU 601)	High 1440 (HD, 4:3)	High (HD, 16:9)
Simple	Pictures: *I, P* Chroma: 4:2:0	Image size[1] Image frequency[2] Bit rate[3]		720 × 576 30 15		
Main	Pictures: *I, P, B* Chroma: 4:2:0	Image size Image frequency Bit rate	325 × 288 30 4	720 × 576 30 15	1440 × 1152 60 100	1920 × 1152 60 80
SNR-scalable	Pictures: *I, P, B* Chroma: 4:2:0	Image size Image frequency Bit rate	325 × 288 30 4, 4[4]	720 × 576 30 15		
Spatially-scalable	Pictures: *I, P, B* Chroma: 4:2:0	Image size Image frequency Bit rate			720 × 576 30 15	
	Enhancement Layer[5]	Image size Image frequency Bit rate			1140 × 1152 60 60, 60[6]	
High[7]	Pictures: *I, P, B* Chroma: 4:2:2	Image size Image frequency Bit rate		720 × 576 30 20	1440 × 1152 60 80	1920 × 1152 60 100
Studio	Pictures: I, P, B Chroma: 4:2:2	Image size Image frequency Bit rate		720 × 608 30 50		

Notes:
[1] Image size specified as samples/line × lines/frame
[2] Image frequency in frames/s
[3] Bit rate in Mbits/s
[4] For *Enhancement Layer 1*
[5] For *Enhancement Layer 1*, except as notes by [6] for *Enhancement Layer 2*
[7] For simplicity, *Enhancement Layers* not specified individually

- The SIF Main Profile/Low Level offers the best picture quality for bit rates below about 5 megabits per second. This provides generally acceptable quality for interactive and multimedia applications. The SIF profile has replaced MPEG-1 in some applications.
- The Main Profile/Main Level grade offers the best picture quality for conventional video system at rates from about 5 to 15 per second.

This provides good quality for broadcast applications, such as play-to-air, where four generations or fewer are required.

- The Studio Profile offers high quality for multiple-generation conventional video applications, such as post-production.
- The High Profile targets HDTV applications.

Studio Profile

Despite the many attributes of MPEG-2, the Main Profile/Main Level remains a less-than-ideal choice for conventional video production because the larger GOP structure makes individual frames hard to access. For this reason the 4:2:2 *Studio Profile* was developed. The Studio Profile expands upon the 4:2:0 sampling scheme of MPEG-1 and MPEG-2. In essence, "standard MPEG" samples the full luminance signal, but ignores half of the chrominance information, specifically the color coordinate on one axis of the color grid. Studio Profile MPEG increases the chrominance sampling to 4:2:2, thereby accounting for both axes on the color grid by sampling every other element. This enhancement provides better replication of the original signal.

The Studio Profile is intended principally for editing applications, where multiple iterations of a given video signal are required or where the signal will be compressed, decompressed, and recompressed several times before it is finally transmitted or otherwise finally displayed.

SMPTE 308M.   SMPTE standard 308M is intended for use in high-definition television production, contribution, and distribution applications [9]. It defines bit streams, including their syntax and semantics, together with the requirements for a complaint decoder for 4:2:2 Studio Profile at High Level. As with the other MPEG standards, 308M does not specify any particular encoder operating parameters.

The MPEG-2 4:2:2 Studio Profile is defined in ISO/IEC13818-2; in SMPTE 308M, only those additional parameters necessary to define the 4:2:2 Studio Profile at High Level are specified. The primary differences are:

1. The upper bounds for sampling density are increased to 1920 samples per line, 1088 lines per frame, and 60 frames per second.
2. The upper bounds for the luminance sample rate is set at 62,668,800 samples per second.
3. The upper bound for bit rates is set at 300 megabits per second.

How MPEG-2 Was Tuned for High Quality of Experience

As an industry, digital television professionals know a lot about quality of experience (QoE) and about losses in real world transmission systems. They have collectively made MPEG-2 encoding choices that provide high QoE, even when the transmission channel is degraded. Their input has been invaluable in making the compression system more robust. Streaming media encoder vendors ought to heed these improvements and incorporate them into their MPEG-4 compression systems, where applicable. This section describes some of the important features of MPEG-2 relied upon by the DTV industry to deliver good pictures to end-users.

The primary application of interest when the MPEG-2 standard was first defined was "true" television broadcast resolution, as specified by ITU-R Rec. 601. This is roughly four times more picture information than the MPEG-1 standard provides. MPEG-2 is a superset, or extension, of MPEG-1. As such, an MPEG-2 decoder also should be able to decode an MPEG-1 stream. This broadcast version adds to the MPEG-1 toolbox provisions for dealing with interlaced video, graceful degradation, and hierarchical coding.

Although MPEG-1 and MPEG-2 each were specified with a particular range of applications and resolutions in mind, the committee's specifications form a set of techniques that support multiple coding options, including picture types and macroblock types. Many variations exist with regard to picture size and bit rates. Also, although MPEG-1 can run at high bit rates and at full ITU-R Rec. 601 resolution, it processes frames, not fields. This fact limits the attainable quality, even at data rates approaching 5 megabits per second.

The MPEG specifications apply only to decoding, not encoding. The ramifications of this approach are:

- Owners of existing decoding software can benefit from future breakthroughs in encoding processing. Furthermore, the suppliers of encoding equipment can differentiate their products by cost, features, encoding quality, and other factors.
- Different schemes can be used in different situations. For example, although *Monday Night Football* must be encoded in real time, a feature film can be encoded in non-real time, allowing for fine-tuning of the parameters via computer or even a human operator.

MPEG-2 Layer Structure

To allow for a simple, yet upgradeable system, MPEG-2 defines only the functional elements—syntax and semantics—of coded streams. Using the same system of *I-*, *P-*, and *B*-frames developed for MPEG-1, MPEG-2 employs a six-layer hierarchical structure that breaks the data into simplified units of information, as listed in Table E.2.

The top *sequence layer* defines the decoder constraints by specifying the context of the video sequence. The sequence-layer data header contains information on picture format and application-specific details. The second level allows for random access to the decoding process by having a periodic series of pictures; it is fundamentally this GOP layer that provides the bi-directional frame prediction. Intraframe-coded (*I*) frames are the entry-point frames, which require no data from other frames in order to reconstruct the image. Between the *I*-frames lie the predictive (*P*) frames, which are derived from analyzing previous frames and performing motion estimation. These *P*-frames require about one third as many bits per frame as *I*-frames. *B*-frames, which lie between two *I*-frames or *P*-frames, are bi-directionally encoded, making use of past and future frames. The *B*-frames require only about one-ninth of the data per frame, compared with *I*-frames.

These different compression ratios for the different frame types leads to different data rates, so that buffers are required at both the encoder output and the decoder input to ensure that the sustained data rate is constant. One difference between MPEG-1 and MPEG-2 is that MPEG-2 allows for a variety of data-buffer sizes, to accommodate different picture dimensions and to prevent buffer under- and overflows.

The data required to decode a single picture is embedded in the *picture layer*, which consists of a number of horizontal *slice layers*, each containing several macroblocks. Each *macroblock layer*, in turn, is made up of a number of individual blocks. The picture undergoes DCT processing, with the slice layer providing a means of synchronization, holding the precise position of the slice within the image frame.

MPEG-2 places the motion vectors into the coded macroblocks for *P*-frames and *B*-frames; these are used to improve the reconstruction of predicted pictures. MPEG-2 supports both field- and frame-based prediction, thus accommodating interlaced signals.

The last layer of MPEG-2's video structure is the *block layer*, which provides the DCT coefficients of either the transformed image information for *I*-frames or the residual prediction error of *B*- and *P*-frames.

Slices

Two or more contiguous macroblocks within the same row are grouped together to form *slices* [10]. The order of the macroblocks within a slice is the same as the conventional television raster scan, being from left to right.

Slices provide a convenient mechanism for limiting the propagation of errors. Because the coded bit stream consists mostly of variable-length code words, any uncorrected transmission errors will cause a decoder to lose its sense of code word alignment. Each slice begins with a slice start code. Because the MPEG code word assignment guarantees that no legal combination of code words can emulate a start code, the slice start code can be used to regain the sense of code-word alignment after an error. Therefore, when an error occurs in the data stream, the decoder can skip to the start of the next slice and resume correct decoding.

The number of slices affects the compression efficiency; partitioning the data stream to have more slices provides for better error recovery, but claims bits that could otherwise be used to improve picture quality.

In the DTV system, the initial macroblock of every horizontal row of macroblocks is also the beginning of a slice, with a possibility of several slices across the row.

Pictures, Groups of Pictures, and Sequences

The primary coding unit of a video sequence is the individual video frame or picture [10]. A video picture consists of the collection of slices, constituting the *active picture area*.

A *video sequence* consists of a collection of two or more consecutive pictures. A video sequence commences with a sequence header and is terminated by an end-of-sequence code in the data stream. A video sequence may contain additional sequence headers. Any video-sequence header can serve as an *entry-point*. An entry point is a point in the coded video bit stream after which a decoder can become properly initialized and correctly parse the bit stream syntax.

Two or more pictures (frames) in sequence may be combined into a GOP to provide boundaries for interframe picture coding and the registration of time code. GOPs are optional within both MPEG-2 and the ATSC DTV system. Figure E.9 illustrates a typical time sequence of video frames.

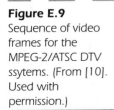

Figure E.9
Sequence of video frames for the MPEG-2/ATSC DTV ssytems. (From [10]. Used with permission.)

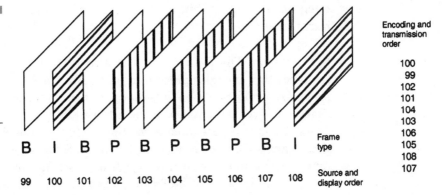

B	I	B	P	B	P	B	P	B	I	Frame type
99	100	101	102	103	104	105	106	107	108	Source and display order

Encoding and transmission order

100
99
102
101
104
103
106
105
108
107

I-Frames. Some elements of the compression process exploit only the spatial redundancy within a single picture (frame or field) [10]. These processes constitute intraframe coding and do not take advantage of the temporal correlation addressed by temporal prediction (interframe) coding. Frames that do not use any interframe coding are referred to as *I*-frames (where "I" denotes *intraframe-coded*). The ATSC video compression system utilizes both intraframe and interframe coding.

The use of periodic *I*-frames facilitates receiver initializations and channel acquisition (for example, when the receiver is turned on or the channel is changed). The decoder also can take advantage of the intraframe-coding mode when uncorrectable channel errors occur. Periodic *I*-frames are also important in streaming media applications, to enable rapid seeking into a media stream, using fast-forward or jump commands, since an *I*-frame entry-point is required to restart the video playback. With motion-compensated prediction, an initial frame must be available at the decoder to start the prediction loop. Therefore, a mechanism must be built into the system so that if the decoder loses synchronization for any reason, it can rapidly reacquire tracking.

The frequency of occurrence of *I*-pictures may vary and is selected at the encoder. This allows consideration to be given to the need for random access and the location of scene cuts in the video sequence. Popular streaming media encoders do not currently allow for the manual placement of I-frames at scene cuts or at other significant points.

P-Frames. *P*-frames, where the temporal prediction is in the forward direction only, allow the exploitation of interframe coding techniques to improve the overall compression efficiency and picture quality [10]. *P*-frames may include portions that are only intraframe-coded.

Each macroblock within a *P*-frame can be either forward predicted or intraframe-coded.

B-Frames. The *B*-frame is a picture type within the coded video sequence that includes prediction from a future frame as well as from a previous frame [10]. The referenced future or previous frames, sometimes called *anchor frames*, are in all cases either *I*- or *P*-frames.

The basis of the *B*-frame prediction is that a video frame is correlated with frames that occur in the past as well as those that occur in the future. Consequently, if a future frame is available to the decoder, a superior prediction can be formed, thus saving bits and improving performance. Some of the consequences of using future frames in the prediction are:

- The *B*-frame cannot be used for predicting future frames.
- The transmission order of frames is different from the displayed order of the frames.
- The encoder and decoder must reorder the video frames, thereby increasing the total latency.

In the example illustrated in Figure E.9, there is one *B*-frame between each pair of *I*- and *P*-frames. Each frame is labeled with both its display order and transmission order. The *I*- and *P*-frames are transmitted out of sequence, so the video decoder has both anchor frames decoded and available for prediction.

B-frames are used for increasing the compression efficiency and perceived picture quality when encoding latency is not an important factor. The use of *B*-frames increases coding efficiency for both interlaced and progressive-scanned material. *B*-frames are included in the DTV system because the increase in compression efficiency is significant, especially for progressive scanning. The same factors mandate the use of *B*-frames in streaming media encoding. The choice of the number of bi-directional pictures between any pair of reference (*I* or *P*) frames can be determined at the encoder.

Motion Estimation. The efficiency of the compression algorithm depends on two things: the creation of an estimate of the image being compressed and subtraction of pixel values of the estimate (or prediction) from the image to be compressed [10]. If the estimate is good, the subtraction will leave a very small residue to be transmitted. In fact, if the predicted estimate were perfect, the difference would be zero for all the pixels in the frame of differences and no new information would need to be sent to the decoder. This condition can be approached for video sequences of essentially still images.

If the estimate is not close to zero for some pixels or many pixels, those differences represent information that needs to be transmitted so that the decoder can reconstruct the correct image. The kinds of image sequences that cause large prediction differences include severe motion and/or sharp details.

Vector Search Algorithm

The video-coding system uses motion-compensated prediction as part of the data-compression process [10]. Thus, macroblock-sized regions in the previously transmitted frames predict macroblocks in the current frame of interest. Motion compensation refers to the fact that the locations of the macroblock-sized regions in the reference frame can be offset to account for local motions. The macroblock offsets are known as *motion vectors*.

The DTV standard does not specify how encoders should determine motion vectors. Streaming media encoders make their own decisions with regard to motion vectors as well. One possible approach is to perform an exhaustive search to identify the vertical and horizontal offsets that minimize the total difference between the offset region in the reference frame and the macroblock in the frame to be coded.

Motion Vector Precision

The estimation of interframe displacement is calculated with half-pixel precision, in both vertical and horizontal dimensions [10]. As a result, the displaced macroblock from the previous frame can be displaced by noninteger displacements and will require interpolation to compute the values of displaced picture elements at locations not in the original array of samples. Estimates for half-pixel locations are computed by averages of adjacent sample values.

Motion Vector Coding

Motion vectors within a slice are differenced, so that the first value for a motion vector is transmitted directly and the following sequence of motion vector differences is sent using variable-length codes (VLC) [10]. Motion vectors are constrained so that all pixels from the motion-compensated prediction region in the reference picture fall within the picture boundaries.

Encoder Prediction Loop

The encoder prediction loop, shown in the simplified block diagram of Figure E.10, is the heart of the video compression system for DTV [10]. The prediction loop contains a prediction function that estimates picture values of the next picture to be encoded in the sequence of successive pictures that constitute the video program. This prediction is based on previous information that is available within the loop, derived from ear-

Figure E.10
Simplified encoder
prediction loop.
(From [10]. Used
with permission.)

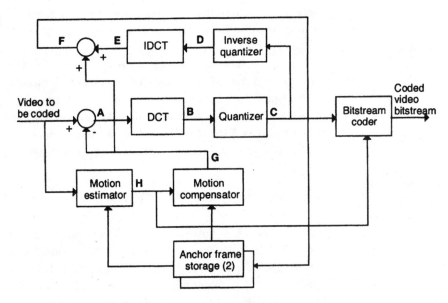

A	Pixel-by-pixel prediction errors
B	Transformed blocks of prediction errors (DCT coefficients)
C	Prediction error DCT coefficients in quantized form
D	Quantized prediction error DCT coefficients in standard form
E	Pixel-by-pixel prediction errors, degraded by quantization
F	Reconstructed pixel values, degraded by quantization
G	Motion compensated predicted pixel values
H	Motion vectors

lier pictures. The transmission of the predicted compressed information works because the same information used to make the prediction is also available at the receiving decoder (barring transmission errors, which are usually infrequent).

The subtraction of the predicted picture values from the new picture to be coded is at the core of predictive coding. The goal is to do such a good job of predicting the new values that the result of the subtraction function at the beginning of the prediction loop is zero or close to zero most of the time.

The prediction differences are computed separately for the luminance and two chrominance components before further processing. As explained in previous discussion of *I*-frames, there are times when prediction is not used, for part of a frame or for an entire frame.

Spatial Transform Block—DCT. The image prediction differences (sometimes referred to as *prediction errors*) are organized into 8 × 8 blocks and a spatial transform applied to the blocks of difference values [10]. In the intraframe case, the spatial transform is applied to the raw, undifferenced picture data. The luminance and two chrominance components are transformed separately. Because the chrominance data is subsampled vertically and horizontally, each 8 × 8 block of chrominance (C_b or C_r) data corresponds to a 16 × 16 macroblock of luminance data, which is not subsampled.

The spatial transform used is the discrete cosine transform. In principle, applying the IDCT (Inverse DCT) to the transformed array would yield exactly the same array as the original. In that sense, transforming the data does not modify the data, but merely represents it in a different form.

The decoder uses the inverse transformation to approximately reconstruct the arrays that were transformed at the encoder, as part of the process of decoding the received compressed data. The approximation in that reconstruction is controlled in advance during the encoding process for the purpose of minimizing the visual effects of coefficient inaccuracies while reducing the quantity of data that needs to be transmitted.

Quantizer. The process of transforming the original data organizes the information in a way that exposes the spatial frequency components of the images or image differences [10]. Using information about the response of the human visual system to different spatial frequencies, the encoder can selectively adjust the precision of transform coefficient representation. The goal is to include as much information about a particular spatial frequency as necessary—and as possible, given the constraints on data transmission—while not using more precision than is needed, based upon visual perception criteria.

For example, in a portion of a picture that is "busy" with a great deal of detail, imprecision in reconstructing spatial high-frequency components in a small region might be masked by the picture's local "business." On the other hand, highly precise representation and reconstruction of the average value or dc term of the DCT block would be important in a smooth area of sky. The dc $F(0,0)$ term of the transformed coefficients represents the average of the original 64 coefficients.

As stated previously, the DCT of each 8×8 block of pixel values produces an 8×8 array of DCT coefficients. The relative precision accorded to each of the 64 DCT coefficients can be selected according to its relative importance in human visual perception. A *quantizer matrix* represents the relative coefficient precision information, which is an 8×8 array of values. Each value in the quantizer matrix represents the coarseness of quantization of the related DCT coefficient.

Two types of quantizer matrix are supported:

- A matrix used for macroblocks that are intraframe-coded
- A matrix used for macroblocks that are non-intraframe-coded

The video coding system defines default values for both the intraframe-quantizer and the non-intraframe-quantizer matrices. Either or both of the quantizer matrices can be overridden at the picture level by the transmission of appropriate arrays of 64 values. Any quantizer matrix overrides stay in effect until the following sequence start code.

The transform coefficients, which represent the bulk of the actual coded video information, are quantized to various degrees of coarseness. As indicated previously, some portions of the picture will be more affected in appearance than others by the loss of precision through coefficient quantization. This phenomenon is exploited by the availability of the quantizer scale factor, which allows the overall level of quantization to vary for each macroblock. Consequently, entire macroblocks that are deemed to be visually less important can be quantized more coarsely, resulting in fewer bits being needed to represent the picture.

For each coefficient other than the dc coefficients of intraframe-coded blocks, the quantizer scale factor is multiplied by the corresponding value in the appropriate quantizer matrix to form the quantizer step size. Quantization of the dc coefficients of intraframe-coded blocks is unaffected by the quantizer scale factor and is governed only by the (0,0) element of the intraframe-quantizer matrix, which is always set to be 8 (ISO/IEC 13818-2).

Entropy Coder.   An important effect of the quantization of transform coefficients is that many coefficients will be rounded to zero after quantization [10]. In fact, a primary method of controlling the encoded data rate is the control of quantization coarseness, because a coarser quantization leads to an increase in the number of zero-value quantized coefficients.

Inverse Quantizer.   At the decoder, the coded coefficients are decoded and an 8 × 8 block of quantized coefficients is reconstructed [10]. Each of these coefficients is *inverse-quantized* according to the prevailing quantizer matrix, quantizer scale, and frame type. The result of inverse quantization is a block of 64 DCT coefficients.

Inverse Spatial Transform Block—IDCT.   The decoded and inverse-quantized coefficients are organized as 8 × 8 blocks of DCT coefficients and the inverse discrete cosine transform is applied to each block [10]. This results in a new array of pixel values, or pixel difference values that correspond to the output of the subtraction at the beginning of the prediction loop. If the prediction loop was in the interframe mode, the values will be pixel differences. If the loop was in the intraframe mode, the inverse transform will produce pixel values directly.

Motion Compensator.   If a portion of the image has not moved, then it is easy to see that a subtraction of the old portion from the new portion of the image will produce zero or nearly zero pixel differences, which is the goal of the prediction [10]. If there has been movement in the portion of the image under consideration, however, the direct pixel-by-pixel differences generally will not be zero and might be statistically very large. The motion in most natural scenes is organized, however, and can be approximately represented locally as a translation in most cases. For this reason, the video coding system allows for *motion-compensated* prediction, whereby macroblock-sized regions in the reference frame may be translated vertically and horizontally with respect to the macroblock being predicted, to compensate for local motion.

The pixel-by-pixel differences between the current macroblock and the motion-compensated prediction are transformed by the DCT and quantized using the composition of the nonintraframe-quantizer matrix and the quantizer scale factor. The quantized coefficients are then coded.

Dual Prime Prediction Mode

The dual prime prediction mode is an alternative "special" prediction mode that is built on field-based motion prediction but requires fewer transmitted motion vectors than conventional field-based prediction [10]. This mode of prediction is available only for interlaced material and only when the encoder configuration does not use *B*-frames. This mode of prediction can be particularly useful for improving encoder efficiency for low-delay applications.

The basis of dual prime prediction is that field-based predictions of both fields on a macroblock are obtained by averaging two separate predictions, which are predicted from the two nearest decoded fields in time. Each of the macroblock fields is predicted separately, although the four vectors (one pair per field) used for prediction all are derived from a single transmitted field-based motion vector. In addition to the single field-based motion vector, a small *differential vector* (limited to vertical and horizontal component values of $+1$, 0, and -1) also is transmitted for each macroblock. Together, these vectors are used to calculate pairs of motion vectors for each macroblock. The first prediction pair is simply the transmitted field-based motion vector. The second prediction vector is obtained by combining the differential vector with a scaled version of the first vector. After both predictions are obtained, a single prediction for each macroblock field is calculated by averaging each pixel in the two original predictions. The final averaged prediction then is subtracted from the macroblock field being encoded.

Adaptive Field/Frame Prediction Mode

Interlaced pictures may be coded in one of two ways: either as two separate fields or as a single frame [10]. When the picture is coded as separate fields, all of the codes for the first field are transmitted as a unit before the codes for the second field. When the picture is coded as a frame, information for both fields is coded for each macroblock.

When frame-based coding is used with interlaced pictures, each macroblock may be selectively coded using either field prediction or frame prediction. When frame prediction is used, a motion vector is applied to a picture region that is made up of both parity fields interleaved together. When field prediction is used, a motion vector is applied to a region made up of scan lines from a single field. Field prediction allows the selection of either parity field to be used as a reference for the field being predicted.

Image Refresh

As discussed previously, a given picture may be sent by describing the differences between it and one or two previously transmitted pictures [10]. For the scheme to work, there must be some way for decoders to become initialized with a valid picture upon tuning into a new channel or stream, or to become reinitialized with a valid picture after experiencing transmission errors. Additionally, it is necessary to limit the number of consecutive predictions that can be performed in a decoder to control the buildup of errors resulting from *IDCT mismatch*.

IDCT mismatch occurs because the video coding system, by design, does not completely specify the results of the IDCT operation. MPEG did not fully specify the results of the IDCT to allow for evolutionary improvements in implementations of this computationally intensive operation. As a result, it is possible for the reconstructed pictures in a decoder to "drift" from those in the encoder if many successive predictions are used, even in the absence of transmission errors. To control the amount of drift, each macroblock is required to be coded without prediction (intraframe-coded) at least once in any 132 consecutive frames.

The process whereby a decoder becomes initialized or reinitialized with valid picture data—without reference to previously transmitted picture information—is termed *image refresh*. Image refresh is accomplished by the use of intraframe-coded macroblocks. The two general classes of image refresh, which can be used either independently or jointly, are:

- Periodic transmission of *I*-frames
- Progressive refresh

Periodic Transmission of I-Frames. One simple approach to image refresh is to periodically code an entire frame using only intraframe coding [10]. In this case, the intracoded frame is typically an *I*-frame. Although prediction is used within the frame, no reference is made to previously transmitted frames. The period between successive intracoded frames may be constant, or it may vary. When a receiver tunes into a new channel where *I*-frame coding is used for image refresh, it may perform the following steps:

- Ignore all data receipt of the first sequence header
- Decode the sequence header and configure circuits based on sequence parameters

- Ignore all data until the next received I-frame
- Commence picture decoding and presentation

When a receiver processes data that contains uncorrectable errors in an *I*- or *P*-frame, there typically will be a propagation of picture errors as a result of predictive coding. Pictures received after the error may be decoded incorrectly until an error-free *I*-frame is received.

Progressive Refresh.   An alternative method for accomplishing image refresh is to encode only a portion of each picture using the intraframe mode [10]. In this case, the intraframe-coded regions of each picture should be chosen in such a way that, over the course of a reasonable number of frames, all macroblocks are coded intraframe at least once. In addition, constraints might be placed on motion-vector values to avoid the possible contamination of refreshed regions through predictions using unrefreshed regions in an uninitialized decoder.

Discrete Cosine Transform

Predictive coding in the MPEG-2 compression algorithm exploits the temporal correlation in the sequence of image frames [10]. Motion compensation is a refinement of that temporal prediction, which allows the coder to account for apparent motions in the image that can be estimated. Aside from temporal prediction, another source of correlation that represents redundancy in the image data is the spatial correlation within an image frame or field. This spatial correlation of images, including parts of images that contain apparent motion, can be accounted for by a spatial transform of the prediction differences. In the intraframe-coding case, where there is by definition no attempt at prediction, the spatial transform applies to the actual picture data. The effect of the spatial transform is to concentrate a large fraction of the signal energy in a few transform coefficients.

To exploit spatial correlation in intraframe and predicted portions of the image, the image-prediction residual pixels are represented by their DCT coefficients. For typical images, a large fraction of the energy is concentrated in a few of these coefficients. This makes it possible to code only a few coefficients without seriously affecting picture quality. The DCT is used because it has good energy-compaction properties and results in real coefficients. Furthermore, numerous fast computational algorithms exist for the implementation of DCT.

Theoretically, a large DCT will outperform a small DCT in terms of coefficient decorrelation and block energy compaction [10]. Better overall performance can be achieved, however, by subdividing the frame into many smaller regions, each of which is individually processed.

If the DCT of the entire frame is computed, the whole frame is treated equally. For a typical image, some regions contain a large amount of detail and other regions contain very little. Exploiting the changing characteristics of different images and different portions of the same image can result in significant improvements in performance. To take advantage of the varying characteristics of the frame over its spatial extent, the frame is partitioned into blocks of 8 × 8 pixels. The blocks then are independently transformed and adaptively processed based on their local characteristics. Partitioning the frame into small blocks before taking the transform not only allows spatially adaptive processing, but also reduces the computational and memory requirements. The partitioning of the signal into small blocks before computing the DCT is referred to as the *block DCT*.

An additional advantage of using the DCT domain is that the DCT coefficients contain information about the spatial frequency content of the block. By utilizing the spatial frequency characteristics of the human visual system, the precision with which the DCT coefficients are transmitted can be in accordance with their perceptual importance. This is achieved through the quantization of these coefficients, as explained in the following sections.

Adaptive Field/Frame DCT. As noted previously, the DCT makes it possible to take advantage of the typically high degree of spatial correlation in natural scenes [10]. When interlaced pictures are coded on a frame basis, however, it is possible that significant amounts of motion result in relatively low spatial correlation in some regions. Allowing the DCTs to be computed either on a field basis or on a frame basis accommodates this situation. The decision to use field- or frame-based DCT is made individually for each macroblock.

Adaptive Quantization. The goal of video compression is to maximize the video quality at a given bit rate and this requires a careful distribution of the limited number of available bits [10]. By exploiting the perceptual irrelevancy and statistical redundancy within the DCT domain representation, an appropriate bit allocation can yield significant improvements in performance. Quantization is performed to reduce the precision of the DCT coefficient values. Through quantization and code word assignment, the actual bit rate compression is achieved. The quantization process is the source of virtually all the loss of information in the compression algorithm. This is an important point, as it simplifies the design process and facilitates fine-tuning of the system.

The degree of subjective picture degradation caused by coefficient quantization tends to depend on the nature of the scenery being coded. Within a given picture, distortions of some regions may be less apparent than in others. The video coding system allows for the level of quantization to be adjusted for each macroblock in order to save bits, where possible, through coarse quantization.

Perceptual Weighting.   The human visual system is not uniformly sensitive to coefficient quantization error [10]. Perceptual weighting of each source of coefficient quantization error is used to increase quantization coarseness, thereby lowering the bit rate. The amount of visible distortion resulting from quantization error for a given coefficient depends on the coefficient number, or frequency, the local brightness in the original image, and the duration of the temporal characteristic of the error. The dc coefficient error results in mean value distortion for the corresponding block of pixels, which can expose block boundaries. This is more visible than higher-frequency coefficient error, which appears as noise or texture.

Displays and the HVS exhibit nonuniform sensitivity to detail as a function of local average brightness. Loss of detail in dark areas of the picture is not as visible as it is in the brighter areas. Another opportunity for bit savings is presented in textured areas of the picture, where high-frequency coefficient error is much less visible than in relatively flat areas. Brightness and texture weighting require analysis of the original image because these areas may be well predicted. Additionally, limiting its duration to one or two frames can easily mask distortion. This effect is most profitably used after scene changes, where the first frame or two can be greatly distorted at normal speed without perceptible artifacts.

When transform coefficients are being quantized, the differing levels of perceptual importance of the various coefficients can be exploited by "allocating the bits" to shape the quantization noise into the perceptually less important areas. This can be accomplished by varying the relative step sizes of the quantizers for the different coefficients. The perceptually important coefficients maybe quantized with a finer step size than the others. For example, low spatial frequency coefficients may be quantized finely and the less important high frequency coefficients may be quantized more coarsely. A simple method to achieve different step sizes is to normalize or weight each coefficient based on its visual importance. All of the normalized coefficients may then be quantized in the same manner, such as rounding to the nearest integer (uniform quantization). Normalization or weighting effectively scales the quantizer from one coefficient to another. The MPEG-2 video compression system utilizes perceptual weighting, wherein the different DCT coefficients are weighted according to a perceptual criterion prior to uniform quantization. Quantizer matrices determine the perceptual weighting. The compression system allows for modifying the quantizer matrices before each picture.

Entropy Coding of Video Data

Quantization creates an efficient, discrete representation for the data to be transmitted [10]. Code word assignment takes the quantized values and produces a digital bit stream for transmission. Hypothetically, the quantized values could be simply represented using uniform- or fixed-length code words. Under this approach, every quantized value would be represented with the same number of bits. As outlined previously in general terms, greater efficiency, in terms of bit rate, can be achieved with entropy coding.

Entropy coding attempts to exploit the statistical properties of the signal to be encoded. A signal, whether it is a pixel value or a transform coefficient, has a certain amount of information, or entropy, based on the probability of the different possible values or events occurring. For example, an event that occurs infrequently conveys much more new information than one that occurs often. The fact that some events occur more frequently than others can be used to reduce the average bit rate.

Huffman Coding.   Huffman coding, which is utilized in the ATSC DTV video compression system, for example, is one of the most common entropy coding schemes [10]. In Huffman coding, a codebook is generated that can approach the minimum average description length (in bits) of events, given the probability distribution of all the events. Events that are more likely to occur are assigned shorter-length code words and those less likely to occur are assigned longer-length code words.

Run Length Coding.   In video compression, most of the transform coefficients frequently are quantized to zero [10]. There may be a few non-zero low-frequency coefficients and a sparse scattering of non-zero high-frequency coefficients, but most of the coefficients typically have been quantized to zero. To exploit this phenomenon, the two-dimensional array of transform coefficients is reformatted and prioritized into a one-dimensional sequence through either a zigzag or alternate scanning process. This results in most of the important non-zero coefficients (in terms of energy and visual perception) being grouped together early in the sequence. They will be followed by long runs of coefficients that are quantized to zero. These zero-value coefficients can be efficiently represented through *run length encoding.*

In run length encoding, the number (run) of consecutive zero coefficients before a non-zero coefficient is encoded, followed by the non-zero coefficient value. The run length and the coefficient value can be entropy-coded, either separately or jointly. The scanning separates most of the zero and the non-zero coefficients into groups, thereby enhancing the efficiency of the run length encoding process. Also, a special *end-of-block* (EOB) marker is used to signify when all the remaining coefficients in the sequence are equal to zero. This approach can be extremely efficient, yielding a significant degree of compression.

In the alternate- or zigzag-scan technique, the array of 64 DCT coefficients is arranged in a one-dimensional vector before run length/amplitude code word assignment. Two different one-dimensional arrangements, or *scan types*, are allowed, generally referred to as *zigzag scan* (shown in Figure E.11a) and *alternate scan* (shown in Figure E.11b). The scan type is specified before coding each picture and is permitted to vary from picture to picture.

Channel Buffer.   Whenever entropy coding is employed, the bit rate produced by the encoder is variable and is a function of the video statistics [10]. Because the bit rate permitted by the transmission system is less than the peak bit rate that may be produced by the variable length coder, a *channel buffer* is necessary at the decoder. This buffering system must be carefully designed. The buffer controller must allow the efficient allocation of bits to encode the video and also ensure that no overflow or underflow occurs.

Buffer control typically involves a feedback mechanism to the compression algorithm whereby the amplitude resolution (quantization) and/or spatial, temporal, and color resolution may be varied in accordance with the instantaneous bit rate requirements. If the bit rate decreases significantly, a finer quantization can be performed to increase it.

Figure E.11
Scanning of coefficient blocks: (a) alternate scanning of coefficients and (b) zigzag scanning of coefficients. (From [10]. Used with permission.)

(a)

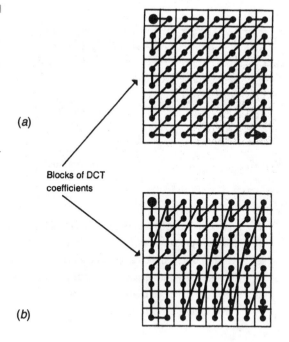

Blocks of DCT
coefficients

(b)

As an example, the ATSC DTV standard specifies a channel buffer size of 8 megabits. The *model buffer* is defined in the DTV video coding system as a reference for manufacturers of both encoders and decoders to ensure interoperability. To prevent overflow or underflow of the model buffer, an encoder may maintain measures of buffer occupancy and scene complexity. When the encoder needs to reduce the number of bits produced, it can do so by increasing the general value of the quantizer scale, which will increase picture degradation. When it is able to produce more bits, it can decrease the quantizer scale, thereby decreasing picture degradation.

Decoder Block Diagram. As shown in Figure E.12, the ATSC DTV video decoder contains elements that invert, or undo, the processing performed in the encoder [10]. The incoming coded video bit stream is placed in the channel buffer and a *variable length decoder* (VLD) removes bits.

The VLD reconstructs 8×8 arrays of quantized DCT coefficients by decoding run length/amplitude codes and appropriately distributing the coefficients according to the scan type used. These coefficients are dequantized and transformed by the IDCT to obtain pixel values or prediction errors.

Figure E.12
ATSC DTV video
system decoder
functional block
diagram. (From [10].
Used with
permission.)

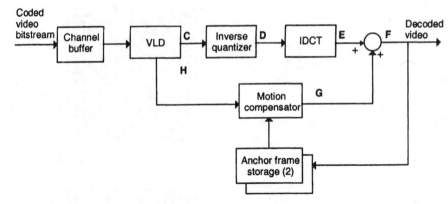

Figure E.12
ATSC DTV video
system decoder
functional block
diagram. (From [10].
Used with
permission.)

C	Prediction error DCT coefficients in quantized form
D	Quantized prediction error DCT coefficients in standard form
E	Pixel-by-pixel prediction errors, degraded by quantization
F	Reconstructed pixel values, degraded by quantization
G	Motion compensated predicted pixel values
H	Motion vectors

In the case of interframe prediction, the decoder uses the received motion vectors to perform the same prediction operation that took place in the encoder. The prediction errors are summed with the results of motion-compensated prediction to produce pixel values.

Spatial and Signal-to-Noise Scalability

Because MPEG-2 was designed in anticipation of the need for handling different picture sizes and resolutions, including standard definition television and high-definition television, provisions were made for a hierarchical split of the picture information into a base layer and two enhancement layers [10]. In this way, standard definition television (SDTV) decoders would not be burdened with the cost of decoding an HDTV signal.

An encoder for this scenario could work as follows. The HDTV signal would be used as the starting point. It would be spatially filtered and subsampled to create a standard resolution image, which then would be MPEG-encoded. The higher definition information could be included in an enhancement layer.

Another use of the hierarchical split would be to provide different picture quality without changing the spatial resolution. An encoder quantizer block could realize both coarse and fine filtering levels. Better error correction could be provided for the more coarse data, so that as signal strength weakened (assuming terrestrial transmission) a step-by-step reduction in the picture signal-to-noise ratio would occur in a way similar to that experienced in broadcast analog signals today. Viewers with poor reception, therefore, would experience a more graceful degradation in picture quality instead of a sudden dropout. These enhancement techniques can also work with streaming, where the base and enhancement layers are transmitted as discrete streams. If the transmission bandwidth reduces, due to network clogging, for example, the enhancement streams could be discarded without the viewer losing the picture, even though the picture quality would visibly degrade.

MPEG-4 Video Compression

MPEG-4 builds upon the video and audio compression tools developed for MPEG-2, extending the concept of a video frame to arbitrary-shaped *video object planes* (VOPs). Thus *I*-frames, *P*-frames, and *B*-frames become *I*-VOPs, *P*-VOPs, and *B*-VOPs. *Groups of Pictures* (GOP) become *Groups of Video* planes (GOV). MPEG-4 also allows for the composition of scenes, comprising a multiplicity of video planes arranged according to a scene description. The scene is rendered in the MPEG-4 decoder for presentation to the user, from a number of *elemental streams*, which represent each video object plane and are synchronized at the receiver in a layer that interfaces the transport mechanisms to the video scene decoder.

MPEG-4 also provides a number of tools to synthesize both audio and video at the decoder, using parametric descriptions. It can combine these synthetic elements with natural elements in a common scene. There are also tools for interactivity, allowing the viewer to change the position of elements in the scene, their own viewpoint, or the acoustic properties of the scene, among others. For the purposes of this discussion, we will describe only those aspects of MPEG-4 that concern the representation of natural video, by way of continuing our examination of the development of key techniques for video compression.

Very Low Bit Rate Video (VLBV)

One of the goals of the MPEG-4 standard was to provide a universal toolkit of video compression techniques that could serve a wide variety of applications, including mobile video conferencing, using very low bandwidth links. This led to the development of the *very low bit rate video* (VLBV) core. The VLBV code allows the coding of video between 5 kilobits and 64 kilobits per second, with low spatial resolution (typically up to CIF resolution) and low frame rates (up to 15 frames per second). VLBV confines itself to conventional rectangular image sequences. MPEG-4 can encode low motion video contents such as talking heads with long predictive sequences to substantially reduce the bitrate. As an example, a SQCIF resolution video sequence (160×120) can be coded at rates between 12 kilobits per second and 28 kilobits per second by using 120-frame, predictive-coded sequences at frame rates of 5 frames per second to 15 frames per second.

Shape Coding

Rectangular-shaped video (called a *texture*) is coded in much the same way as it is in MPEG-2, except that the dimensions of the rectangle can be arbitrarily chosen. The way arbitrary-shaped video objects are coded makes use of an alpha channel. The actual video object is defined over a rectangular area, called a *mask*. The mask is sized to accommodate the greatest horizontal and vertical extent of the video, to the nearest 16 pixel multiple. The alpha channel denotes whether or not the underlying video pixels are visible. Shapes may be coded as binary or gray-scale data. Binary shapes are the simplest, with the alpha data merely controlling whether a pixel is visible or transparent. Unfortunately, this leads to edge effects that violate the Nyquist sampling theorem, and can lead to unwanted visible edge aliasing artifacts. Grayscale shapes, in contrast, permit a smooth, bandwidth-controlled transition between object and background. This alpha blending technique makes arbitrary-shaped video objects appear more realistic. Binary shapes find application for simpler objects, where edge aliasing is not an issue. Burning a rectangular picture into a video session scene would be an example. Coding of *binary alpha blocks* is trivial for wholly transparent or opaque pixels. However, edge pixels representing the boundary of the shape are coded using techniques derived from IBM's patented *arithmetic coding*, an entropy encoding technique, which, unlike Huffman coding, self-adjusts its codeword choices according to the statistics of the input entropy that it keeps.

Context-Based Arithmetic Encoding. The algorithm used to encode alpha channel bits with partial transparency is called *context-based arithmetic encoding* (CAE). MPEG-4 extends the algorithm to include the use of motion estimation. As with regular arithmetic encoding, coding is based on a continuously updated probability estimate for each incoming symbol. In the basic intraCAE, the probability estimate is computed from ten pixels, above and to the left of the pixel being encoded. For interCAE, where motion estimation and prediction are used, pixels from the current video object plane (VOP) and some from the reference VOP (an *I*-VOP or *P*-VOP) form the basis for the probability estimate.

Grayscale shapes are represented in a similar way to luminance signals. The pixels are usually quantized to 8-bit resolution, with 0 representing total transparency and 255 representing total opacity. The image plane, as with all MPEG video, is divided into macroblocks. Macroblocks falling entirely outside the boundary of the object are marked as "all zero" (i.e., fully transparent) and those wholly within the boundary of the object marked as "all 255" (i.e., fully opaque). Macroblocks that contain any part of the shape's boundary are coded in a similar way to the underlying video texture, using motion-compensated DCT coding.

Texture Coding

Texture coding is analogous to conventional moving image coding in MPEG-2. The coding method builds upon earlier work from the JPEG and MPEG standards. In all MPEG-4 profiles, except for the studio profiles, 4:2:0 YUV video representations are used to describe textures. Although a rectangular area represents the extent of the video object, however, only those macroblocks in the image made visible by the shape signal are actually encoded. In other words, texture coding is performed only for those macroblocks wholly or partially within the boundary of the arbitrary shape.

I-VOPs are coded in much the same way as MPEG-2 *I*-frames, except efficiency has been improved by using a more efficient predictor. In MPEG-4, the predictor measures horizontal and vertical brightness gradients and predicts the dc coefficient value from the block above or the block to the left, in the direction of the lesser gradient.

We know from our discussion of MPEG-2 that correlation helps in compression, because correlation identifies image redundancy. Image areas of similar texture have similar ac coefficients, after DCT. The most significant coefficients are the lower order ones, which contain the greatest proportion of the image's energy. These are the non-zero coefficients in the first row and/or column of the coefficient matrix for the DCT-coded macroblock and these coefficients are quantized least aggressively, since quantization errors in these coefficients are the most visible. In MPEG-4, the ac coefficients of either the first row or column are predicted from those of the block immediately above, the block to the immediate left, or the block diagonally above-left.

We also described, in our discussion of MPEG-2, how coefficients are scanned (zigzag or alternate) and variable length encoded. MPEG-4 improves upon both scanning and variable length coding. The method chosen for coefficient readout is determined by the dc prediction. When there is no dc prediction, the MPEG-2 zigzag scanning technique is used. If the dc coefficient was predicted from the block to the left, *alternate-vertical* scanning is used (a scanning system biased toward reading out the vertical coefficients first). Unsurprisingly, if the dc coefficient was predicted from the block above, *alternate-horizontal* scanning is used and the horizontal coefficients are read out first.

To improve the efficiency of variable length encoding, two different VLC tables are provided in MPEG-4. The quantization level determines the choice of table. The VLC codes are reversible, for error resilience. If there is an error in the bit stream, decoding can continue up to the error. Data after the error may be decoded by starting at the end of the block and decoding the VLC codes in reverse until the error is reached.

Boundary Coding

When blocks straddle part of the arbitrary-shaped object's boundary, an interesting problem arises. The pixels that lie outside the boundary of the image have to have some value, yet ideally the choice of these pixel values, which form no part of the visible image, shouldn't add coefficient energy to the overall block, or there will be a marked effect on the DCT process. Choosing to make these pixels black, for example, would be a poor choice. To avoid annoying the DCT transformation process, the block must be padded with pixel values that do the least harm. In MPEG-4, all pixels that are not part of the image are given a value equal to the average value of all of the pixels that are part of the image (note that the pixels that are visible remain unchanged during this operation). The padding is refined by stepping through the pixels that are outside the visible object and performing a correction based on the average value of any neighbors that lie within the object. This process of writing "dummy values" to those pixels that will not be visible minimizes the energy of the coefficients when the block is DCT transformed.

Coding Arbitrary-Shaped Video Objects

With arbitrary-shaped video objects, blocks are either wholly transparent, wholly opaque, or are part of the boundary of the VOP. In the first case, no texture coding is required. In the second, the texture may be intracoded or predicted with motion compensation, as with MPEG-2 macroblocks.

For those blocks that form part of the VOP boundary, the shape will either be described as binary (every pixel is either transparent or opaque) or grayscale (supporting pixels with partial transparency). In this case, both the shape and the texture must be coded. The shape (alpha-channel) and texture (video image data) can be intracoded or predictive coded. When coding is predictive, the simplest case is when the *motion vector difference* (MVD) is zero (i.e., motion vector prediction is perfect). In this case, the motion vector points to an exact match for the block, so the coding simply requires a "skip" code. If the match is imperfect, residuals may need to be transmitted if the match is not sufficiently good.

Sprites

MPEG-4 has another interesting video object, useful for backgrounds. In computer games, the background against which all the action takes place is much larger than the viewable frame size. All the game's characters are superimposed on this background and more of the background is revealed as the characters move around the game space (as the "camera" following the characters' pans). MPEG-4 allows such a sprite to be transmitted once and then updates to the view of the section of sprite to be sent as cropping and warping information.

Transmitting the entire sprite before anything else can happen can lead to unacceptable start-up latency, so MPEG-4 allows the sprite to be progressively transmitted and reconstructed at the decoder as needed.

Progressive encoding means that the sprite could consist of a 360-degree panoramic view, but only a low-resolution version of part of the panorama need be transmitted at first to minimize delay. Because sprites are always static images, they are coded as *I*-VOPs.

Advanced Coding Extensions (ACE)

MPEG-4 Version 2 introduced three new tools to improve coding efficiency for video objects, collectively known as the *advanced coding extensions* (ACE). These yield improvements in coding efficiency of up to 50%, compared to MPEG-4 Version 1 (and therefore, MPEG-2, to a first approximation). The new tools are discussed in the following sections.

Global Motion Compensation. Global Motion Compensation allows the overall motion of the video object to be coded with a very few parameters. If the object remains stationary, or if it moves in its entirety, there is no need to indicate the motion of each component block individually.

Quarter Pel Motion Compensation. The improved resolution of motion vectors substantially reduces prediction errors and hence the need to transmit residuals. MPEG-2 used half pel motion compensation.

Shape Adaptive DCT. Shape-adaptive DCT may improve the coding efficiency of boundary blocks. Instead of blindly applying a DCT to an 8 × 8 block (64 pixels), one-dimensional DCT is applied, first vertically, then horizontally, but only to pixels that belong to the visible object (called *active pixels*).

To commence the process, each column of the block is examined for active pixels. For any column that contains active pixels, the pixels are "top justified" (moved to the top of the column) and a DCT of dimension 1 × (number of active pixels in column) is performed. When the vertical transforms are complete, each row in the block is examined for active pixels. For any row containing active pixels, the pixels are "left justified" and a DCT of dimension 1 × (number of active pixels in row) is performed. This results in fewer coefficients to be transmitted.

Because the contour of the object is transmitted separately by shape coding, the pixels can be moved back into the right places in the decoder.

Fine Grain Scalability

Fine grain scalability is also known as signal-to-noise scalability. What happens in the encoder is that the DCT coefficient matrix can be "sliced up" and sent as a base layer and multiple enhancement layers, which can be transmitted or not according to available bandwidth. The decoder can use the base layer and one or more enhancement layers to reconstruct the image.

The way the DCT matrix of coefficients is sliced is by bit plane. If there are 4 bits in the coefficient representations, the base layer may simply take the most significant bit, the first enhancement layer the next most significant bit, and so on. Additionally, low frequency coefficients in the matrix can be artificially weighted, prior to bit slicing, to ensure the base layer contains more of the low frequency components needed for image reconstruction. Note also that with fine grain scalability, downstream relay routers can take DCT coefficients and bit slice them. If the base layer took the top 4 bits and the enhancement layer took the remaining 4 bits of an 8-bit quantized matrix, for example, the router could further slice the enhancement layer into two layers.

Error Robustness

MPEG-4 addresses applications that attempt to transmit video over lossy networks, such as wireless cellular networks, for example. To address these applications, MPEG-4 provides three categories of tool to improve error robustness. These categories are not unique to MPEG-4, but have been used by researchers working on general video error resilience. The three main categories are discussed next.

Resynchronization.   Resynchronization tools attempt to enable resynchronization between the decoder and the bit stream after a residual error or errors have been detected. In general, the data between the synchronization point prior to the error and the first point where synchronization is reestablished is discarded. Resynchronization tools attempt to minimize the amount of data discarded by the decoder, so that downstream data recovery and error concealment tools can mask the effect of the data loss.

MPEG-4 uses a packet approach, similar to the *group of blocks* (GOB) structure used by ITU standards H.261 and H.263. In these standards, a GOB is defined as one or more rows of macroblocks. The discussion of slices earlier in this appendix showed how they could be used to resynchronize bit streams. GOBs and slices are the same thing. GOBs are delineated by a *GOB start code*, which is distinct from a picture start code and allows the decoder to resynchronize to the bit stream before the next picture. The GOB approach places start codes spatially. With variable rate encoding, this means that resynchronization markers will be unevenly spaced throughout the bit stream. As a consequence, high motion areas of the picture will be more prone to data loss, which will also be more difficult to conceal.

The video packet approach adopted by MPEG-4, in contrast, provides periodic resynchronization markers throughout the bit stream. Marker placement is not based on the number of macroblocks, but on the number of bits contained in that video packet. This marker is distinguishable from all possible VLC codewords as well as the VOP start code. The video packet header contains all the information needed to resynchronize the decoder. It is also possible to link header extensions to provide additional resynchronization information, in case the VOP header has been corrupted.

When MPEG-4 error resilience tools are in use, some of the compression efficiency tools must be modified. For example, all predictively coded information must be constrained within a video packet, to prevent the propagation of errors.

MPEG-4 has also adopted a second resynchronization method, called *fixed interval synchronization*. This method requires that the VOP start code and the video packet start codes (resynchronization markers) appear only at legal fixed interval locations in the bit stream. This prevents the decoder from incorrectly interpreting a corrupted bit stream as a VOP start code, since with fixed interval synchronization, the decoder only needs to search for a VOP start code at the beginning of each fixed interval.

Data Recovery.   After synchronization has been re-established, it is necessary to recover data that would otherwise be lost. MPEG-4 provides tools, which are not simply error correcting codes, but instead are techniques that encode the data in a fundamentally error resilient manner. One of these tools is the *reversible variable length code* (RVLC). In this approach, the variable length codewords are designed so that they can be read both in the forward and reverse directions, as was mentioned earlier in the discussion of texture coding. In essence, the resynchronization marker may be used as a starting point to read the bit stream in reverse order, up to the error in the stream that caused synchronization to be lost. Without reversible codes, the data from the error occurrence to the resynchronization point would ordinarily have been discarded as useless by the decoder, even though many valid symbols may have been received, simply because, while running out of synchronization, the decoder has no way of recognizing valid symbols.

Error Concealment. Error concealment acts as a very important last line of defense in a robust, error-tolerant video codec. The effectiveness of error concealment strategies is highly dependent on the performance of the resynchronization scheme. If the resynchronization tools can effectively isolate and localize the fault in the bit stream, then the error concealment problem becomes much more tractable. In low bit rate, low delay applications, the simple expedient of copying blocks from the previous decoded VOP provides perfectly usable results.

To further enhance concealment capabilities, MPEG-4 provides an additional error-resilient mode that improves the ability of the decoder to localize an error. This approach uses data partitioning, separating the motion and the texture. A second resynchronization marker is inserted between the motion and texture information, so that if texture information is lost due to error, motion information is used to conceal these errors by motion-compensating the previous decoded VOP.

Concatenation

When television production and post-production was young and analog video was the only available option, many of the stages in production involved playing video, in real time, from one machine to another, re-recording or processing it at the next stage in production. Each processing and re-recording stage added a generation loss. When television production became digital, this workflow was essentially preserved in aspic. Many production stages, even today, still involve playing back baseband video in real time, between processing stations. Even though the compressed data could be transferred over a high speed LAN much faster than real time playback, long-established production habits and backward-compatibility concerns constrain the data flow to take place the way grandma used to do it (assuming, of course, that grandma was a video producer). This is an important problem because if the analog production workflow is slavishly mimicked in a digital production plant, using lossy compression encoders and decoders at every production stage interface, the video data goes through a number of compress/ decompress cycles, with each transfer potentially losing more of the information. The losses are especially acute if the pixel values in the video information change because of processing or if aspect ratio conversion takes place. Information loss is worse at low bit rates or high compression ratios; information, once lost, cannot be recovered.

Compression and decompression in the same format is not usually called concatenation. However, when different compression schemes are used, concatenation artifacts are likely. The more generations in the production process, the more artifacts can potentially be introduced. Some of these artifacts are insignificant and barely noticeable, whereas others are considerable and objectionable. In a typical streaming media production chain, it is common for the cameras to capture material in DV format, for the post-production plant to be wired for MPEG-2, and for the final stream to be encoded into, say, MPEG-4.

Concatenation artifacts can be mitigated in two ways: reducing the number of compress/decompress cycles in the production chain and working with higher quality or losslessly compressed images. Working with high quality, lightly compressed data upstream, unfortunately, costs more, since all of the equipment has to be capable of handling vast amounts of data. The electronics must work faster, storage has to be larger and higher bandwidth networks and interfaces are needed.

For streaming media production, a balance must be struck between the cost of production and the amount of degradation due to concatenation artifacts that is tolerable. The saving grace is that typical compress/decompress cycle losses do not typically degrade image quality as badly as low-band analog tape generation loss, so it is possible to start with relatively modest image quality at the camera and still produce an acceptable presentation at the end user. On the other side of the equation, there is a point of diminishing returns where working with ever higher bit rate images provides only marginal improvements in image quality and end-user quality of experience.

Transcoding from one streaming media delivery format to another, especially if the streams have high compression ratios or low bit rates, is probably never going to produce satisfactory results, simply because the popular formats treat their video signals and compression algorithms in different ways and concatenation effects will, in all likelihood, prevent the output from being usable. Transcoding one delivery stream format into another is, therefore, not recommended. Using a domestic DVD as source, with its MPEG-2 encoding, is probably going to produce marginal streaming media quality as well, because of concatenation effects.

In general, it is not possible to hide artifacts by up-sampling from a low-quality, highly compressed stream to a higher quality one. Concatenation artifacts persist and propagate. Hence, there is a need to archive streaming media masters using high quality, losslessly compressed formats, ideally. This will allow the media assets to be reformatted and repurposed in the future. Also, the overall quality of the production chain is only as good as the worst quality compression system in the chain. If any upstream stage in the process is very lossy, things can only get worse downstream.

Typical MPEG Artifacts

Although each type of program material consists of a unique set of video characteristics, some generalizations concerning the artifacts that can be expected with MPEG-based compression systems can be made [13]. The artifacts are determined in large part by the algorithm implementations used by specific MPEG encoding vendors. Possible artifacts include the following:

- **Block effects**—These may be seen when the eye tracks a fast-moving, detailed object across the screen. The blocky grid appears to remain fixed while the object moves beneath it. This effect also may be seen during dissolves and fades. It typically is caused by poor motion estimation and/or insufficient allocation of bits in the coder.
- **Mosquito noise**—This artifact may be seen at the edges of text, logos, and other sharply defined objects. The sharp edges cause high-frequency DCT terms, which are coarsely quantized and spread spatially when transformed back into the pixel domain.
- **Dirty window**—This condition appears as streaking noise that remains stationary while objects move beneath it. In this case, the encoder may not be sending sufficient bits to code the residual (prediction) error in the P- and B-frames.
- **Wavy noise**—This artifact often is seen during slow pans across highly detailed scenes, such as a crowd in a stadium. The coarsely quantized high-frequency terms resulting from such images can cause reconstruction errors to modulate spatially as details shift within the DCT blocks.

It follows, then, that certain types of motion do not fit the MPEG linear translation model particularly well and are, therefore, problematic. These types of motion include:

- Zooms
- Rotations
- Transparent and/or translucent moving objects
- Dissolves containing moving objects

Furthermore, certain types of image elements cannot be predicted well. These image elements include:

- Shadows
- Changes in brightness resulting from fade-ins and fade-outs
- Highly detailed regions
- Noise effects
- Additive noise

Efforts continue to minimize coding artifacts. Success lies in the skill of the system designers in adjusting the many operating parameters of a video encoder. One of the strengths of the MPEG standard is that it allows—and even encourages—diversity and innovation in encoder design.

SMPTE RP202

SMPTE Recommended Practice 202 is an important step in the world of digital video production. Equipment conforming to this practice will minimize concatenation artifacts by optimizing macroblock alignment [14]. As MPEG compression becomes pervasive, multiple compression and decompression cycles will inevitably occur. Concatenation of codecs may be needed for production, post-production, transcoding, and format conversion. Any time video transitions to or from the coefficient domain of MPEG are performed, care must be exercised on alignment of the video, both horizontally and vertically, as it is coded from the raster format or decoded and placed in the raster format.

The first problem is shifting the video horizontally and vertically. Over multiple compression and decompression cycles, this could substantially distort the image. Less obvious, but just as important, is the need for macroblock alignment to reduce artifacts between encoders and decoders from various equipment vendors. If concatenated encoders do not share common macroblock boundaries, then additional quantization noise, motion estimation errors, and poor mode decisions may result. Likewise, encoding decisions that may be carried through the production and post-production process with recoding data present will rely upon macroblock alignment. Decoders must also exercise caution in the placement of the active video in the scanning format so that the downstream encoder does not receive an offset image.

With these issues in mind, RP202 specifies the spatial alignment for MPEG-2 video encoders and decoders. Both standard definition and high definition video formats for production, distribution, and emission systems are addressed. Table E.4 gives the recommended coding ranges for MPEG-2 encoders and decoders. Although not specifically addressed in the practice, MPEG-4 streaming media encoders that take input from a MPEG-2 production plant could also benefit from macroblock and spatial alignment.

TABLE E.4

Recommended MPEG-2 Coding Ranges for Various Video Fromats (After [14])

Format	Resolution pels x lines	Coded pels	Coded lines			MPEG-2 profile and level
			Field 1	Field 2	Frame	
480I	720 × 480	0–719	23–262	286–525		MP @ML
480P	720 × 480	0–719			46–525	MP @ HL
512I	720 × 512	0–719	7–262	270–525		422P @ ML
512P	720 × 512	0–719			14–525	422P @ HL
576I	720 × 576	0–719	23–310	336–623		MP @ ML
608I	720 × 608	0–719	7–310	320–623		422P @ ML
720P	1280 × 720	0–1279			26–745	MP @ HL
720P	1280 × 720	0–1279			26–745	422P @ HL
1080I	1920 × 1088[1]	0–1919	21–560	584–1123		MP @ HL
1080I	1920 × 1088[1]	0–1919	21–560	584–1123		422P @ HL
1080P	1920 × 1088[1]	0–1919			42–1121	MP @ HL
1080P	1920 × 1088[1]	0–1919			42–1121	422P @ HL

[1]The active image only occupies the first 1080 lines.

Digital Audio Data Compression

As with video, high on the list of priorities for the professional audio industry is to refine and extend the range of digital equipment capable of the capture, storage, post-production, exchange, distribution, and transmission of high-quality audio, be it mono, stereo, or 5.1 channel surround sound [15]. This demand is being driven by end-users, broadcasters, filmmakers, and recording industry professionals alike, who are moving rapidly towards a "tapeless" environment. Indeed, high-end recording professionals have had the option of completely tapeless production since the late 1980s. Over the last two decades, there have been advancements in DSP technology that have supported research engineers in their endeavors to produce the necessary hardware, particularly in the field of digital audio data compression or—as it is referred to—*bit rate reduction*. There exist a number of real-time or, in reality, near instantaneous compression coding algorithms. These can significantly lower the circuit bandwidth and storage requirements for the transmission, distribution, and exchange of high-quality audio. From an applications point of view, the real drivers for better audio compression techniques are portability of media, instantaneous access, reduced storage cost and/or performance requirements, and the ability to own or access a greater variety of audio assets inexpensively.

The introduction of the *compact disk* (CD) digital audio format in 1983 set a quality benchmark that the manufacturers of subsequent professional audio equipment strive to match or improve. In actual fact, professional digital audio manipulation tools, such as the Fairlight CMI, existed prior to 1983, but the expense of digital memory and the lack of availability and cost of high quality analog anti-aliasing filters, high-precision digital to analog converters and, in particular, high precision analog to digital converters served to hold developments back. Indeed, analog operational amplifiers, the workhorse of the professional audio equipment industry up to that time, were barely able to meet the performance demands of CD audio quality. It was only with the introduction of the CD consumer format that these critical electronic parts were produced in significant numbers and achieved commodity pricing. The discerning consumer now expects the same audio quality from radio and television receivers and streaming media players as they have enjoyed from CD players for almost two decades. This leaves the broadcaster and media streamer with an enormous challenge.

PCM versus Compression

It can be an expensive and complex technical exercise to fully implement a linear *pulse code modulation* (PCM) infrastructure, except over very short distances and within studio areas [15]. To demonstrate the advantages of distributing compressed digital audio over wireless or wired systems and networks, consider once again the CD format as reference. A stereo CD transfers information (data) at 1.411 megabits per second, which would require a circuit with bandwidth of approximately 700 kHz to avoid distortion of the digital signal. In practice, additional bits are added to the signal for channel coding, synchronization, and error correction; this increases the bandwidth demands yet again. 1.5 MHz is the commonly quoted bandwidth figure for a circuit capable of carrying a CD or similarly encoded linear PCM digital stereo signal. This can be compared with the 20 kHz needed for each of two circuits to distribute the same stereo audio in analog format, a 75-fold increase in bandwidth requirements. (Note that 1.5 MHz poses few challenges to today's vanilla office-grade digital networks, such as 100BaseT and Gigabit Ethernet, for example, but the point is made. Broadband Internet connections of 500 kilobits per second would not be capable of carrying a 1.411-megabit signal, as a case in point).

Audio Bit Rate Reduction

In general, analog audio transmission requires fixed input and output bandwidths [16]. This condition implies that in a real-time compression system, the quality, bandwidth, and distortion/noise level of both the original and the decoded output should not be *subjectively* different, thus giving the appearance of a lossless and real-time process.

In a technical sense, all practical real-time bit rate reduction systems can be referred to as "lossy." In other words, the digital audio signal at the output is not identical to the input signal data stream. However, some compression algorithms are, for all intents and purposes, lossless; they lose as little as 2% of the original signal. Others remove approximately 80% of the original signal.

Redundancy and Irrelevancy. A complex audio signal contains a great deal of information, some of which, because the human ear cannot hear it, is deemed irrelevant [16]. The same signal, depending on its complexity, also contains information that is highly predictable and, therefore, can be made redundant.

Redundancy, measurable and quantifiable, can be removed in the coder and replaced in the decoder; this process often is referred to as *statistical compression*. *Irrelevancy*, on the other hand, referred to as *perceptual coding*, once removed from the signal cannot be replaced and is lost, irretrievably. This is entirely a subjective process, with each proprietary algorithm using a different psychoacoustic model.

Critically perceived signals, such as pure tones, are high in redundancy and low in irrelevancy. They compress quite easily; almost totally a statistical compression process. Conversely, noncritically perceived signals, such as complex audio or noisy signals, are low in redundancy and high in irrelevancy. These compress easily in the perceptual coder, but with the total loss of all the irrelevancy content.

Human Auditory System (Psychoacoustics).   The sensitivity of the human ear is biased toward the lower end of the audible frequency spectrum, around 3 kHz [16]. At 50 Hz, the bottom end of the spectrum and 17 kHz at the top end, the sensitivity of the ear is down by approximately 50 dB relative to its sensitivity at 3 kHz (Figure E.13). Additionally, very few audio signals—music- or speech-based—carry fundamental frequencies above 4 kHz. The designers of the *predictive* range of compression algorithms take advantage of these characteristics of the ear, the structure of audible sounds and the redundancy content of the PCM signal to reduce bit rate.

Another well-known feature of the hearing process is that loud sounds mask out quieter sounds at a similar or nearby frequency. This compares with the action of an automatic gain control, turning the gain

Figure E.13
Generalized frequency response of the human ear. Note how the PCM process captures signals that the ear cannot distinguish. (From [16]. Used with permission.)

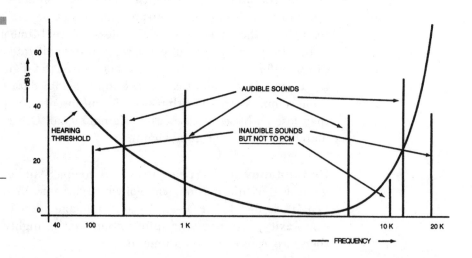

down when subjected to loud sounds, thus making quieter sounds less likely to be heard. For example, as illustrated in Figure E.14, if we assume a 1 kHz tone at a level of 70 dBu, levels of greater than 40 dBu at 750 Hz and 2 kHz would be required for those frequencies to be heard. The ear also exercises a degree of temporal masking, being exceptionally tolerant of sharp transient sounds.

It is by mimicking these additional psychoacoustic features of the human ear and identifying the irrelevancy content of the input signal that the *transform* range of low bit rate algorithms operate, adopting the principle that if the ear is unable to hear the sound then there is no point in transmitting it in the first place.

Figure E.14
Example of the masking effect of a high-level sound. (From [16]. Used with permission.)

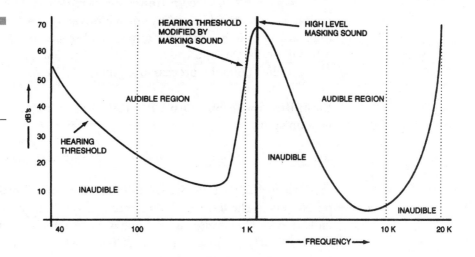

Quantization. Quantization is the process of converting an analog signal to its representative digital format or, as in the case with compression, the requantizing of an already converted signal [16]. This process is the limiting of a finite level measurement of a signal sample to a specific present integer value. This means that the *actual* level of the sample may be greater or smaller than the preset *reference* level it is being compared with. The difference between these two levels, called the *quantization error*, is compounded in the decoded signal as *quantization noise*.

Quantization noise, therefore, will be injected into the audio signal after each A/D and D/A conversion, the level of that noise being governed by the bit allocation associated with the coding process (i.e., the number of bits allocated to represent the level of each sample taken of the analog signal). For linear PCM, the bit allocation is commonly 16. The level of each audio sample, therefore, will be compared with one of 2^{16} or 65,536 discrete levels or steps.

Compression or bit rate reduction of the PCM signal leads to the requantizing of an already quantized signal, which will unavoidably inject further quantization noise. It always has been good operating practice to restrict the number of A/D and D/A conversions in an audio chain. Nothing has changed in this regard and now the number of compression stages also should be kept to a minimum. Additionally, the bit rates of these stages should be set as high as practical; put another way, the compression ratio should be as low as possible.

Sooner or later, after a finite number of A/D, D/A conversions and passes of compression coding, of whatever type—the accumulation of quantization noise and other unpredictable signal degradations eventually will break through the noise/signal threshold, be interpreted as part of the audio signal, be processed as such, and be heard by the listener.

Sampling Frequency and Bit Rate.   The bit rate of a digital signal is defined by:

Sampling frequency × bit resolution × number of audio channels

The rules regarding the selection of a sampling frequency are based on Nyquist's theorem [16]. This ensures that, in particular, the lower sideband of the sampling frequency does not fold into the base-band audio. Objectionable and audible aliasing effects would occur if the two bands were to overlap. In practice, the sampling rate is set slightly above twice the highest audible frequency, which makes filter designs less complex and less expensive. (In fact, the setting of the sampling frequency at slightly more than twice the audio base-bandwidth is a historical artifact, since most conversion systems today employ oversampling, which enables very precise definition of the audio pass-band through digital anti-alias filtering at very low cost. When anti-aliasing filters were analog, they were both imprecise and expensive.)

In the case of a stereo CD with the audio signal having been sampled at 44.1 kHz, this sampling rate produces audio bandwidths of approximately 20 kHz for each channel. The resulting audio bit rate = 44.1 kHz × 16 × 2 = 1.411 megabits per second, as discussed previously.

Prediction and Transform Algorithms

Most audio compression systems are based upon one of two basic technologies [16]:

- Predictive or *adaptive differential* PCM (ADPCM) time-domain coding
- Transform or *adaptive* PCM (APCM) frequency-domain coding

It is in their approaches to dealing with the redundancy and irrelevancy of the PCM signal that these techniques differ.

The time domain or *prediction* approach includes G.722, which has been a universal standard since the mid-70s and was joined by a proprietary algorithm, apt-X100. Both these algorithms deal mainly with redundancy.

The frequency domain or *transform* method adopted by a number of algorithms deal in irrelevancy, adopting psychoacoustic masking techniques to identify and remove those unwanted sounds. This range of algorithms include the industry standards ISO.MPEG-1 Layers 1, 2, and 3; apt-Q; MUSICAM; Dolby AC-2 and AC-3, and others.

Subband Coding.   Without exception, all of the algorithms mentioned in the previous section process the PCM signal by splitting it into a number of frequency subbands, in one case as few as two (G.722) or as many as 1024 (apt-Q) [15]. MPEG-1 Layer 1, with 4:1 compression, has 32 frequency subbands and is the system found in the now rare Digital Compact Cassette (DCC). The MiniDisc ATRAC proprietary algorithm at 5:1 has a more flexible multi-subband approach, which is dependent on the complexity of the audio signal.

Subband coding enables the frequency domain redundancies within the audio signals to be exploited. This permits a reduction in the coded bit rate, compared to PCM, for a given signal fidelity. Spectral redundancies are also present as a result of the signal energies in the various frequency bands being unequal at any instant of time. By altering the bit allocation for each subband, either by dynamically adapting it according to the energy of the contained signal or by fixing it for each subband, the quantization noise can be reduced across all bands. This process compares favorably with the noise characteristics of a PCM coder performing at the same overall bit rate.

Subband Gain. On its own, subband coding, incorporating PCM in each band, is capable of providing a performance improvement or gain compared with that of full band PCM coding, both being fed with the same complex, constant level signal [15]. The improvement is defined as subband gain and is the ratio of the variations in quantization errors generated in each case while both are operating at the same transmission rate. The gain increases as the number of subbands increase and with the complexity of the input signal. However, the implementation of the algorithm also becomes more difficult and complex.

Quantization noise generated during the coding process is constrained within each subband and cannot interfere with any other band. The advantage of this approach is that the masking by each of the subband dominant signals is much more effective because of the reduction in the noise bandwidth. Figure E.15 charts subband gain as a function of the number of subbands for four essentially stationary, but differing, complex audio signals.

Figure E.15
Variation of subband gain as a function of the number of subbands. (From [16]. Used with permission.)

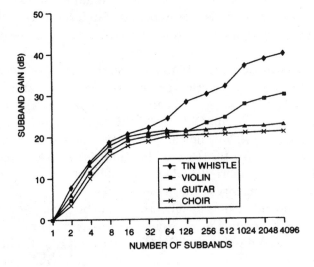

In practical implementations of compression codecs, several factors tend to limit the number of subbands employed. The primary considerations include:

- The level variation of normal audio signals leading to an averaging of the energy across bands and a subsequent reduction in the coding gain
- The coding or processing delay introduced by additional subbands
- The overall computational complexity of the system

The two key issues in the analysis of a subband framework are:

- Determining the likely improvement associated with additional subbands
- Determining the relationships between subband gain, the number of subbands, and the response of the filter bank used to create those subbands

APCM Coding. The APCM processor acts in a similar fashion to an automatic gain control system, continually making adjustments in response to the dynamics—at all frequencies—of the incoming audio signal [15]. Transform coding takes a time block of signal, analyzes it for frequency and energy, and identifies irrelevant content. Again, to exploit the spectral response of the ear, the frequency spectrum of the signal is divided into a number of subbands and the most important criteria are coded with a bias toward the more sensitive low frequencies. At the same time, through the use of psychoacoustic masking techniques, those frequencies, which (it is assumed) will be masked by the ear, are also identified and removed. The data generated, therefore, describes the frequency content and the energy level at those frequencies, with more bits being allocated to the higher-energy frequencies than those with lower energy.

The larger the time block of the signal being analyzed, the better the frequency resolution and the greater the amount of irrelevancy identified. The penalty, however, is an increase in coding delay and a decrease in temporal resolution. A balance has been struck with advances in perceptual coding techniques and psychoacoustic modeling leading to increased efficiency. It is reported in that, with this approach to compression, some 80% of the input audio can be removed with acceptable results [16].

The hybrid arrangement of working with time-domain subbands and simultaneously carrying out a spectral analysis can be achieved by using a *dynamic bit allocation* process for each subband. This subband APCM approach is found in the popular range of software-based MUSICAM, Dolby AC-2, and ISO/MPEG-1 Layers 1 and 2 algorithms. Layer 3, a more complex method of coding and operating at much lower bit rates, is, in essence, a combination of the best functions of MUSICAM and ASPEC, another adaptive transform algorithm. Table E.5 lists the primary operational parameters for these systems.

Additionally, some of these systems exploit the significant redundancy between stereo channels by using a technique known as *joint stereo coding.* After the common information between left and right channels of a stereo signal has been identified, it is coded only once, thus reducing the bit rate demands yet again.

Each of the subbands has a defined *masking threshold.* The output data from each of the filtered subbands is requantized with just enough bit resolution to maintain adequate headroom between the quantization noise and the masking threshold for each band. In more complex coders (e.g., ISO/MPEG-1 Layer 3), those subbands with the greater need for increased masking threshold separation use any spare bit capacity. The maintenance of these signal-to-masking threshold ratios is crucial if further compression is contemplated for any post-production or transmission process.

TABLE E.5

Operational Parameters of Subband APCM Algorithm (After [16])

Coding systems	Compression ratio	Subbands	Bit rate, kbits/s	A to A delay, ms[1]	Audio bandwidth, kHz
Dobly AC-2	6:1	256	256	45	20
ISO Layer 1	4:1	32	384	19	20
ISO Layer 2	Variable	32	192–256	>40	20
ISO Layer 3	12:1	576	128	>80	20
MUSICAM	Variable	32	128–384	>35	20

[1]The total system delay (encoder-to-decoder) of the coding system.

Processing and Propagation Delay

As noted previously, the current range of popular compression algorithms operates, for all intents and purposes, in real-time [15]. However, this process does of necessity introduce some measurable delay into the audio chain. All algorithms take a finite time to analyze the incoming signal, which can range from a few milliseconds to tens and even hundreds of milliseconds. The amount of processing delay will be crucial if the equipment is to be used in an interactive or two-way application. As a rule of thumb, more than 20 ms of delay in a two-way audio exchange is problematic. Propagation delay in satellite and long terrestrial circuits is a fact of life. A two-way hook up over a 1000 km, full duplex, telecom digital link has a propagation delay of 3 ms in each direction. This is comparable to having a conversation with someone standing one meter away. It is obvious that even over a very short distance, the use of a codec with a long processing delay characteristic will have a dramatic effect on operation.

Bit Rate and Compression Ratio

The ITU has recommended the following bit rates when incorporating data compression in an audio chain [15]:

- 128 kilobits per second per mono channel (256 kilobits per second for stereo) as the minimum bit rate for any stage if further compression is anticipated or required.
- 192 kilobits per second per mono channel (384 kilobits per second for stereo) as the minimum bit rate for the first stage of compression in a complex audio chain.

These markers place a 4:1 compression ratio at the "safe" end in the scale. However, more aggressive compression ratios, currently up to a nominal 20:1, are available. Keep in mind, though, that low bit rate, high-level compression can lead to problems many stages of compression are required or anticipated.

With successive stages of compression, either or both the noise floor and the audio bandwidth will be set by the stage operating at the lowest bit rate. It is, therefore, worth emphasizing that after these platform have been set by a low bit rate stage, they cannot be subsequently improved by using a following stage operating at a higher bit rate.

Bit Rate Mismatch.   A stage of compression may well be followed in the audio chain by another digital stage, either of compression or linear, but more importantly, operating at a different sampling frequency [15]. If a D/A conversion is to be avoided, a sample rate converter must be used. This can be a stand-alone unit or it may already be installed as a hardware or software module in existing equipment. Where a following stage of compression is operating at the same sampling frequency but a different compression ratio, the bit resolution will change by default.

If the stages have the same sampling frequencies, a direct PCM or AES/EBU digital link can be made, thus avoiding the conversion to the analog domain.

Editing Compressed Data

The linear PCM waveform associated with standard audio workstations is only useful if decoded [15]. The resolution of the compressed data may or may not be adequate to allow direct editing of the audio signal. The minimum audio sample that can be removed or edited from a transform-coded signal will be determined by the size of the time block of the PCM signal being analyzed. The larger the time block, the more difficult the editing of the compressed data becomes.

Audio Compression Schemes Important to Streaming

Subband APCM coding has found numerous applications in the professional audio industry and the streaming media industry. We will examine ISO/MPEG-1 Layer 2 (MUSICAM by another name) and ISO/MPEG-1 Layer 3 (the popular MP3 format used by such applications as Winamp, MusicMatch, Napster, and leading streaming media formats). We will also examine Dolby AC-3, which is used in DVD players and the ATSC DTV system, and finally, the MPEG AAC system favored by MPEG-4 streaming media implementations.

ISO/MPEG-1 Layer 2 (MUSICAM). This algorithm differs from Layer 1 by adopting more accurate quantizing procedures and by additionally removing redundancy and irrelevancy on the generated scale factors [15]. The ISO/MPEG-1 Layer 2 scheme operates on a block of 1152 PCM samples, which at 48 kHz sampling represents a 24 ms time block of the input audio signal. Simplified block diagrams of the encoding/decoding systems are given in Figure E.16.

The incoming linear PCM signal block is divided into 32 equally spaced subbands using a polyphase analysis filter bank (Figure E.16a). At 48 kHz sampling, this equates to the bandwidth of each subband being 750 Hz. The bit allocation for the requantizing of these subband samples is then dynamically controlled by information derived from analyzing the audio signal, measured against a preset psychoacoustic model.

The filter bank, which displays manageable delay and minimal complexity, optimally adapts each block of audio to balance between the effects of temporal masking and inaudible pre-echoes.

Figure E.16
ISO/MPEG-1 Layer 2
system: (a) enoder
block diagram, (b)
decoder block
diagram. (After [15].)

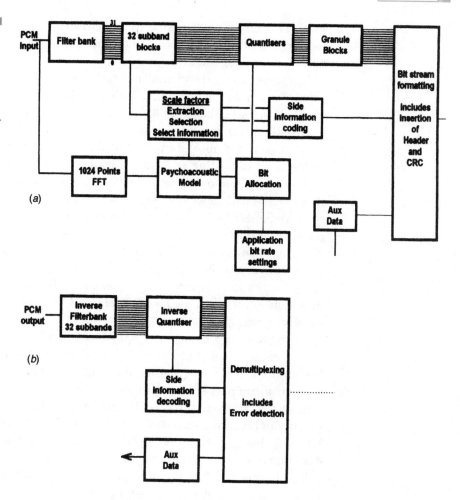

The PCM signal is also fed to a *fast Fourier transform* (FFT) running in parallel with the filter bank. The aural sensitivities of the human auditory system are exploited by using this FFT process to detect the differences between the wanted and unwanted sounds and the quantization noise already present in the signal and then to adjust the signal-to-mask thresholds, conforming to a preset perceptual model.

This psychoacoustic model is only found in the coder, thus making the decoder less complex and permitting the freedom to exploit future improvements in coder design. The actual number of levels for each quantizer is determined by the bit allocation. This is arrived at by setting the *signal-to-mask ratio* (SMR) parameter, defined as the difference between the minimum masking threshold and the maximum signal level. This minimum masking threshold is calculated using the psychoacoustic model and provides a reference noise level of "just noticeable" noise for each subband.

In the decoder, after demultiplexing and deciphering of the audio and side information data, a dual-synthesis filter bank reconstructs the linear PCM signal in blocks of 32 output samples (Figure E-16b).

A scale factor is determined for each 12-subband sample block. The maximum of the absolute values of these 12 samples generates a *scale factor* word consisting of 6 bits, a range of 63 different levels. Because each frame of audio data in Layer 2 corresponds to 36 subband samples, this process will generate 3 scale factors per frame. However, exploiting some redundancy in the data can reduce the transmitted data rate for these scale factors. Three successive subband scale factors are analyzed and a pattern is determined. This pattern, which is obviously related to the nature of the audio signal, will decide whether one, two, or all three scale factors are required. The decision will be communicated by the insertion of an additional *scale factor select information* (SCFSI) data word of 2 bits.

In the case of a fairly stationary tonal-type sound, there will be very little change in the scale factors and only the largest one of the three is transmitted; the corresponding data rate will be $(6 + 2)$ or 8 bits. However, in a complex sound with rapid changes in content, the transmission of two or even three scale factors may be required, producing a maximum bit rate demand of $(6 + 6 + 6 + 2)$ or 20 bits. Compared with Layer 1, this method of coding the scale factors reduces the allocation of data bits required for them by half.

The number of data bits allocated to the overall bit pool is limited or fixed by the data rate parameters. These parameters are set out by a combination of sampling frequency, compression ratio and, where applicable, the transmission medium. In the case of 20 kHz stereo being transmitted over ISDN, for example, the maximum data rate is 384 kilobits per second, sampling at 48 kHz, with a compression ratio of 4:1.

After the number of side information bits required for scale factors, bit allocation codes, CRC (Cyclic Redundancy Check), and other functions have been determined, the remaining bits left in the pool are used in the recoding of the audio subband samples. The allocation of bits for the audio is determined by calculating the SMR, via the FFT, for each of the 12 subband sample blocks. The bit allocation algorithm then selects one of the 15 available quantizers with a range such that the overall bit rate limitations are met and the quantization noise is masked as far as possible. If the composition of the audio signal is such that there are not enough bits in the pool to adequately code the subband samples, then the quantizers are adjusted down to a best-fit solution with (hopefully) minimum damage to the decoded audio at the output.

If the signal block being processed lies in the lower one-third of the 32 frequency subbands, a 4-bit code word is simultaneously generated to identify the selected quantizer; this word is, again, carried as side information in the main data frame. A 3-bit word would be generated for processing in the mid-frequency subbands and a 2-bit word for the higher frequency subbands. When the audio analysis demands it, this allows for *at least* 15, 7, and 3 quantization levels, respectively, in each of the three spectrum groupings. However, each quantizer can, if required, cover from 3 to 65,535 levels; additionally, if no signal is detected then no quantization takes place.

As with the scale factor data, some further redundancy can be exploited, which increases the efficiency of the quantizing process. For the lowest quantizer ranges (i.e., 3, 5, and 9 levels), three successive subband sample blocks are grouped into a "granule" and this, in turn, is defined by only one code word. This is particularly effective in the higher frequency subbands where the quantizer ranges are invariably set at the lower end of the scale.

Error detection information can be relayed to the decoder by inserting a 16-bit CRC word in each data frame. This parity check word allows for the detection of up to three single bit errors or a group of errors up to 16 bits in length. A codec incorporating an error concealment regime can either mute the signal in the presence of errors or replace the impaired data with a previous, error free, data frame. The typical data frame structure for ISO/MPEG-1 Layer 2 audio is given in Figure E.17.

The compression ratios for ISO/MPEG-1 Layer 2 typically fall between 6:1 and 8:1. This corresponds to bit rates of 256 to 192 kilobits per second for a stereo signal.

Figure E.17
ISO/MPEG-1 Layer 2 data frame structure. (After [15].)

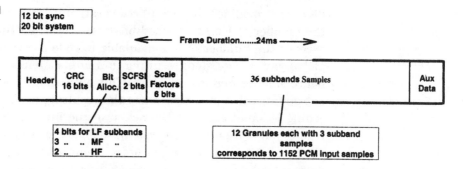

ISO/MPEG-1 Layer 3 Compression (MP3). ISO/MPEG-1 Layer 3 audio, also known as the infamous MP3 format, builds upon the Layer 2 codec, but has compression ratios that fall between 10:1 and 12:1, corresponding to 128 and 112 kilobits per second respectively, for a stereo audio signal. It adds a number of "advanced features" to Layer 2. The frequency resolution is 18 times higher, which allows a Layer 3 encoder to better adapt the quantization noise to the masking threshold. A *modified discrete cosine transform* (MDCT) is used to increase the frequency resolution. Layer 3 uses entropy coding (like MPEG video) to further reduce redundancy. It also uses a bit reservoir (like MPEG video) to suppress artifacts in critical moments. Finally, Layer 3 may use more advanced joint-stereo coding methods. Joint stereo techniques are used where available bit rate is limited, but stereo signals are desired.

The filter bank used in Layer 3 is a hybrid filter bank, consisting of a polyphase filter bank (as in Layer 2) and the MDCT. The hybrid form was chosen to maintain backward compatibility with Layers 1 and 2. A Layer 3 decoder should be able to decode audio encoded using the lower layers.

A system of two nested iteration loops is used for quantization and coding in the Layer 3 encoder. These are the inner iteration *rate* loop and the outer iteration *noise control/distortion* loop. Quantization is done using a power-law quantizer, where larger values are automatically coded with less accuracy than smaller values. Some noise shaping is thus already built into the quantization process, in much the same way as α-law or A-law converters use nonuniform quantization to achieve improvements in signal to noise ratio.

The quantized values are entropy-coded using Huffman codes, which are lossless. This is called *noiseless coding* because no noise is added to the audio signal.

The two nested iteration loops find the optimum gain and scale factors for a given block, bit rate, and perceptual model output using an *analysis-by-synthesis* technique. The inner loop is called the rate loop because it modifies the overall coder rate until it is small enough. When the number of Huffman-coded bits resulting from the coding operation exceeds the number of bits available to code a given block of data, the global gain is adjusted, resulting in a larger quantization step size and hence smaller quantized values. This operation is repeated with different quantization step sizes until the resulting bit demand for Huffman coding is small enough. The action of the inner loop is analogous to a traditional noise reduction system, such as those made famous by Dolby, since gain is manipulated in order to optimize the effective signal-to-noise ratio, as would happen using a broadband audio level compressor.

The outer loop shapes the quantization noise according to the masking threshold for each subband. The system starts with a default scale factor of 1.0 for each subband. If the quantization noise in a given band exceeds the allowable noise, determined by the masking threshold supplied by the perceptual model, the scale factor for the band is adjusted to reduce the quantization noise. Since smaller quantization noise requires a larger number of quantization steps (and thus, a higher bit rate), the rate adjustment loop has to be repeated every time new scale factors are computed. In other words, the rate loop is nested within the noise control loop. The outer noise control loop executes until the actual noise, computed as the difference between the original spectrum values and the quantized spectrum values, is below the masking threshold for every subband.

Compression techniques closely related to ISO/MPEG-1 Layer 3 audio compression find application in a number of popular streaming media formats.

MPEG-2 AAC. Of particular note is MPEG-2 *advanced audio coding* (AAC), a highly advanced perceptual code, used initially for digital radio applications. The AAC code improves on previous techniques to increase coding efficiency. For example, an AAC system operating at 96 kilobits per second produces the same subjective sound quality as ISO/MPEG-1 Layer 2 operating at 192 kilobits per second; a 2:1 reduction in bit rate. The driving force to develop AAC was the quest for an efficient coding method for surround signals, like the five-channel systems (left, right, center, left-surround, right-surround) used in cinemas today. There have been algorithms for these signals in MPEG for quite a while, but optimum efficiency was not reached, for technical and historical reasons. The set aim, therefore, was a considerable decrease in the bit rate necessary to represent surround sound program material.

There are three main modes (Profiles) in the AAC standard:

- **Main**—Used when processing power and especially memory are readily available.
- **Low complexity (LC)**—Used when processing cycles and memory use are constrained.
- **Scaleable sampling rate (SSR)**—Appropriate when a *scalable decoder* is required. A scalable decoder can be designed to support different levels of audio quality from a common bit stream; for example, having both high- and low-cost implementations to support higher and lower audio qualities, respectively.

Different profiles trade off encoding complexity for audio quality at a given bit rate. For example, at 128 kilobits per second, the Main profile AAC code has a more complex encoder structure than the LC AAC code at the same bit rate, but provides better audio quality as a result.

Like all perceptual coding schemes, MPEG-2 AAC makes use of the signal masking properties of the human auditory system in order to reduce the amount of data. In so doing, the quantization noise is distributed to frequency bands in such a way that it is masked by the aggregate signal (in other words, it remains inaudible). Even though the basic structure of this coding method hardly differs from its predecessors, there are some new aspects worth noting.

A block diagram of the AAC system general structure is given in Figure E.18. The blocks in the drawing are referred to as "tools" that the coding algorithm uses to compress the digital audio signal. While many of these tools exist in most audio perceptual coders, two are unique to AAC: the *temporal noise shaper* (TNS) and the *filterbank* tool. The TNS uses a backward adaptive prediction process to remove redundancy between the frequency channels that are created by the filterbank tool. A true novelty in the area of time/frequency coding schemes, the TNS shapes the distribution of quantization noise in the time domain by prediction in the frequency domain. Prediction is a technique commonly used in speech coding systems. AAC benefits from the fact that many types of audio signal are easy to predict. Voice signals, in particular, experience considerable improvement through TNS.

Figure E.18

Functional block diagram of the MPEG-2 AAC coding system.

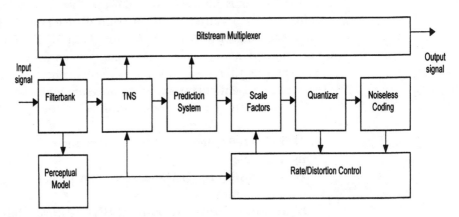

MPEG-2 AAC provides the capability of up to 48 main audio channels, with sampling frequencies between 8 kHz and 96 kHz, 16 low frequency effects channels, 16 overdub/multilingual channels, and 10 data streams. By comparison, ISO/MPEG-1 Layer 1 provides two channels and Layer 2 provides 5.1 channels (maximum). AAC is not backward compatible with the Layer 1 and Layer 2 codes, since MPEG-2 AAC uses a different type of transform than MPEG-1 audio coding. In contrast to the hybrid filter bank of ISO/MPEG Audio Layer 3, chosen for reasons of compatibility but displaying certain structural weaknesses, MPEG-2 AAC uses a plain *modified discrete cosine transform* (MDCT). Together with the increased window length (2048 instead of 1152 lines per transformation) the MDCT outperforms the filter banks of previous coding methods. By allowing finer control of quantization resolution through the choice of MDCT, the given bit rate can be used more efficiently.

MPEG-2 AAC also improves upon the three-layered MPEG-1 audio coding tools by providing a *low sample rate extension* to address very low bit rate applications with limited bandwidth requirements. Sampling frequencies of 16, 22.05 and 24 kHz have been added to the 32, 44.1 and 48 kHz sampling frequencies specified in the MPEG-1 audio coding layers and the bit rate now extends down to 8 kilobits per second.

AAC is an important tool in the coding of natural audio in MPEG-4. With its superior sound quality for a given bit rate, or lower bit rate for a given sound quality, applications that rely on MPEG-1 Layer 3 compression, such as the popular portable MP3 players available today, will begin to adopt AAC in preference to earlier compression standards. In particular, the advantages for streaming media delivery of surround sound experiences, using limited bandwidth, will make AAC an attractive choice for MPEG-4 compliant streaming media applications.

Dolby AC-3 5.1 Channel Surround Sound. Dolby AC-3 is a proprietary multichannel perceptual coding technique that predates MPEG-2 AAC. It does many of the same things as AAC; namely deliver surround sound to end-users. It is mentioned here for historical perspective and because the system finds application in DVD, ATSC DTV, and D-cinema. Although Dolby helped in the development of AAC, they have also actively and consistently promoted their own AC-3 technology. As with AAC, AC-3 allows bit-rate and number of channels to be tailored to particular applications. It combines high quality sounds with good data efficiency, delivering multichannel sound at a lower data rate than is needed for just one channel on a CD.

Dolby AC-3 had its origins in developing a digital multichannel sound delivery system for feature film presentation in movie theaters, when four channels of sound were matrixed onto two optical tracks. When Dolby Stereo films, as they are known, are transferred to two channel video formats, like VHS, the four-channel encoding survives intact. Consumers began constructing "home theaters", with surround sound systems, to get a similar sound field to that which they would have experienced watching films in a theater from their videotapes.

The actual development of the AC-3 coding system was in response to a requirement for a four or more discrete channel audio solution for the high definition television (HDTV) system. Like the Dolby Digital film format, AC-3 provides separate channels for left, right, and center speakers at the front; two surround speakers at the sides; and a subwoofer at the listener's option.

AC-3 was designed to provide the following features:

- Data identifying each program's original production format—mono, stereo, matrixed, or discrete surround—can be sent to eliminate confusion at playback or reception.
- Program material can be coded when it is originally mixed so that subjectively constant, dialog-keyed loudness is maintained as the listener switches between program sources. No alteration of program dynamics is involved, only playback volume.
- Decoders can be designed to provide optimum mix-downs from multichannel programming, such as a matrix-encoded two-track mix for analog Dolby Surround decoding, a conventional stereo mix, or even a mono mix.
- When programs with wide dynamic range, such as movie soundtracks, are played at low volume, the system can apply appropriate compression to preserve low-level content. The degree of compression can be made to vary according to need.
- The listener can program the AC-3 decoder (known commercially as a Dolby Surround Digital decoder) to route nondirectional low bass only to those channels in the system that have wide range speakers or subwoofers.

Dolby AC-3 was the first perceptual coding standard designed specifically to code multichannel digital audio. It benefited from the development of two other perceptual coding systems, Dolby AC-1 and AC-2, and from the development of what are in essence *analog* perceptual coding systems: the full catalog of Dolby professional and consumer noise reduction systems. Indeed, Dolby Laboratories' experience with audio noise reduction was useful in developing AC-3's effective data rate reduction, since the fewer the bits used to describe an audio signal, the greater the noise—and Dolby had made a decent living from reducing noise.

The Australian Broadcasting Corporation first used Dolby AC-l in 1985 for DBS applications. Partly because of its low decoder cost, AC-l has since been adopted for other DBS services, satellite communication networks, and digital "cable radio" systems; the data rate is 220–325 kilobits per second per channel depending on application. A refined form of adaptive delta modulation (ADM), the data stream contains information not on the absolute value of the audio signal, but on the change in value from sample to sample. Techniques adapted from Dolby noise reduction, such as continually varying step-size and pre-emphasis, improve on basic ADM performance.

Dolby AC-2 uses advanced adaptive transform coding for professional audio applications; its data rate being 128 or 192 kilobits per second per audio channel. Frequency-domain signal processing in a multiplicity of narrow bands took full advantage of noise masking, resulting in data rate reduction combined with high signal transparency.

Dolby noise reduction, an important precursor to the development of AC-3, works by lowering the noise when no audio signal is present, while allowing strong audio signals to cover or mask the noise at other times. Thus, it takes advantage of the psychoacoustic phenomenon known as *auditory masking*. Even when audio signals are present in some parts of the spectrum, Dolby NR reduces the noise in the other parts so the noise remains imperceptible. This is because audio signals can only mask noise that occurs at nearby frequencies.

AC-3 was designed to take advantage of human auditory masking, in common with all other perceptual audio coding schemes. It divides the audio spectrum of each channel into narrow frequency bands of different sizes, optimized with respect to the frequency selectivity of human hearing. This makes it possible to sharply filter coding noise so that it is forced to stay very close in frequency to the frequency components of the audio signal being coded. Reducing or eliminating coding noise, wherever there are no audio signals to mask it, can subjectively preserve the sound quality of the original signal. In this respect, a perceptual coding system like AC-3 is essentially a form of very selective noise reduction.

In AC-3, bits are distributed among the filter bands as needed by the particular frequency spectrum or dynamic nature of the program, as with AAC. A built-in model of auditory masking allows the coder to alter its frequency selectivity (as well as time resolution) to make sure that a sufficient number of bits are used to describe the audio signal in each band, thus ensuring noise is fully masked. AC-3 also decides how the bits are distributed among the various channels from a common bit pool. This technique allows channels with greater frequency content to demand more data than sparsely occupied channels, for example, or strong sounds in one channel to provide masking for noise in other channels.

AC-3's masking model and its shared bit pool are key factors in its spectrum efficiency. Furthermore, where other coding systems use considerable (and precious) data to carry instructions to their decoders, AC-3 can use proportionally more of the transmitted data to represent audio, which arguably means higher sound quality.

AC-3 can process at least 20-bit dynamic range digital audio signals over a frequency range from 20 Hz to 20 kHz. The bass effects channel covers 20 to 120 Hz. Sampling rates of 32, 44.1 and 48 kHz are supported. Data rates in AC-3 coding range from as low as 32 kilobits per second, for a single mono channel, up to 640 kilobits per second, covering a wide range of applications. Typical applications include 384 kilobits per second for 5.1-channel Dolby Surround Digital consumer formats, and 192 kilobits per second for two-channel audio distribution.

With AAC and AC-3 being similar in what they can do for the end-user, it is hard to see why one system should be preferred over the other. According to supporters of the MPEG-2 AAC standard, being a more recently developed, elaborate, and complex compression algorithm than the older AC-3, it provides superior compression efficiency and subjectively better sound quality. AAC and AC-3 are both transform coders, but AAC uses a filterbank with a finer frequency resolution that enables superior signal compression. AAC also uses a number of new tools, such as temporal noise shaping, backward adaptive linear prediction, joint stereo coding techniques, and Huffman coding of quantized components, each of which provides additional audio compression capability. Furthermore, AAC is much more flexible than AC-3, in that AAC supports a wide range of sampling rates and bit rates, from one to 48 audio channels, a multitude of low frequency enhancement channels, multilanguage capability, and support for embedded data streams. However, AC-3 has been cemented into the ATSC DTV standard and finds application in the highly successful domestic DVD standard, among other applications, so it isn't likely to vanish anytime soon.

References

1. Lakhani, Gopal: "Video Compression Techniques and Standards," *The Electronics Handbook*, Jerry C. Whitaker (ed.), CRC Press, Boca Raton, Fla., pp. 1273-1282, 1996.
2. Solari, Steve. J.: *Digital Video and Audio Compression*, McGraw-Hill, New York, 1997.

3. Netravali, A.N., and B.G. Haskell: *Digital Pictures, Representation and Compression*, Plenum Press, 1988.

4. Gilge, M.: "Region-Oriented Transform Coding in Picture Communication," *VDI-Verlag, Advancement Report*, Series 10, 1990.

5. DeWith, P. H. N.: "Motion-Adaptive Intraframe Transform Coding of Video Signals," *Philips J. Res.*, vol. 44, pp. 345-364, 1989.

6. Isnardi, M., and T. Smith: "MPEG Tutorial," *Proceedings of the Advanced Television Summit*, Intertec Publishing, Overland Park, Kan., 1996.

7. Nelson, Lee J.: "Video Compression," *Broadcast Engineering*, Intertec Publishing, Overland Park, Kan., p. 42, October 1995.

8. Arvind, R., et al.: "Images and Video Coding Standards," *AT&T Technical J.*, p. 86, 1993.

9. SMPTE 308M, "MPEG-2 4:2:2 Profile at High Level," SMPTE, White Plains, N.Y., 1998.

10. ATSC, "Guide to the Use of the ATSC Digital Television Standard," Advanced Television Systems Committee, Washington, D.C., doc. A/54, Oct. 4, 1995.

11. "IEEE Standard Specifications for the Implementation of 8 x 8 Inverse Discrete Cosine Transform," std. 1180-1990, Dec. 6, 1990.

12. Nelson, Lee J.: "Video Compression," *Broadcast Engineering*, Intertec Publishing, Overland Park, Kan., pp. 42-46, October 1995.

13. Smith, Terry: "MPEG-2 Systems: A Tutorial Overview," Transition to Digital Conference, *Broadcast Engineering*, Intertec Publishing, Overland Park, Kan., Nov. 21, 1996.

14. SMPTE Recommended Practice: RP 202 (Proposed), "Video Alignment for MPEG-2 Coding," SMPTE, White Plains, N.Y., 1999.

15. Wylie, Fred: "Audio Compression Technologies" in *NAB Engineering Handbook*, 9th ed., Jerry C. Whitaker (ed.), National Association of Broadcasters, Washington, D.C., 1998.

16. Wylie, Fred: " Audio Compression Techniques," *The Electronics Handbook*, Jerry C. Whitaker (ed.), CRC Press, Boca Raton, Fla., pp. 1260-1272, 1996.

17. Lyman, Stephen: "A Multichannel Audio Infrastructure Based on Dolby E Coding," *Proceedings of the NAB Broadcast Engineering Conference*, National Association of Broadcasters, Washingtom, D.C., 1999.

18. Terry, K. B., and S. B. Lyman: "Dolby E—A New Audio Distribution Format for Digital Broadcast Applications," *International Broadcasting Convention Proceedings*, IBC, London, England, pp. 204-209, September 1999.

Bibliography

Bennett, Christopher: "Three MPEG Myths," *Proceedings of the 1996 NAB Broadcast Engineering Conference*, National Association of Broadcasters, Washington, D.C., pp 129-136, 1996.

Bonomi, Mauro: "The Art and Science of Digital Video Compression," *NAB Broadcast Engineering Conference Proceedings*, National Association of Broadcasters, Washington, D.C., pp. 7-14, 1995.

Brandenburg, K., and Gerhard Stoll: "ISO-MPEG-1 Audio: A Generic Standard for Coding of High Quality Digital Audio," *92nd AES Convention Proceedings*, Audio Engineering Society, New York, N.Y., 1992, revised 1994.

Dare, Peter: "The Future of Networking," *Broadcast Engineering*, Intertec Publishing, Overland Park, Kan., p. 36, April 1996.

Fibush, David K.: "Testing MPEG-Compressed Signals," *Broadcast Engineering*, Overland Park, Kan., pp. 76-86, February 1996.

Freed, Ken: "Video Compression," *Broadcast Engineering*, Overland Park, Kan., pp. 46-77, January 1997.

IEEE Standard Dictionary of Electrical and Electronic Terms, ANSI/IEEE Standard 100-1984, Institute of Electrical and Electronics Engineers, New York, 1984.

Jones, Ken: "The Television LAN," *Proceedings of the 1995NAB Engineering Conference*, National Association of Broadcasters, Washington, D.C., p. 168, April 1995.

Smyth, Stephen: "Digital Audio Data Compression," *Broadcast Engineering*, Intertec Publishing, Overland Park, Kan., February 1992.

Stallings, William: *ISDN and Broadband ISDN*, 2nd Ed., MacMillan, New York.

Taylor, P.: "Broadcast Quality and Compression," *Broadcast Engineering*, Intertec Publishing, Overland Park, Kan., p. 46, October 1995.

Whitaker, Jerry C. and Harold Winard (eds.): *The Information Age Dictionary*, Intertec Publishing/Bellcore, Overland Park, Kan., 1992.

Web Resources

http://www.iis.fhg.de/amm/techinf/layer3/index.html
http://www.iis.fhg.de/amm/techinf/faq.html
http://mpeg.telecomitalialab.com/faq/mp2-aud/mp2-aud.htm
http://smirnoff.nncity.ru/smirnoff/music/AAC-FAQ.htm
http://www.dolby.com/tech/entirmpc.html

GLOSSARY

.NET—Microsoft's distributed applications environment and initiative.

2RDD—Multimedia Rights Data Dictionary created by the DOI Foundation.

3G—Third generation; referring to the next generation of cellular phone network technology.

802.11—IEEE standards designation for wireless networking in a personal area network (a.k.a. WiFi).

802.15—IEEE standards designation for wireless networking optimized for media transport (a.k.a. WiMedia).

802.16—IEEE standards designation for wireless networking over a metropolitan area network (a.k.a. WirelessMAN).

AAC (Advanced Audio Coding)—An improvement on the popular MP3 audio coding scheme.

ACELP (Algebraic Code Excited Linear Prediction)—A technique for compressing speech-quality audio.

ActiveX—Microsoft's technology for embedding self-registering software objects in Web pages and other programs.

ADSL (Asynchronous Digital Subscriber Line)—A way to transmit data to the home at high speed over existing telephone company copper wires.

AGP (Accelerated Graphics Port)—A hardware bus for rapidly moving graphics data from a central processing unit to a graphics adapter.

AIFF (Apple Interchange File Format)—A file container for digital media content.

AMI-C (Automotive Multimedia Interface Collaboration)—A body working to standardize data buses in automobiles.

ANSI (American National Standards Institute)—The primary organization for fostering the development of technology standards in the United States.

AOL (America Online)—A global Internet service provider.

ASF (Advanced Streaming Format)—Microsoft's proprietary streaming media file format, which interleaves audio and video data.

ASX (Advanced Stream Redirector)—A file format used with Microsoft Windows Media Technologies.

ATM (Asynchronous Transfer Mode)—A networking technology used by telecommunications companies to ensure virtual switched circuit quality of service on digital networks.

ATVEF (Advanced Television Enhancement Forum)—An industry body to promote the use of standards-based interactivity with digital television.

AVI (Audio Video Interleave)—Microsoft's file format for capturing raw audio and video in an interleaved fashion.

AVO (Audio Visual Object)—The discrete element in an MPEG-4 scene presentation.

AVR (Audio Visual Research)—A sound file format on the 680x0-based Atari ST computers.

BCH (Bose-Chaudhuri-Hocquenghem)—A forward error-correction algorithm, used in control channels for cellular TDMA in the US, for example.

BIFS (Binary Format for Scenes)—An MPEG-4 scene description format.

BiM (Binary Metadata)—Used in MPEG-7 to create binary descriptions about media.

Bit—(Binary Digit); the smallest discrete unit of information.

BMP—Bitmap file extension for files that hold image information as a raw bitmap.

C#—Microsoft's newest programming language.

C++—A programming language.

CAD—Computer Aided Design.

CARP—Cache Array Routing Protocol.

CATV—Community Access TeleVision, also known as Cable TV.

CBT (Core Based Trees)—A sparse mode multicast routing protocol.

CCTV—Closed Circuit Television.

CD—Compact Disk.

CDMA (Code Division Multiple Access)—A cellular telephone technique for carrying multiple channels on a wireless network.

CDN (Content Delivery Network)—A network designed specifically to reliably deliver content.

CD-R—Compact Disk Recordable.

CLEC (Competitive Local Exchange Carrier)—One of the companies that were supposed to gain access to the phone companies' local loops in order to offer competitive broadband service to consumers.

Compact PCI (Compact Peripheral Component Interconnect)—A hardware standard to create processing units, typically used in the telecommunications industry.

CPU (Central Processing Unit)—Also known as a microprocessor—the beating heart of every PC.

CR-LDP (Constraint-based Routing Label Distribution Protocol)—A routing protocol for traffic engineering on the Internet, used to ensure quality of service.

D-Cinema (Digital Cinema)—Technology to enable cinema quality presentation without the use of film.

DCT (Discrete Cosine Transform)—A mathematical device used in video compression algorithms.

DDL (Description Definition Language)—Used in MPEG-7 to define description schemes.

DHTML (Dynamic Hyper Text Mark-up Language)—Web page tags that cause browsers to animate the content.

DID (Digital Item Declaration)—MPEG-21 declaration of the existence of a digital content item.

DII&D (Digital Item Identification and Description)—MPEG-21 unique identifier and description for digital content items.

DirectShow—Microsoft's application programming interface for audio and video playback and manipulation; a subset of the DirectX suite.

DirectX—Microsoft's suite of technologies for multimedia rendering on Windows platforms.

DMAT (Digital Music Access Technology)—Secure Digital Music Initiative's trademark for products that are compliant with SDMI specifications.

DMIF (Delivery Multimedia Integration Framework)—MPEG-4 technology for receiving and synchronizing elemental streams delivered over several networks.

DNS (Domain Name System)—The way that Internet domain names are located and translated into Internet protocol addresses.

DOI (Digital Object Identifier)—A digital equivalent to the ISBN book cataloging system for digital media assets.

DoS (Denial of Service)—An attack in which system resources are maliciously used up, keeping legitimate users from obtaining normal service.

DPRL (Digital Property Rights Language)—Xerox PARC digital media commerce language.

DRM (Digital Rights Management)—Generic term for technologies that protect digital copyrights and for Microsoft's own rights-management product.

DSL (Digital Subscriber Line)—A method of transmitting digital data over standard copper twisted-pair telephone wires.

DTV (Digital Television)—A way of broadcasting television programming digitally.

DV (Digital Video)—Tape and compression format for digital video camcorders.

DVB (Digital Video Broadcast)—Technical standard for digital television transmission.

DVD (Digital Versatile Disk, a.k.a. Digital Video Disk)—Popular format for the distribution of digital media content.

DVD-Audio—A carrier for digital audio, using DVD technology and standards.

DVD-RAM—A recordable version of the DVD.

DVMRP—Distance Vector Multicast Routing Protocol.

DWDM (Dense Wave Division Multiplexing)—A method of using a single optical fiber to carry much more digital data.

DWT (Discrete Wavelet Transform)—A mathematical device used in video compression algorithms.

EBU—European Broadcasting Union.

ENG—Electronic News Gathering.

EPSF—Extensible Proxy Services Framework.

Exabit—10^{18} bits.

Ezine—Electronic version of a fanzine; a magazine published by fans on niche subjects.

FEC (Forward Error Correction)—A technique that allows perfect reconstruction of data transmitted via a lossy channel, by spreading the information over several data packets with redundancy.

FGS (Fine Granular Scalability)—MPEG-4 method of building up picture detail successively, from separate elemental streams.

FITL (Fiber in the Loop)—Refers to the presence of fiber optic carriers in the local loop, which connects individual consumers to the telephone network.

Flash—Macromedia's interactive multimedia authoring tool.

FlexMux (Flexible Multiplex)—Part of the MPEG-4 architecture; it allows multiple elemental streams to be flexibly synchronized.

FM (Frequency Modulation)—A popular analog radio transmission system.

Fps (Frames Per Second)—A measurement of film or video temporal resolution.

FreeBSD—BSD stands for Berkeley Software Distribution. FreeBSD is a free, open-source version of the BSD UNIX operating system developed at Berkeley University.

FTP (File Transfer Protocol)—A protocol for transporting files over the Internet.

FTTC (Fiber to the Curb)—A fiber optical connection serving a group of houses.

FTTH (Fiber to the Home)—A direct fiber optical connection to the home.

FTTN (Fiber to the Neighborhood)—An optical fiber connection serving a neighborhood.

GB (Gigabyte)—A byte is 8 bits and a gigabyte is 10^9 bytes.

GEO (Geosynchronous Earth Orbit)—Satellites that maintain the same apparent position above the earth.

GIF (Graphics Interchange Format)—A file format for compressing graphic images losslessly.

GPS (Global Positioning System)—A system for locating any point on earth, based on signals from a network of specialized location satellites.

GSM (Global System for Mobile communication)—Used for cellular telephony.

GUI—Graphical User Interface.

H.263—An ITU video compression standard.

HDCD (High Definition Compact Disk)—Microsoft's proprietary system for releasing higher-resolution music CDs which are compatible with the existing CD format.

HDSL—High bit-rate DSL.

HDTV—High Definition Television.

HFC (Hybrid Fiber/Coax)—A broadband connection technology.

HSV (Hue Saturation Value)—A way of describing colors with these parameters.

HTML (Hyper Text Markup Language)—The language of Web page authoring.

HTML+Time—Microsoft's version of a markup language that directs the browser to perform dynamic effects and transitions; now incorporated in SMIL 2.0.

HTTP (Hyper Text Transport Protocol)—How the elements of a Web page are transported from server to browser.

ICAP (Internet Content Adaptation Protocol)—A protocol to allow edge servers to perform value-added services.

ICCP (Inter Cache Cooperation Protocol)—An extension to HTTP and ICP, which allows purging of cached objects, tracing of HTTP requests through a sequence of proxies, and the removal of URLs from ICP replies.

ICMP—Internet Control Message Protocol.

ICP (Internet Cache Protocol)—A UDP-based protocol used for locating instances of cached responses in neighbor caches.

ICQ—Not an acronym but meant to suggest "I Seek You"; a popular Internet chat and instant messaging service.

IDRM (Internet Digital Rights Management)—An organization sponsored by the IETF to standardize digital rights management on the Internet.

IEC (International Electrotechnical Commission)—Produces international standards on electrical and electronic matters.

IEEE (Institution of Electrical and Electronic Engineers)—The body that sponsors electrical technical standards, particularly for wireless networking.

IETF (Internet Engineering Task Force)—The technical body that standardizes and develops protocols for the Internet.

IFPI—International Federation of the Phonographic Industry.

IGMP—Internet Group Management Protocol.

ILEC (Incumbent Local Exchange Carriers)—Also known as local phone companies.

IP (Internet Protocol)—The method by which packets of data are transported over the Internet.

IPMP (Intellectual Property Management and Protection)—MPEG-21 method for digital rights management.

IRML (Intermediary Rule Mark-up Language)—An XML-based language used to describe service-specific execution rules on an edge server.

ISDN (Integrated Services Digital Network)—The digital connection to the home provided by the phone company.

IS-IS—Intermediate System to Intermediate System protocol.

IS-IS/TE—Intermediate System to Intermediate System with Traffic Engineering.

ISMA—Internet Streaming Media Alliance.

ISO—International Organization for Standardization.

ISP (Internet Service Provider)—Any company that provides access to the Internet.

ISV (Independent Software Vendor)—An author of application software.

IT—Information Technology.

ITU—International Telecommunications Union.

IVDS-DTV—Interactive Video and Data Services delivered using digital television carriers.

IVR (Interactive Voice Response)—Systems used to respond automatically to voice commands, often used by call centers.

JPEG (Joint Picture Experts Group)—A standardized method for still-image compression.

JPEG2000—Next generation of still image compression.

Kilobit—10^3 bits.

LAN—Local Area Network.

LC-RTP—Loss Collection Real Time Protocol.

LDP (Label Distribution Protocol)—Used to ensure network quality of service.

LEO (Low Earth Orbit)—A satellite network being deployed to transport network data.

Linux—UNIX-like operating system originally written by Linus Torvalds, now an open-source system.

LMDS (Local Multipoint Distribution System)—A broadband network.

LSP (Label Switched Path)—A packet-network equivalent of a virtual circuit, used to guarantee quality of service.

Lumen—A unit of light intensity, generally used as a guide to the brightness of digital projectors.

MBone (Multicast Backbone)—An experimental Internet infrastructure established for multicast transmission.

Megabit—10^6 bits.

MHEG-5—Multimedia and Hypermedia Experts Group standard for interactive digital media.

MHP—Multimedia Home Platform, which seeks to converge broadcast and the Internet in a common application-programming interface, mainly for set-top boxes.

MIDI—Musical Instrument Digital Interface.

M-JPEG (Motion Joint Picture Experts Group)—A video compression format consisting of sequences of JPEG images.

MMDS (Multipoint Multichannel Distribution Services)—A wireless broadband delivery technology.

MMS (Microsoft Media Streaming protocol)—A derivative of the Real Time Protocol.

Mmusic (Multiparty Multimedia Session Control)—An IETF working group responsible for multicast protocols.

MOSPF (Multicast Open Shortest Path First)—A multicast routing protocol.

MOST (Media Oriented Systems Transfer)—An automotive standard for in-car data transmission.

MOV—Apple QuickTime Movie format; container file format for digital media.

MP3—Shorthand for a file format that contains MPEG-2 Layer 3-encoded audio.

MPAA—Motion Picture Association of America.

MPEG—Motion Picture Experts Group.

MPEG-1—Video and audio compression standard.

MPEG-2—Improved video and audio compression standard.

MPEG-4—Standard for multimedia compression and transmission.

MPEG-7—Standard for digital media description

MPEG-21—Standard for digital media transactions.

MPLS (Multi-Protocol Label Switching)—A technique for tag switching in Internet routers, first proposed by Cisco Systems.

MSN (Microsoft Network)—A competitor to AOL.

Multicast—A technique for broadcasting Internet data packets to multiple receivers.

MVDS (Microwave Video Distribution Systems)—Also known as cellular TV.

Napster—A program used to share files peer to peer. Consumers used it primarily for swapping digital music files (illegally).

NECP (Network Element Control Protocol)—A lightweight protocol for signaling between servers and network elements that forward traffic to them, primarily to perform load balancing.

NTSC (National Television Standards Committee)—A standard for analog television broadcasting in the US.

NVOD (Near Video On Demand)—A video on demand system that has longer latency, often because material must be sourced from deep archive, as opposed to being available from online storage.

OC48—2.45 gigabits per second Optical Carrier; a fiber optic backbone.

OC768—40 gigabits per second, state-of-the-art fiber optical carrier.

OECD—Organization for Economic Cooperation and Development.

OpenDML—Open Digital Media file format extensions to the AVI file format, originally proposed by Matrox.

OPES (Open Pluggable Web Services)—Allows the plug-in inclusion of code which runs on an edge server.

ORDL (Open Digital Rights Language)—A vocabulary for the expression of terms and conditions over digital content.

OS X—Apple's operating system 10.

OSPF (Open Shortest Path First)—A routing algorithm and protocol.

OSPF/TE—Open Shortest Path First with Traffic Engineering.

P2P (Peer to Peer)—A method of sharing digital media.

PAL (Phase Alternating Line)—An analog broadcast television standard in much of Europe.

PC—Personal Computer.

PDA (Personal Digital Assistant)—A mobile computing device.

Perl—A scripting language used primarily for text manipulation.

Petabit—10^{15} bits.

PICT—An image file format developed by Apple.

PIM-DM—Protocol Independent Multicast–Dense Mode.

PIM-SM—Protocol Independent Multicast–Sparse Mode.

Pixel—A picture element; a dot on the screen.

PNA (Progressive Networks Audio)—A legacy streaming protocol.

PNG (Portable Network Graphics)—An image file format similar to GIF.

POP (Point of Presence)—An Internet access point.

PVR (Personal Video Recorder)—A device for recording television programs to hard disk.

QoE (Quality of Experience)—The overall impression formed by a user of a service.

QoS (Quality of Service)—The timely delivery of data packets, without loss.

QXGA (Quadruple Extended Graphics Array)—A graphics adapter capable of producing images of 2048 × 1536 pixels.

RADSL (Rate Adaptive DSL)—A version of ADSL where modems test the line at startup and adapt their operating speed to the fastest speed the line can handle.

RAM—RealNetwork's redirection file format, analogous to Microsoft's ASX.

RBN (Real Broadcast Network)—A streaming media hosting service.

RDT (Real Data Transport)—A RealNetworks proprietary alternative to RTP.

RGB (Red Green Blue)—A description of the color of a pixel.

RIAA—Recording Industry Association of America.

RIFF (Resource Interchange File Format)—Microsoft's proprietary family of digital media file formats.

RIP—Routing Information Protocol.

RITL (Radio in the Loop)—A wireless broadband connectivity solution.

RM (Real Media)—RealNetworks' streaming media file format.

RMA (Real Media Audio)—Previous name of RealSystem.

RMP (Real Metadata Package)—RealNetworks' metadata file format, for descriptions of secure content.

RMS (Real Media Secure)—The encrypted version of a streaming media file, used in media commerce applications.

RSVP (Resource Reservation Protocol)—A method for guaranteeing quality of service.

RSVP-TE—Resource Reservation Protocol with Traffic Engineering.

RTCP (Real Time Control Protocol)—Used with RTP to ensure synchronization.

RTP (Real Time Protocol)—Time-stamped data packets used to transport streaming media.

RTSP (Real Time Streaming Protocol)—A control protocol for streaming media, to enable pause, fast forward, rewind, etc.

RUP (Resource Update Protocol)—The generic method of updating content as it passes through a content delivery network.

SAMI (Synchronized Accessible Media Interchange)—Microsoft's closed captioning technology for streaming media.

SAML—Security Assertion Mark-up Language.

SAP (Session Announcement Protocol)—A method for announcing a multicast session to clients.

SCCP—Simple Conference Control Protocol.

S-curve—Describes technology lifecycles and adoption rates.

SDAP—Session Directory Announcement Protocol.

SDK—Software Development Kit.

SDMI—Secure Digital Music Initiative.

SDP—Session Description Protocol.

SDSL (Symmetric Digital Subscriber Line)—A broadband connection in which upload and download speeds are equivalent.

SECAM (Sequential Couleur Avec Memoire)—French analog television broadcast system.

Shockwave—Macromedia's open interactive graphics file format.

SIP—Session Initiation Protocol.

SLA—Service Level Agreement.

SMIL—Synchronized Multimedia Integration Language.

SMPTE—Society of Motion Picture and Television Engineers.

SMS—Short Message Service.

Solaris—Sun Microsystems' operating system.

SONET/SDH (Synchronous Optical Network/Synchronous Digital Hierarchy)—Optical trunking standards.

SSL (Secure Sockets Layer)—A way of sending encrypted data packets over the Internet.

SureStream—RealNetworks' proprietary technology for maintaining stream playback continuity despite network congestion.

SVG (Scalable Vector Graphics)—An image file format.

TCP (Transport Control Protocol)—A method for "best-effort" delivery of data packets in the Internet.

TCP/IP—Transport Control Protocol/Internet Protocol.

Terabit—10^{12} bits.

TFT (Thin Film Transistor)—A display technology commonly used on laptop computers.

TOS (Type of Service)—Indicates required quality level for each IP data packet.

TransMux (Transport Multiplex)—An MPEG-4 architectural feature allowing synchronized delivery of elemental streams via a variety of simultaneous network technologies.

TTL (Time To Live)—A parameter used to influence routing decisions for each packet of IP data.

UDP (User Datagram Protocol)—A simple, non-error-checked data transport protocol.

UMTS (Universal Mobile Telecommunication System)—A third-generation cellular telephone network, also known as WCDMA.

Unicast—Point-to-point transfer of data.

UNIX—An operating system.

URI—Universal Resource Identifier.

URL (Universal Resource Locator)—Also known as a Web site address.

VB Script (Visual Basic Script)—A scripting language by Microsoft.

VC (Virtual Circuit)—A method of reserving network resources in ATM networks, so that quality of service is guaranteed.

VCR—Video Cassette Recorder.

VDSL—Very high-bit rate DSL.

VHS (Vertical Helix Scan)—A very popular analog videocassette format.

VLBV (Very Low Bit rate Video)—MPEG-4 compression techniques to render video streamed over low-bandwidth channels.

VLSI (Very Large Scale Integration)—A silicon chip fabrication technology that is able to pack a large amount of circuitry onto a single die.

VOD—Video On Demand.

VoIP (Voice Over Internet Protocol)—A rival to the phone system's switched-circuit telephony that is often much cheaper to use.

VPN (Virtual Private Network)—A private data network that makes use of the public telephone network, maintaining privacy through the use of a tunneling protocol and security procedures.

VR—Virtual Reality.

VRML—Virtual Reality Mark-up Language.

VSAT (Very Small Aperture Terminal)—A satellite technology that allows reception using very small satellite dishes.

VXML (Voice Extensible Mark-up Language)—A technology that allows a user to interact with the Internet through voice recognition technology, with a voice browser and/or telephone.

WAN—Wide Area Network.

WAV (Wave)—A Microsoft file format for audio.

WAX—Windows audio redirection file format.

WBEM/WMI—Web Based Enterprise Management/Windows Management Instrumentation.

WCCP (Web Cache Coordination/Communication/Control Protocol)—Primarily for redirecting requests away from the origin server and to the edge cache.

WCDMA (Wideband Code Division Multiple Access)—A third-generation cellular wireless technology.

WCIP (Web Cache Invalidation Protocol)—Methods to guarantee cache consistency and synchronization with the content on the origin server.

Web TV—Microsoft-sponsored system for accessing the Internet using a television set-top box.

WebDVD—DVD that contains links to Web sites and content, allowing high-quality video to be combined with up-to-the-minute content.

WEBI (Web Intermediaries)—The IETF working group concerned with edge network protocols.

WiFi—IEEE wireless personal area networking technology.

WiMedia—IEEE wireless local media transport networking technology.

WirelessMAN—IEEE wireless metropolitan area networking technology.

WLL (Wireless Local Loop)—A broadband connectivity technology that solves the last-mile problem using a wireless interconnection.

WM—Windows Media file format.

WMA—Windows Media Audio file format.

WMF (Windows Media Format)—Microsoft's proprietary streaming media format.

WMV—Windows Media Video file format.

WPAD (Web Proxy Auto Discovery)—A mechanism that enables Web browsers and other clients automatically to locate an appropriate proxy cache within their domain.

WREC (Web Replication and Caching)—An IETF working group.

X3D (Extensible 3D)—An XML-based successor to VRML.

XACML (Extensible Access Control Mark-up Language)—A specification in XML, for expressing policies for information access over the Internet.

XDSL—A generic way of referring to the family of DSL technologies.

XMCL—Extensible Media Commerce Language.

XML—Extensible Mark-up Language.

XOR (Exclusive OR)—A logical function used in cryptography and forward error-correction codes.

XrML—Extensible Rights Mark-up Language.

YUV (Luminance Bandwidth Chrominance)—A method of describing colors of pixels.

SOURCES

Introduction

Crossing the Chasm, Geoffrey A. Moore and Regis McKenna, Harper Business Books, 1999, ISBN 0–066620023

The Medium

What Is Streaming Media?
Understanding Streaming Media: A guide to Webcasting Rich Media over the Internet, White paper, Al Kovalick, Pinnacle Systems, http://www.pinnaclesys.com

The ABC's of Streaming Video: An Executive Brief Covering the What, How and Why's, Imerge Consulting Group White paper, http://www.imergeconsulting.com

Take the Streaming Tour, Gary Bryant, http://www.broadbandindustrynews.com/html/whatsstreaming.shtml

Ad Disputes Tune Web Radio Out, John Borland, CNET News.com, April 11, 2001, http://news.cnet.com/news/0-1005-200-5575140.html

Regulators Leave Personalized Web Radio Question Up in the Air, John Borland, CNET News.com, December 8, 2000, http://news.cnet.com/news/0-1005-200-4062888.html?tag=rltdnws

RIAA, Webcasters Stand Off Over Licensing Pacts, John Borland, CNET News.com, September 26, 2000, http://news.cnet.com/news/0-1005-200-2870221.html?tag=rltdnws

Clear Channel Using Web-Only Adds to Resume Radio Webcasts, http://www.digitalmediawire.com/archives_061801.html

Why the Backlash? Anger toward [the Internet] justified or not, depends on your POV, Stephen Schleicher, http://www.digitalwebcast.com/2001/08_aug/editorials/backlash.htm

I'm A Saboteur, Daniel Pink, Fast Company Issue 40, p 242, http://www.fastcompany.com/online/40/wf_gatto.html

Spotmagic Corporate Press Pack, circa December 2000, Spotmagic Inc., 1700 California Street, Suite 430, SF, CA 94109, http://www.spotmagic.com

Wireless Video To Link Speeding Ambulances with Hospitals, Brian McDonough, August 24, 2001, http://wireless.newsfactor.com/perl/printer/13061

Mobile Video: Ready or Not, Here It Comes?, Brian McDonough, August 27, 2001, http://wireless.newsfactor.com/perl/printer/13114

What Is VoIP?, http://www.innomedia.com/ip_telephony/voip/index.htm

About Videoshare, http://www.allcam.com/about/default.asp

X10 Ninja Pan and Tilt Camera Kit, http://www.x10.com/products/
   x10_vk74a.htm

X10 Cameras, http://www.x10.com/products/cameras.htm

'Pop-Under' Ads Click Off Web Surfers, Michael Park, July 6, 2001, Fox News,
   http://www.foxnews.com/story/0,2933,28826,00.html

The Technology Behind www.AndieAndMike.org, http://www.andieandmike.org/
   howitworks.htm

Clay Brothers' Video Rocketry, http://www.dph.com/vidroc/Vidroc_intro.html

Sharp Unveils MPEG-4 Camcorder, David Tanaka, November 22, 1999,
   http://www.canadacomputes.com/v3/print/1,1019,1754,00.html

Defining Virtual Reality: Dimensions Determining Telepresence, Jonathan
   Steuer, October 15, 1993, http://cyborganic.com/People/jonathan/Academia/
   Papers/Web/defining-vr1.html

Can DVD Bring the World Together? Producers Soup Up the Internet with the
   Video Horsepower of a Shiny, Spinning Disc, Bryant Frazer, AV Video Multimedia Producer, http://216.246.51.100/2001/05_may/features/may2001/
   dvd_report/dvd_1.htm

How Does Streaming Media Work?

Streaming Media Technology, White paper, Steve Cresswell, June 2000,
   Smart421, http://www.smart421.com

Harry Nyquist, http://www.geocities.com/bioelectrochemistry/nyquist.htm

Claude Shannon 1916-2001, http://www.research.att.com/~njas/doc/ces5.html

Sampling, 50 Years After Shannon, Proceedings of the IEEE, Vol. 88 No. 4,
   April 2000

Analog to Digital Conversion: Sampling, http://www.cs.ucl.ac.uk/staff/J.Crow-
   croft/mmbook/book/node96.html

How Analog-Digital Recording Works, Marshal Brain, http://www.howstuff-
   works.com/analog-digital.htm/printable

Trevor Marshall article on Sampling, Byte magazine, http://www.byte.com/
   print/documentID=702

Worldcom, Using Siemens/Optisphere Equipment, Places 3.2 Terabits of Traffic
   Per Second on Existing Network Fiber, March 13, 2001, http://www.icn.
   siemens.com/icn/news/2001/01031301.html

Telecom Startup Thinks It Can Lick the Last-Mile Problem, Vishesh Kumar,
   The Industry Standard, July 1, 2001, http://www.thestandard.com/article/0,
   1902,27569,00.html

Basic Concepts of WCDMA Radio Access Network, White paper,
   http://www.ericsson.com

This is UMTS, http://www.ericsson.com/wcdma/umts

The UMTS Forum: Future Service Revenue Opportunities in 3G,
   http://www.ericsson.com/wcdma/news/art_umts_forum.shtml

UMTS Forum Position Paper No. 1, August 2001, http://www.umts-forum.org

Report No. 9, The UMTS Third Generation Market, Structuring the Service
   Revenue Opportunities, September 2000, http://www.umts-forum.org

The Last Mile: Where Telecommunications Traffic Slows to a Crawl, Flournoy
   and Scott, ITS Projects, http://www.tcomschool.ohiou.edu/its_pgs/lastmi.html

How Telecommunications' Last Mile Problem Relates to Broadcasters: Proposal
   to National Association of Broadcasters, Don Flournoy, ITS Projects -
   http://www.tcomschool.ohiou.edu/its_pgs/nab.html

Windows Media Player for Handheld PC FAQ, http://www.microsoft.com/win-
   dows/windowsmedia/en/software/handheld/faq.asp

Windows Media Player 7.1 for Pocket PC FAQ, http://www.microsoft.com/windows/windowsmedia/en/software/pocket/faq.asp

Windows Media Player for Palm-size PC FAQ, http://www.microsoft.com/windows/windowsmedia/en/software/palmsize/faq.asp

Windows Media Player 7.1 FAQ, http://www.microsoft.com/windows/windowsmedia/en/software/V7/V7FAQ.asp

Windows Media Rights Manager Flow, http://www.microsoft.com/windows/windowsmedia/wm7/drm/architecture.asp

Things to Consider Before Issuing Licenses with Windows Media Rights Manager 1, Andrea Pruneda, Microsoft Streaming Media Division, December 1999, http://msdn.microsoft.com/library/en-us/dnwmt/html/Generallicensing.asp?frame=true

Intelligent Streaming, Bill Birney, Microsoft Corporation, White paper, October 2000, http://www.microsoft.com/library/en-us/dnwmt/html/IntStreaming.asp?frame=true

Features of DRM, http://www.microsoft.com/windows/windowsmedia/wm7/drm/features.asp

Windows Media 7.1 Walkthrough, http://www.microsoft.com/windows/windowsmedia

Archived chats and chat transcripts, http://msdn.microsoft.com/chats/recent.asp

Microsoft Studios Case Study, http://www.microsoft.com/windows/windowsmedia/en/content_provider/casestudies/msstudios.asp

Virgin Megastores Case Study, http://www.microsoft.com/windows/windowsmedia/en/content_provider/casestudies/virgin.asp

Windows Media Technologies Roadmap, http://www.microsoft.com/windows/windowsmedia/en/fatures/roadmap.asp

Windows Media Services FAQ, http://www.microsoft.com/windows/windowsmedia/en/support/faq_strm.asp

Digital Broadcast Manager FAQ, http://www.microsoft.com/windows/windowsmedia/en/technologies/dbm/faq.asp

Serving for Windows Media Services: The Basics, October 2000, http://msdn.microsoft.com/library/en-us/dnwmt/html/dnwmt_serving.asp?frame=true

Windows Media Encoder 7.1 Tools, http://www.microsoft.com/windows/windowsmedia/technologies/resource/encoder.asp

Windows Media Services Tools, http://www.microsoft.com/windows/windowsmedia/technologies/resource/services.asp

eTesting Labs Study Finds Nearly Three Times as Many People Tested Prefer Quality of Windows Media Video 8 Over RealVideo 8, http://www.microsoft.com/presspass/press/2001/Apr01/04-23v8testPR.asp

Windows Media Technologies FAQ, http://www.microsoft.com/windows/windowsmedia/en/support/faq.asp

Windows Media Shrinks File Sizes, Cecily Barnes, CNET News.com, March 28, 2001, http://news.cnet.com/news/0-1005-200-5338600.html?tag=rltdnws

Windows Media "Corona," http://www.microsoft.com/windows/windowsmedia/thirdgen/default.asp

"Corona" Platform, DVD, e-mail posting to streamingmedia.com Streaming Media Business Models mailing list, David Caulton, Competitive Analysis and Reviews, SME Microsoft Announcements, Microsoft Digital Media Division, December 11, 2001

Windows Media Services in Windows .NET Server: Fast Stream, http://www.microsoft.com/windows/windowsmedia/services/net/fast_streaming.asp

Windows Media Services in Windows .NET Server: Dynamic Content Delivery, http://www.microsoft.com/windows/windowsmedia/services/net/dynamic_content.asp

Windows Media Services in Windows .NET Server: Extensible Platform, http://www.microsoft.com/windows/windowsmedia/services/net/extensible_platform.asp

Windows Media Services in Windows .NET Server: Industrial Strength, http://www.microsoft.com/windows/windowsmedia/services/net/industrial_strength.asp

Windows Media Services in Windows .NET Server: Scenarios, http://www.microsoft.com/windows/windowsmedia/services/net/scenarios.asp

Introducing Windows Media, http://www.microsoft.com/windows/windowsmedia/en/overview/default.asp

Enterprise, http://www.microsoft.com/windows/windowsmedia/en/business/default.asp

Benefits, http://www.microsoft.com/windows/windowsmedia/en/business/benefits.asp

Features, http://www.microsoft.com/windows/windowsmedia/en/features/default.asp

Intelligent Streaming, http://www.microsoft.com/windows/windowsmedia/en/features/intellistream/default.asp

Windows Media On-Demand Producer, http://www.microsoft.com/windows/windowsmedia/en/technologies/tools/ondemand.asp

Superior Audio Quality, http://www.microsoft.com/windows/windowsmedia/en/features/audio/default.asp

Dare to Compare!, http://www.microsoft.com/windows/windowsmedia/en/compare/default.asp

Windows Media Format Tools, http://www.microsoft.com/windows/windowsmedia/technologies/resource/format.asp

Tour of Microsoft Windows Media Tools, http://msdn.microsoft.com/library/en-us/dnwmt/html/TourWMTools.asp?frame=true

Windows Media Encoder 7.1 FAQ, http://www.microsoft.com/windows/windowsmedia/en/wm7/encoder/faq.asp

Windows Media Technologies, http://www.microsoft.com/windows/windowsmedia/en/wm7/default.asp

Welcome to Windows Media Player 7.1, http://www.microsoft.com/windows/windowsmedia/en/software/Playerv7.asp

Windows Media Format 7, http://www.microsoft.com/windows/windowsmedia/en/wm7/Format.asp

Windows Media Encoder 7.1, http://www.microsoft.com/windows/windowsmedia/en/wm7/encoder.asp

Windows Media Rights Manager, http://www.microsoft.com/windows/windowsmedia/en/wm7/drm.asp

Windows Media 7.1 SDK, http://msdn.microsoft.com/library/en-us/dnwmt/html/wmsdk.asp?frame=true

Windows Media Services, http://www.microsoft.com/windows/windowsmedia/en/technologies/services.asp

Codecs 101 for Windows Media Technologies, http://msdn.microsoft.com/library/en-us/dnwmt/html/codecs.asp?frame=true

About HDCD, http://www.hdcd.com/about/index.html

What is HDCD?, http://www.hdcd.com/about/whatisHDCD.html

QuickTime Heads Toward Open Standards, David Nagel, http://www.digitalwebcast.com/2001/12_dec/features/applequicktimelive011217.htm

Fast-start vs. Streaming, http://www.apple.com/quicktime/products/tutorials/httpvsrtsp.html

Fast Start Movies, http://www.apple.com/quicktime/products/tutorials/faststart.html

Real-Time Transport Protocol, http://www.apple.com/quicktime/products/tutorials/rtp.htm

Server Movies, http://developer.apple.com/techpubs/quicktime/qtdevdocs/REF/Streaming.b.htm

Composite Streaming and Non-Streaming Tracks, http://developer.apple.com/techpubs/quicktime/qtdevdocs/REF/Streaming.d.htm

Sorenson Broadcaster 1.1 FAQ, http://www.sorenson.com/web/support/faq/sb_faq.jsp

Sorenson Video 3 and MPEG-4 Codecs, White paper, http://www.sorenson.com

Sorenson Vcast FAQ, http://www.sorenson.com/web/support/faq/vc_faq.jsp

Sorenson Video Professional Edition 3, http://www.sorenson.com/web/support/faq/sv_faq.jsp

QuickTime FAQ, http://www.apple.com/quicktime/products/qt/qt_faq.html

QuickTime Specifications, http://www.apple.com/quicktime/specifications.html

QuickTime Components, http://www.apple.co,/quicktime/products/qt/components.html

QuickTime Streaming Server, http://www.apple.com/quicktime/products/qtss

QuickTime Streaming Server FAQ, http://www.apple.com/quicktime/products/qtss/qtssfaq.html

QuickTime Tutorials, http://www.apple.com/quicktime/products/tutorials

QuickTime Services, http://www.apple.com/quicktime/products/services

An Introduction to QuickTime, http://developer.apple.com/quicktime/qttutorial/overview.html

An Overview of the Delivery Multimedia Integration Framework for Broadband Networks, Jean-Francois Huard and George Tselikis, IEEE Communications Surveys, Fourth Quarter 1999 Vol. 2 No. 4, http://www.comsoc.org/pubs/surveys

E-Vue, http://www.e-vue.com

IVast Products, http://www.ivast.com/products.html

WebCine Encoder, http://www.mpeg-4.philips.com/products/encoder/index.asp

WebCine Server, http://www.mpeg-4.philips.com/products/server/index.asp

WebCine Player, http://www.mpeg-4.philips.com/products/player/index.asp

MPEG-4: A System Designer's View, Robert Bleidt, EE Times, November 12, 2001 http://www.eetimes.com/story/OEG20011112S0043

MPEG-4: Looks Great!, Jan Ozer, June 18, 2001, http://www.extremetech.com/print_article/0,3428,a%253D3780,00.asp

MPEG-4: A Multimedia Standard for the Third Millenium, Stefano Battista, Franc Casolino, and Claudio Lande, IEEE Multimedia Magazine, http://computer.org/multimedia/articles/mpeg4_1.htm

MPEG-4 Coding Standard, http://www.comm.toronto.edu/~karen/projects/15.MPEG4

Things Called MPEG-4 That Aren't MPEG-4, http://www.dv.com/magazine/2001/0501/waggoner0501.html

MPEG and Multimedia Communications, Leonardo Chiariglione, CSELT, Turin, Italy, http://www.cselt.stet.it/ufv/leonardo/paper/isce96.htm

Frequently Asked Questions about MPEG-7, http://www.darmstadt.gmd.de/mobile/MPEG7/FAQ.html

Overview of the MPEG-4 Standard, Rob Koenen, http://www.cselt.it/mpeg/standards/mpeg-4/mpeg-4.htm

MPEG-7 Behind the Scenes, Jane Hunter, D-Lib Magazine, Vol. 5 No. 9, September 1999 http://www.dlib.org/dlib/september99/hunter/09hunter.html

Overview of the MPEG-7 Standard, Jose Martinez, http://mpeg.telecomitalial-ab.com/standards/mpeg-7/mpeg-7.htm

MPEG-4 Multimedia for Our Time, Rob Koenen, IEEE Spectrum, Vol. 36 No. 2, February 1999, http://www.cselt.it/mpeg/documents/koenen/mpeg-4.htm

MPEG-21 Overview, Jan Bormans and Keith Hill, http://mpeg.telecomitalialab.com/standards/mpeg-21/mpeg-21.htm

MPEG-4 Boosts Multimedia, http://www.zdnet.co.uk/itweek/brief/2001/09/internet

MPEG-4 Very Low Bit Rate Video, http://wwwam.hhi.de/mpeg-video/papers/sikora/vlbv.htm

Benefits of MPEG-4, http://www.ingenient.com/mpeg4.htm

ISMA Members, http://www.isma.tv/html/about/ourmembers.shtml

Internet Streaming Media Alliance Formed to Accelerate Adoption of Standards, Laura Nugent, December 12, 2000, http://ism-alliance.org/cgi-bin/phtm/prlist.pl?9988

Video Compression: The Big Squeeze, Adrian Pennington, Broadband magazine, June 15, 2001 p. 14-15

Recent Advances in Video Compression, http://www.cs.ru.ac.za/Honours/mmcourse/compression/mpeg/eth/intro.html

MPEG-4 Video Standards for Multimedia Applications, http://wwwam.hhi.de/mpeg-video/papers/sikora/ISCAS/IEEE-ISCAS.htm

Multimedia Technology Review Paper, Sylvia Willie and Hohnny Gui, May 23, 1997, http://www2.fit.qut.edu.au/frill/willie/MMST_papers/johnny%20gui%2001970895/john

Content Networking and Edge Services: Leveraging the Internet for Profit, stardust.com White paper, http://www.stardust.com/cdn

Content Delivery Network, A Reference Guide, Matthew Liste, April 2001, Thrupoint, Inc., http://www.thrupoint.com

The Cisco Content Networking Architecture, http://www.cisco.com/warp/public/779/largeent/learn/technologies/content_networking

Start-up Companies Plan to Speed Up the Net, John Borland and Corey Grice, CNET News.com, August 5, 1999, http://news.cnet.com/news/0-1004-200-345730.html?tag=rltdnws

Akamai Aims to End Web Waits, John Borland, CNET News.com, June 15, 1999, http://news.cnet.com/news/0-1004-200-343683.html?tag=rltdnws

Net Jams Hinder Faster Connections, Jeff Pelline, CNET News.com, October 22, 1997, http://news.cnet.com/news/0-1004-200-323300.html?tag=rltdnws

Akamai Success Rides On the Need for Faster Net, John Borland, CNET News.com, November 2, 1999, http://news.cnet.com/news/0-1006-200-1426854.html?tag=bplst

Networking Firms Aim to End Web Wait, Wylie Wong, CNET News.com, November 2, 1999, http://news.cnet.com/news/0-1006-200-1427214.html?tag=bplst

Digital Island Global Delivery Service Level Agreements, http://www.digitalisland.net/services/sla.shtml

Hosted Applications Key to Voice, John Weald, September 6, 2001, http://www.planetanalog.com/story/OEG20010906S0078

Firms Work to Kill Net Congestion, John Borland, CNET News.com, October 29, 1999, http://news.cnet.com/news/0-1006-200-1425329.html?tag=bplst

Cisco CDN Software Service Provider Edition 1.0, White paper, http://www.cisco.com

Akamai: Solutions: Content Delivery Services: Streaming Media, http://www.akamai.com/html/en/sv/streaming_media.html

Akamai Solutions: Streaming Applications, http://www.akamai.com/html/en/sv/streaming_applications.html

Akamai Solutions: Content Delivery Services: Global Traffic Management, http://www.akamai.com/html/en/sv/global_traffic_mgmnt.html

Customize Web Content with EdgeSuite, http://www.akamai.com/html/en/sv/content_targeting.html

Akamai Solutions: Content Delivery Services: Business Intelligence, http://www.akamai.com/html/en/sv/business_intelligence.html

Akamai Technology: Core Technologies, http://www.akamai.com/html/en/tc/core_tech.html

Cache Array Routing Protocol v1.0, Vinod Valloppillil, Microsoft Corp, Keith Ross, University of Pennsylvania, Internet Draft, February 26, 1998, http://icp.ircache.net/carp.txt

Cache Digests Frequently Asked Questions, Niall Doherty, http://squid-cache.org/Doc/FAQ/FAQ-16.html

Is the Internet Heading for a Cache Crunch?, Russell Baird Tewksbury, http://www.isoc.org/inet98/proceedings/1k/1k_2.htm

Caching Support in Standards-based RTSP/RTP Servers, Frederick, Geagan, Kellner, Periyannan, IETF Internet Draft, March 10, 2000, http://www.web-cache.com/Writings/Internet-Drafts/draft-periyannan-rtsp-caching-01.txt

Edge Services: The Next Wave of Computing, Scott Mace, Position Paper, http://www.stardust.com

Streaming Media with CacheFlow, White paper, November 2000, http://www.cacheflow.com

ICAP FAQ, http://www.i-cap.org/faq

Edge Side Includes, http://www.esi.org

Web Caching-Related Protocols and Standards, http://www.web-cache.com/Writings/protocols-standards.html

Hyper Text Caching Protocol, Paul Vixie, February 1999, http://icp.ircache.net/htcp.txt

Application of Internet Cache Protocol, Claffy Wessels, September 1997, http://icp.ircache.net/rfc2187.txt

Network Cache Performance Measurements, White paper, September 3, 1998, CacheFlow Inc., http://www.cacheflow.com

Monitoring the Performance of a Cache for Broadband Customers, Warfield, Mueller, Sember, Telstra Australia, http://www.isoc.org/inet98/proceedings/1k/1k_3.htm

Web Caching Meshes: Hit or Miss?, Verschuren, de Jong, Melve, http://www.isoc.org/inet98/proceedings/1k/1k_1.htm

EDGE-FX Cache Streaming Media Extension, http://www.f5networks.com/f5products/edgefx/edgefxsme.html

Intelligence at the Edge of Broadband, Greg Jones, http://www.broadbandindustrynews.com/html/bl-bbgateways_jones.shtml

Going Edge to Edge: Making the World Wide Web Truly Worldwide, Orblynx White paper, http://www.broadbandindustrynews.com/html/wp_orblynx.shtml

Stream Caching: The Cost Effective Solution for Speeding Up and Improving the Internet Experience, Entera White paper, http://www.broadbandindustrynews.com/html/wp_streamcaching_entera.shtml

Intelligence at the Edge of Broadband, Greg Jones, http://www.broadbandindustrynews.com/html/bl-bbgateways_jones.shtml

Enterprise: Delivering Content to the Edge, White paper, Bill Birney, Microsoft Digital Media Division, June 2001, http://www.microsoft.com/windows/windowsmedia

Internet QoS: the Big Picture, XiPeng Xiao and Lionel M. Ni, White paper, Department of Computer Science, Michigan State University, East Lansing, MI, IEEE Network Magazine, March/April 1999 pp 8-18

RSVP, http://searchsystemsmanagement.techtarget.com/sDefinition/0,,sid20_gci214274,00.html

Resource Reservation Protocol, http://www.cisco.com/univercd/cc/td/doc/cisintwk/ito_doc/rsvp.htm

Short Bibliography on Traffic Control and QoS in IP Networks, Olivier Bonaventure, Infonet group, University of Namur, Belgium, November 23, 2000, http://www.infonet.fundp.ac.be?TC/biblio.pdf

Multiprotocol Label Switching (MPLS) Traffic Engineering, http://www.cisco.com/univercd/cc/td/doc/product/software/ios120/120newft/120limit/120s/120s5/mpls_te.htm

Guidelines for Tunneling RTSP/RTP over HTTP, Emblaze Systems, 3GPP TSG-SA WG4 (Codec Working Group), 3GPP SA4 Meeting #19, Tokyo, Japan, December 3-7, 2001, Tdoc S4 (01)0606

Tutorial: MPLS Label Distribution and Signaling, Rick Gallaher, November 1, 2001, http://www.convergedigest.com/tutorials/mpls2/page1.htm

MPLS, http://www.webopedia.com/TERM/M/MPLS.html

Multiprotocol Lambda Switching Comes Together, Luc Ceuppens, http://lw.pennnet.com/Articles/article_display.cfm?Section=Articles&ARTICLE_ID=77930&Publication_ID=13

Generalized Multiprotocol Label Switching: An Overview of Signaling Enhancements and Recovery Techniques, Banerjee, Drake, Lang, Turner, Awduche, Berger, Kompella, Rekhter, IEEE Communications Magazine, July 2001

RTSP Frequently Asked Questions, http://www.rtsp.org/2001/faq.html

Some Frequently Asked Questions about RTP, http://www.cs.columbia.edu/~hgs/rtp/faq.html

Internetworking Equipment Design: MPLS, IPv6 bring integrated services closer, Gary Hemminger, EE Times, November 27, 2001, http://www.commsdesign.com/story/OEG20011127S0017

RTP Control Protocol, RTCP, http://www.freesoft.org/CIE/RFC/1889/13.htm

RealSystem Media Commerce Suite, Technical White Paper, see http://www.realnetworks.com

Total Solutions for Education, http://www.realnetworks.com/education/index.html?src=noref,rnhmpg_080301,rnhmln

XMCL Initiative, http://www.xmcl.org/initiative.html

Introducing RealOne Player, http://www.real.com

RealOne Player Authoring Overview, http://www.real.com

Introduction to RealSystem iQ Production, http://www.realnetworks.com

RealJukebox FAQ: Constant and Variable Bit Rate Encoding, http://service.real.com/help/faq/rjbvrfaq.html

Microsoft Windows Media Services and RealNetworks RealSystem, Feature Comparison, May 2001, http://www.approach/digitalmedia

RTSP Interoperability With RealSystem Server 8, RealSystem iQ White paper, December 7, 2000, http://www.realnetworks.com

Live Broadcast Distribution With RealSystem Server 8, RealSystem iQ White paper, December 7, 2000, http://www.realnetworks.com

Streaming QuickTime With RealSystem Server 8, RealSystem iQ Technical Blueprint Series, December 7, 2000, http://www.realnetworks.com

RealOne Platform, http://www.realnetworks.com/solutions/ecosystem/realone.html

RealNetworks Moves to MPEG-4, Jose Alvear, December 10, 2001, streamingmedia.com online, http://www.streamingmedia.com

Tutorial: Corporate Streaming with RealSystem iQ, Larry Bouthillier, streamingmedia.com online, http://www.streamingmedia.com/backend/tutorials/view.asp?tutorial_id=145&clean=y

RealSystem Proxy 8 Overview, RealSystem iQ White paper, December 15, 2000, http://www.realnetworks.com

RealSystem Server Best Practices, White paper, June 15, 2001, http://www.realnetworks.com

Introduction to Digital Media Delivery, http://www.realnetworks.com/realsystem/whatis_realsystem.html

Getting Started with Streaming Media, http://www.realnetworks.com/devzone/feature/archive_features/get_started_faq.html

RealSystem Media Commerce Suite, http://www.realnetworks.com/realsystem/mediacommerce.html

What is RealSystem iQ?, http://www.realnetworks.com/realsystem/whatis_rsiq.html

RealSystem iQ Helps Your Business, http://www.realnetworks.com/realsystem/benefits.html

RealSystem iQ Resources, http://www.realnetworks.com/realsystem/tech_overview.html

What's New With RealSystem iQ?, http://www.realnetworks.com/solutions/ecosystem/realsystem.html

RealSystem Server Product Line Comparison, http://www.realnetworks.com/products/servers.comparison.html

RealSystem Server Intranet, http://www.realnetworks.com/products/servers/intranet/index.html

Envivio Products, http://www.envivio.com/solutions/products.html

Serving Suggestion, Alison Campbell, October 22, 2001, streamingmedia.com online, http://europe.streamingmedia.com/r/printerfriendly.asp?id=7977

ZD Labs Study Reveals Windows Media Scalability Offers Over 9,000 Concurrent Video Streams and 99.9999999 Percent Packet Delivery Reliability, White Paper, http://www.microsoft.com/windows/windowsmedia/en/compare/ZDLabs.asp

Streaming Methods: Web Server vs. Streaming Media Server, White paper, http://www.microsoft.com/windows/windowsmedia/en/compare/webservvstreamsrv.asp

Frequently Asked Questions on the Multicast Backbone (MBone), http://www.cs.columbia.edu/~hgs/internet/mbone-faq.html

Higher Level Protocols Used With IP Multicast, IP Multicast Initiative White paper, Vicki Johnson and Marjory Johnson, http://www.ipmulticast.com/community/whitepapers/highprot.html

How IP Multicast Works, IP Multicast Initiative White paper, Vicki Johnson and Marjory Johnson, http://www.ipmulticast.com/community/whitepapers/howipmcworks.html

Synchronized Multimedia Integration Language (SMIL 2.0), http://www.w3.org/TR/smil20

SMIL Standard Feature Comparison, http://www.oratrix.com/Products/CompareSMIL?bstate=Compare

Understanding SAMI 1.0, http://www.msdn.microsoft.com/library/en-us/dnacc/html/atg_samiarticle.asp?frame=true

VTrails Product Overview, http://www.vtrails.com/product

VTrails Technology Overview, http://www.vtrails.com/technology/overview2.html

Sharing the Streaming Burden, Tania Hershman, September 16, 2000, Wired News Online, http://www.wired.com/news/print/0,1294,38677,00.html

Groove Product Backgrounder, White paper, http://www.groove.net

Why Peer-to-Peer?, White paper, http://www.groove.net

Groove and .NET, White paper, http://www.groove.net

Owning Your Content In a Digital World, White paper, Widevine Technologies Inc., http://www.widevine.com

The Pre-Encryption Fallacy, White paper, Widevine Technologies Inc, see http://www.widevine.com

How To Steal Streaming Media, White paper, Widevine Technologies Inc., http://www.widevine.com

The Widevine Cypher Cryptosystem, White paper, Widevine Technologies Inc., http://www.widevine.com

Why Was Streaming Media Invented?

Real Time Audio and Video Playback on the Web, Bob Godwin-Jones, Language Learning and Technology, Vol. 1 No. 1 July 1997 pp. 5–8, http://polyglot.cal.msu.edu.llt/vol1num1/emerging.html

Video on the World Wide Web, http://www.videonics.com/videos/about-web-video.html

The Digital Media Association, http://www.digimedia.org

Video Over The Internet, http://www.rad.com/networks/1996/video/video.htm

Streaming, http://www.webopedia.com/TERM/s/streaming.html

IRB 203 Xing Stream Works MP3, http://internetradio.about.com/library/weekly/aa022699.htm?pid=2815&cob=home

Streaming Media and the Internet, Gary James, http://www.allencomm.com

Companies Rethink Corporate Travel, http://www.storagenet2001.com/document_lr.asp?doc_id=8434

Meet You at the Virtual Meeting, December 1999, Softwaremag.com, http://www.softwaremag.com/archive/1999dec/VirtualMeeting.html

IP/TC Delivers Video Communications to the Desktop, Demonstration CD Release 2.0, Cisco Systems Inc. 1995-1998, http://www.cisco.com/iptv

Cisco IP/TV 3400 Series Servers, http://www.cisco.com/warp/public/cc/pd/mxsv/iptv3400/index.shtl

New Developments in Digital Video, Bob Godwin-Jones, Language Learning & Technology, Vol. 2 No. July 1, 1998 pp. 11–13, http://polyglot.cal.msu.edu/llt/vol2num1/emerging

Microsoft Video for Windows, Desktop Video AGOCG Report, http://www.agocg.ac.uk/reports/mmedia/25/repor_15.htm

NetShow Video Server, Terry Cornall, January 15, 1998, http://www.ctie.monash.edu.au/emerge/multimedia/links/nshowvs.htm

The New Media Players: An Exploration of the Current State of Web Video, Nels Johnson, http://www.dv.com/magazine/1998/1198/johnson1198.html

ServerWatch's Review of Microsoft NetShow, http://serverwatch.internet.com/review/av-netwshow.html

Microsoft NetShow 3.0, Larry Seltzer, August 14, 1998, http://www.zdnet.com/pcmag/secondlooks/s1980813.html

Microsoft NetShow 3.0, Jan Ozer, July 30, 1998, http://www.zdnet.com/pcmag/firstlooks/9807/f980707a.html

RealNetworks vs. Microsoft, Larry Seltzer, http://www.zdnet.com/pcmag/pclabs/inside/in980727.htm

Microsoft Corp.: Microsoft NetShow, http://www.zdnet.com/pcmag/features/stream/svidr2.htm

Microsoft NetShow 2.0 Technical Whitepaper, Microsoft Corporation, July 1997, http://msdn.microsoft.com/archive/en-us/dnarnetsh../msdn_ns20wpfinal.asp

RealSystem G2 (Beta), Jan Ozer, July 24, 1998, http://www.zdnet.com/pcmag/firstlooks/9807/f980724a.html

RealServer 7.0x and Earlier Documentation, http://service.real.com/help/library/earlier_servers.html

Why is Streaming Media Better?

Streaming Companies Push Toward TV Quality, Patricia Jacobus, CNET News.com, December 12, 2000, http://news.cnet.com/news/0-1005-200-4104631.html?tag=rltdnws

Commentary: TV-Quality Video Coming to Web, Eventually, Lou Latham, CNET News.com, December 12, 2000, http://news.cnet.com/news/1,10000,0-1005-201-4118193-0.html

Madonna Webcast Shows Internet is Not TV, Gwendolyn Mariano, CNET News.com, November 28, 2000, http://news.cnet.com/news/0-1005-200-3890055.html?tag=rltdnws

Sony Joins MHP Bandwagon, Junko Yoshida, EE Times, August 24, 2001, http://www.eetimes.com/story/OEG20010824S0089

Open Platform Presages Interactive Digital TV in Europe, Junko Yoshida, EE Times, August 24, 2001, http://www.edtn.com/story/lead/ OEG20010824S0060-T

CacheVision Unveils Open Video Recorder Platform, Junko Yoshida, EE Times, August 23, 2001, http://www.eetimes.com/story/OEG20010823S0068

Interactive Streaming: The Creative Foundation of More Compelling, More Competitive Web Sites, Jon Leland, http://www.mediamall.com

Servers in Streaming Applications, http://www.realitywave.com/technology-paper-servers.asp

Who's Selling Streaming Ads?, December 7, 2001, http://www.mediapost.com/dtls_dsp_news.cfm?newsI=102253

Tiananmen Square 1989, The Declassified History, Jeffrey Richelson and Michael Evans, June 1, 1999, http://www.gwu.edu/~nsarchiv/NSAEBB/NSAEBB16

Virage Streaming Video Seminar, http://www.virage.com

Who Is Driving Streaming Media's Innovation?

IDM Motion Wavelets Video Codec: Description of Basic Principles and Coding Examples, http://www.idm.ru/wavelets.htm

Internet Streaming Media Alliance Releases ISMA 1.0, Janet Feddish, October 2, 2001, http://ism-alliance.org/cgi-bin/phtm/prlist.pl?9986

Quality Comparison: RealVideo, Windows Media Video and Sorenson Video, Francesco Schiavon, November 20, 2001, streamingmedia.com online, http://www.streamingmedia.com/r/printerfriendly.asp?id=8058

MPEG-4 Products and Services, http://www.m4if.org/products/index.html

New Technologies Promise Streaming Rebirth, John Borland, CNET News.com, April 16, 2001, http://news.cnet.com/news/0-1005-200-5630041.html

Kasenna, Software for Enabling Video-on-Demand, http://www.kasenna.com

Welcome to Oratrix Development, http://www.oratrix.com

Proposal for ITU-T H.26L: A Wavelet-Based Video Coding Scheme Using OBMC and Image Warping Prediction, Heising, Marpe, Cycon, Petukhov, http://invinet.hhi.de/itu/sld001.htm

Multimedia Communications: Coding, Systems and Networking, Tsuhan Chen, tsuhan@ece.cmu.edu

JPEG2000: The Upcoming Still Image Compression Standard, Skodras, Christopoulos and Ebrahimi, Proceedings of the 11th Portuguese Conference on Pattern Recognition, May 11-12, 2000, pp. 359-366

A Study of JPEG2000 Still Image Coding Versus Other Standards, Diego Santa-Cruz and Touradj Ebrahimi, Proceedings of the X European Signal Processing Conference, Tampere, Finland, September 5–8, 2000 Vol. 2 pp. 673-676, http://ltswww.epfl.ch/~dsanta

JPEG2000 Wavelet Compression Spec Approved, R. Colin Johnson, EE Times, December 29, 1999, http://www.eetimes.com/story/OEG19991228S0028

An Analytical Study of JPEG 2000 Functionalities, Santa-Cruz, Ebrahami, Askelof, Larsson and Christopoulos, Proceedings of SPIE, Vol. 4115 of the 45th annual SPIE meeting, Applications of Digital Image Processing XXIII, http://ltswww.epfl.ch/~dsanta

Emerging Graphics File Formats, Sue Chastain, http://graphicsoft.about.com/cs/formatsnew

Vector Format for Zooming, http://www.vfzoom.com

What's Wrong with Streaming Media?

Microsoft Previews Windows Media for DVD Players, Rick Merritt, EE Times, December 1, 2001, http://www.eetimes.com/story/OEG20011211S0054

Net Video Not Yet Ready for Prime Time, John Borland, CNET News.com, February 5, 1999, http://news.cnet.com/news/0-1004-200-338361.html?tag=rltdnws

Streaming Technology: Still Rough Waters, David Strom, November 9, 1998, http://www.computerworld.com/cwi/Printer_Friendly_Version/0,1212,NAV47_STO34294,00.html

Information and Content Exchange, http://www.icestandard.org

Information and Content Exchange Implementation Cookbook: Getting Started with Web Syndication, Souzis, Popkin, Khoury and Hunt, http://www.idealliance.org

Spec Sprawl Derails Home Net Consensus, Junko Yoshida, EE Times, December 14, 2001, http://www.eetimes.com/story/OEG20011214S0090

Automotive Groups Mend Rift, Charles Murray, EE Times, December 14, 2001, http://www.eetimes.com/story/OEG20011214S0035

High Performance Wireless Home Networks: An Ultra-Wideband Solution, Fantasma Networks Inc. White paper, March 19, 2001, http://www.fantasma.net

Assuring QoE on Next Generation Networks, White paper, http://www.empirix.com

Measuring Streaming Quality: An Overview of Methodology Behind the Keynote Scale, White paper, http://www.keynote.com

DoS Attack Hits New York Times, Robyn Weisman, October 31, 2001, http://www.newsfactor.com/perl/printer/14500

Trends in Denial of Service Attack Technology, Kevin Houle and George Weaver, CERT Coordination Center White paper, Carnegie Mellon University, October 2001, http://www.cert.org

DoS Attacks: Easier to Launch, Harder to Fight, Jay Lyman, November 6, 2001, http://www/newsfactor.com/perl/printer/14593

Synchronous Packet Streaming White Paper, http://www.packetstream.com/whitepaper.html

QoS Protocols and Architectures, White paper, http://www.stardust.com/qos/whitepapers/protocols.htm

Network Bandwidth Considerations, White paper, see http://www.webex.com

Quality of Service Considerations In a Multimedia Home Network, White paper, February 5, 2001, http://www.magisnetworks.com

Streaming Security Scares, Alison Campbell, November 5, 2001, streamingmedia.com online, http://europe.streamingmedia.com/r/printerfriendly.asp?id=8024

Virage Solutions Product Brochure, http://www.virage.com

The Reality of Wireless Broadband, Randy Roberson, http://www.broadbandindustrynews.com/html/bl_tantivity.html

Planet of the Apps, Jason Thompson, October 29, 2001, streamingmedia.com online, http://www.streamingmedia.com/r/printerfriendly.asp?id=7994

When Will Streaming Media Be Ready for Prime Time?

Broadband Future: Reassessing Design, George Mattathil, http://www.broadbandindustrynews.com/html/mattathil7.shtml

The Case for Streaming Media, George Mattathil, http://www.broadbandindustrynews.com/html/mattathil6.shtml

Regulation, Deregulation, Regulation, George Mattathil, http://www.broadbandindustrynews.com/html/mattathil_5.shtml

Broadband Around the World, George Mattathil, http://www.broadbandindustrynews.com/html/mattathil_4.shtml

Examining Telecommunications Infrastructure, George Mattathil, http://www.broadbandindustrynews.com/html/mattathil_3.shtml

IP Syndrome, George Mattathil, http://www.broadbandindustrynews.com/html/mattathil_1.shtml

VoIP, What's The Hang-up?, Chris Martin, http://www.broadbandindustrynews.com/html/f_empowertel1.shtml

Broadband, Where are We?, Jay Lyman, http://www.broadbandindustrynews.com/html/f_lyman1.shtml

The Future of Broadband, John Pickens, http://www.broadbandindustrynews.com/html/f_future_of_com21.shtml

UK Target is Hopeless: Only Government Action Can Make Deadline Achievable, Sarah Arnott, Computing magazine, November 29, 2001, p. 1

Why Rights Management Is Wrong (and What To Do Instead), Mark Manasse, January 21, 2001, http://gatekeeper.dec.com/pub/DEC/SRC/technical-notes/SRC-2001-002.html

DRM: The (Possible) Downside for End Users, April 29, 2000, http://xml.coverpages.org/DRM-the-downside.html

Intellectual Property Management and Protection in MPEG Standards, Rob Koenen, January 2001, http://mpeg.telecomitalialab.com/standards/ipmp

Windows Media Rights Manager FAQ, http://www.microsoft.com/windows/windowsmedia/en/wm7/DRM/FAQ.asp

Open Digital Rights Language Initiative, http://ordl.net

MPEG Rights Expression Language, Robin Cover, November 30, 2001, http://www.oasis-open.org/cover/mpegRights.html

Extensible Media Commerce Language Initiative To Create Standard Business Rules for Digital Media Market, June 20, 2001, http://xml.coverpages.org/ni2001-06-20-a.html

Welcome to XMCL.org, http://www.xmcl.org

Extensible Media Commerce Language (XMCL), Robin Cover, November 21, 2001, http://xml.coverpages.org/xmcl.html

The Digital Object Identifier System: Digital Technology Meets Content Management, Norman Paskin, http://www.doi.org/sun_pap2.html

XrML FAQ, http://www.xrml.org/faq.htm

XrML: The Technology Standard for Trusted Systems in the eContent Marketplace, White paper, http://www.xrml.org/white_paper.htm

Extensible Rights Markup Language (XrML), Robin Cover, June 23, 2001, http://xml.coverpages.org/xrml.html

Digital Rights Management for Electronic Publishing, Rosenblatt, Ram, Moynahan, Mutter, http://seminars.seyboldreports.com/seminars/2000_san_franscisco/26/26_transcript.html

Digital Rights Management, http://www.giantstepsmts.com/drm.htm

Secure Digital Music Initiative Portable Device Specification Version 1.0, http://www.sdmi.org

EFF "Intellectual Property: Digital Rights Management (DRM) Systems & Copy-Protection Schemes" Archive, http://www.eff.org/pub/Intellectual_property/DRM

Microsoft DRM Offerings, http://www.microsoft.com/windows/windowsmedia/wm7/drm/offerings.asp

Features of DRM, http://www.microsoft.com/windows/windowsmedia/wm7/drm/features.asp

Microsoft DRM New Security Features, http://www.microsoft.com/windows/windowsmedia/wm7/drm/newin7.asp

Digital Rights Management for Microsoft Windows Media Technologies: The Proven DRM Platform for Secure Digital Media E-Commerce, White paper, http://www.microsoft.com/windows/windowsmedia

Understanding Secure Audio Path, White paper, http://www.microsoft.com/windows/windowsmedia

Security Overview of Windows Media Rights Manager, Microsoft Digital Media Division, September 2001, http://www.microsoft.com/windows/windowsmedia

Digital Rights Management FAQ, http://www.microsoft.com/windows/windowsmedia/WM7/DRM/FAQ.asp

Vodaphone Fails 3G Test, September 7, 2001, http://news.bbc.co.uk/hi/english/business/newsid_1530000/1530091.stm

DoCoMo Launches Live Video for Mobile Phones, Jay Wrolstad and Dan McDonough Jr., August 28, 2001, http://wireless.newsfactor.com/perl/printer/13148

Enhancing Communications with Rich Media, Graham Seabrook, http://www.broadbandindustrynews.com/html/bl_ridgeway_seabrook.shtml

In Search of Streaming Audio, Video and Software Standards, Steig Westerberg, http://www.broadbandindustrynews.com/html/bl_stream_theory_westerburg.shtml

MPEG4 Video Floods Internet, Special Report, Nikkei Electronics Asia: January 2000, http://www.nikeibp.com/nea/2000_jan/specrep

Streaming Media Gets Down to Business, Stephen Russell, http://www.broadbandindustrynews.com/html/f_russell.shtml

Taalee Semantic Engine Press Kit, circa December 2000, http://www.taalee.com

The Singingfish.com Multimedia Search Engine, Product Brochure, circa December 2000, http://www.singingfish.com

The Audience

Who Will Watch?

Broadband Access for All, Reza Ahy, http://www.broadbandindustrynews.com/html/bl_apertonet.shtml

Broadband: Community Builder or The Great Isolator, John Brothers, http://www.broadbandindustrynews.com/html/bl_incanta.shtml

Webcast Audience Over 50 Percent of Internet Users: Recent Study Reveals over 70 Million People Stream, Jose Alvear, August 23, 2001, streamingmedia.com online, http://www.streamingmedia.com/article.asp?id=7835

Fashion Show Outdoes Clinton Testimony, Jeff Pelline, CNET News.com, February 4, 1999, http://news.cnet.com/news/0-1007-200-338301.html?tag=bplst

Web Surfers Attracted to Streamed Content, Gwendolyn Mariano, CNET News.com, December 12, 2000, http://news.cnet.com/news/0-1005-200-4118543.html?tag=bplst

Peer-to-Peer Readies for Takeoff, Joyce Routson, November 29, 2001, streamingmedia.com online, http://europe.streamingmedia.com/r/printerfreindly.asp?id=8087

Executive Broadcast Solution, White paper, June 2001, http://www.approach.com/digitalmedia

Mercedes-Benz USA Case Study, http://www.misrosoft.com/windows/windowsmedia/en/archive/casestudies/mbusa/default.asp

How Will We Watch Streaming Media?

Streaming Beyond the PC: Opportunities in Digital Media for Non-PC Devices, A Streaming Media Research Report, The Carmel Group, http://www.streamingmedia.com

Creating the Networked Home, Andy Trott, Paul Entwistle, White paper, http://www.broadbandindustrynews.com/html/wp_networkedhome.shtml

Optimizing Your Network for Streaming Services, Arthur Rabinovitz, Optibase White paper, http://www.broadbandindustrynews.com/html/optibase1.shtml

Residential Gateways: Single Device Connects Multiple Broadband Access and Home Networking Technologies, Amit Dhir, http://www.broadbandindustrynews.com/html/wp_res.shtml

Bluetooth's Place in the Broadband Home, Rolf Johansson, White paper, http://www.broadbandindustrynews.com/html/f_ericsson_bluetooth.shtml

National Semiconductor Powers Set-Top Boxes for e.Biscom's FastWeb Service, Trish Gessner, Solveig Loesch, March 22, 2001, http://www.national.com/news/item/0,1735,623,00.html

Set Top Boxes and Information Appliance Device Driver Development, http://www.intelligraphics.com/settop_multimedia.html

LC-RTP (Loss Collection RTP): Reliability for Video Caching in the Internet, Zink, Jonas, Griwodz, Steinmetz, http://www.kom.e-technik.tu-darmstadt.de/publications/abstracts/ZJGS00-1.html

Media Streaming and Conferencing Solution for Cellular Networks, Dror Gill, July 2000, White paper, http://www.zapex.co.il/tech_mpeg4_1.shtml

Standards for Multimedia Streaming and Communication over Wireless Networks, Dror Gill, July 2000, White paper, http://www.zapex.co.il/tech_mpeg4_2.shtml

MPEG-4 Over Wireless Networks, Ian Thornton, ARM Ltd, White paper, http://www.arm.com

Get Set for Set-Top Boxes, Mark Shapiro, May 15, 2000, http://www.technocopia.com/gadgets-20000515-settopboxes.html

Mobile Video, Part 2: Who Wants It?, Brian McDonough, August 29, 2001, http://wireless.newsfactor.com/perl/printer/13182

Extreme Retailing: A Market No One Expected, Scott Lehane, HDTV News, http://web-star.com/hdtv/enterprise.html

Intel Moves Into Wireless Home Networking, Dan McDonough, Jr., August 30, 2001, http://wireless.newsfactor.com/perl/printer/13227

Wireless Networking: Does Infrared Have a Chance?, Tim McDonald, August 30 2001, http://wireless.newsfactor.com/perl/printer/13228

Mobile Video, Part 3: Playing for Keeps, Brian McDonough, August 31, 2001, http://wireless.newsfactor.com/perl/printer/13256

RealNetworks/Symbian Launch Mobile Video: Companies Get World Commercial First with Nokia 9210 Clips, Mark Mayne, July 25, 2001, streamingmedia.com online, http://www.streamingmedia.com/article.asp?id=7732

The X-Box, Mark Pesce, Wired magazine May 2001, p. 139

Tiny Takami Plasma, Review, Personal Computer World Magazine July 2001, pp. 66-67

Wireless Tour Guide Tags Along with Sightseers, Dan McDonough, Jr., August 27, 2001, http://wireless.newsfactor.com/perl/printer/13104

When Will We Watch?

How Embedded Computing Will Enable The Pervasive Internet, White paper, http://www.windriver.com/corporate/html/smart_white.html

Report: E-tail Prosperity Is in the Details, Keith Regan, September 10, 2001, http://wireless.newsfactor.com/perl/printer/13432

Windows Media Technologies 7 in Enterprises, http://www.microsoft.com/windows/windowsmedia/en/business/enterprises.asp

Enterprises Using Windows Media, http://www.microsoft.com/windows.windowsmedia/en/business/others.asp

Streaming Software Over the 'Net, Ross Avner, November 26, 2001, streamingmedia.com online, http://europe.streamingmedia.com/r/printerfriendly.asp?id=8069

Processor Vendors Tout Mobile Video Encoding Capabilities, Anthony Cataldo, EE Times, November 28, 2001, http://www.eetimes.com/story/OEG20011128S0026

Why Watch Streaming Media?

What Hooks Online Consumers?, Kimberly Hill, CRM Daily.com, August 31, 2001, http://www.ecommercetimes.com/perl/printer/13260

Hollywood Studios To Launch On-Demand Movie Service: Biggest Effort to Date for On-Demand Digital Distribution by the Movie Industry, Jason Thompson, August 16, 2001, streamingmedia.com online, http://www.streamingmedia.com/article.asp?id=7804

Beeb Chooses Real IQ, Jason Thompson, August 14, 2001, streamingmedia.com online, http://www.streamingmedia.com/article.asp?id=7798

Maslow's Hierarchy of Needs, http://web.utk.edu/~gwynne/maslow.htm

Maslow's Hierarchy of Needs, http://www.connect.net/georgen/maslow.htm

Maslow's Hierarchy of Needs, William Huitt, January 10 1998, http://www.valdosta.edu/whuitt/col/regsys/maslow.html

What Will We Watch?

Video on Demand: Who's Holding the Ace?, Laura Ierfino and Milan Sallaba, October 21, 2001, streamingmedia.com online, http://europe.streamingmedia. com/r/printerfriendly.asp?id=7976

Rockwell Streams Well, Max Bloom, August 27, 2001, streamingmedia.com online, http://www.streamingmedia.com/article.asp?id=7780

Film Applications of Windows Media Technologies, http://www.microsoft.com/ windows/windowsmedia/en/content_provider/film/default.asp

Broadcast Applications of Windows Media Technologies, http://www. microsoft.com/windows/windowsmedia/en/content_provider/broadcast/default. asp

Music Applications of Windows Media Technologies, http://www.microsoft.com/ windows/windowsmedia/en/music/default.asp

Digital Plan for Individual Learning, December 10, 2001, http://news.bbc.co.uk/ hi/english/education/newsid_1701000/1710781.stm

Emerging Construction Technologies: VizStream 3D Streaming Technology, http://www.new-technologies.org/ECT/Internet/vizstream.htm

Metastream Streaming 3D, http://www.lhdesign.com/portfolio/14_stream/ 14_stream3d.html

VRML, Virtual Reality and Streaming 3D, http://www.opengl.org/users/ apps_hardware/applications/vrml.html

Macromedia Director 8.5 Streaming 3D Cast Members, http://www.macromedia. com/support/director/3d_lingo/streaming_3d/streaming_3d.htm

Beyond the Bubble: The Future of Film Distribution on the Net, Joe Tripician, http://www.istreamtv.com/film_distrib.htm

The Reign of Content in E-Commerce, Mark Vigoroso, October 19, 2001, http://ecommercetimes.com/perl/printer/14114

The Business

How Will Anyone Make Money with Streaming Media?

Enterprise Streaming: Return on Investment, A Streaming Media Research Report, August 2001, http://www.streamingmedia.com

Chicago Sun Times Tests Streaming Video Ad: Technology Doesn't Require Player or Plug-in, Carl Sullivan, August 24, 2001, Editor and Publisher Online, http://209.11.43.220/editorandpublisher/headlines/article_display.jsp? vnu_contnent_id=1022860

Lowering the Cost of Training: The Cisco IP/TV Network Streaming Video Solution, Cisco Systems White paper, http://www.cisco.com

Affordable Infrastructure for Stream Playback in the Internet, Carsten Griwodz, Alex Jonas, Michael Zink, Technical Report TR-KOM-1999-07, Darmstadt University, http://www.kom-e.teknik.tu-darmstadt.de

J.D. Edwards: A Streaming Media ROI Case Study, Microsoft Windows Media Technologies White paper, http://www.microsoft.com/windows/windowsmedia/ en/archive/casestudies/JDEdwards

DSL Speeds Employee Education, Lowers Delivery Costs, Monsong Chen, Info-Value Computing Inc., http://www.broadcastpapers.com/education-print.htm

Mercedes-Benz Shifts into Streaming Gear, Max Bloom, November 12, 2001, streamingmedia.com online, http://europe.streamingmedia.com/r/ printerfriendly.asp?id=8035

Microsoft Windows Media Technologies Rapid Economic Justification White paper, Produced by Approach Inc, October 2001, http://www.approach.com

Kendra, An Introduction, Daniel Harris, August 22, 2001, Kendra Initiative 2001, http://www.kendra.org.uk/documents/kendra-an-introduction-draft-current.html

Revenue Generation Opportunities With Digital Media, Bill Birney, Jim Travis, Microsoft Digital Media Division, December 2000, http://www.microsoft.com/windows/windowsmedia

Internet in Media Time, Stacy Lawrence, May 1, 2001, The Industry Standard, http://www.thestandard.com/article/0,1902,14571,00.html

Microsoft Windows Media Technologies, Advertising, http://www.microsoft.com/windows/windowsmedia/en/wm7/advertise.asp

Using Windows Media Technologies for Advertising on the Internet, Cherylene Simonetti, Bill Birney, Jim Travis, Microsoft Digital Media Division, May 2001, http://msdn.microsoft.com/library/en-us/dnwmt/html/AdsWM.asp?frame=true

Get Rich Quick!, Alex Gibbons, Akamai, November 30 2001, Streamingmedia.com, http://europe.streamingmedia.com/r/printerfriendly. asp?id=8088

The Business Case for Broadband, Joseph Varello, http://www.broadbandindustrynews.com/html/bl_vorello_everest.shtml

Streaming Media in the Enterprise, Cost Per Stream Analysis, May 2001, White paper, http://www.approach.com/digitalmedia

Cahner's In-Stat Group: Corporate Streaming Media Use to Grow, August 15, 2001, http://www.nua.com/surveys/index.cgi?f=VS&art_id=905357085&rel=true

The Secret Lies of Broadband, Niki Scevak, July 3, 2001, http://www.australia.internet.com/r/article/jsp/sid/10651

Digital Fountain Inc., Media Backgrounder, http://www.digitalfountain.com

Streaming Fountain Product Overview, http://www.digitalfountain.com/products/streamingFountain.htm

Complex Math Makes Downloads and Streaming More Efficient, Jason Meserve, Network World Fusion, April 17, 2001, http://www.nwfusion.com/news/2001/0417math.html

Digital Fountain Looks to Bolster IP Multicast Networks, Jason Meserve, Network World, February 14, 2000, http://www.nwfusion.com/news/2000/0214infra.html

Digital Fountain Technology Overview White paper, http://www.digitalfountain.com

Accessing Multiple Mirror Sites in Parallel: Using Tornado Codes to Speed Up Downloads, John Byers, Michael Luby, Michael Mitzenmacher, http://www.digitalfountain.com

FLID-DL: Congestion Control for Layered Multicast, John Byers, Michael Frumin, Gavin Horn, Michael Luby, Michael Mitzenmacher, Alex Roetter, William Shaver, http://www.digitalfountain.com

Videophone Comes of Age; Companies Reap the Rewards of New Technology, Hilary Douglas, October 24, 2001, http://wireless.newsfactor.com/perl/printer/14375

Who Will Make Money?

AOL Time Warner Executive Upbeat: Company and Internet Have Bright Futures, Co-Chief Operating Officer Says, Alec Klein, December 12, 2001, Washington Post, http://www.washingtonpost.com/ac2/wp-dyn/A28813-2001Dec11?language=printer

Study Confirms Streaming Video's Largest One-Month Gain Ever: According to AccuStream iMedia Research, September 2001 Was a Blockbuster Month for Streaming, Charlie White, http://www.digitalwebcast.com/2001/12_dec/features/cw_accustream_research.htm

Get the Picture? Newcomers Venture into Still-fuzzy Video on Demand Market, Karen Brown, November 2000, Broadband Week Magazine, http://www.broadbandweek.com/news/0011/print/0011_apps_vod.htm

The Day Digital Music Stopped Making Sense: Road to Consumer Success Not Led By The Usual Suspects, Andrew Cullen, http://www.broadbandindustrynews.com.html/f_digitalmusic.shtml

The Case for Corporate Streaming, White paper, Robert Skip Yourd, http://www.broadbandindustrynews.com/html/wp_2netfx.shtml

The State of Streaming Media in Business Today, Bill Torneo, http://www.broadbandindustrynews.com/html/bl_tellsoft.shtml

Beyond MP3, Gary Marshall, .net magazine, February 2002, pp. 30–37

AOL, Sony Announce Far-Reaching Broadband Pact, Lisa Gill, November 13, 2001, http://www.newsfactor.com/perl/printer/14750

Tutorial: Making Brand Launches Sizzle, Christine Perey, October 20, 2001, http://europe.streamingmedia.com/r/printerfriendly.asp?id=7975

The Switch to Holography, Alex Saltman, Wired Magazine May 2001, p. 80

The n-Dimensional Superswitch, Josh McHugh, Wired Magazine May 2001, pp. 88–98

When Will Streaming Media Make Money?

Money, Money, Money, Is a Turn Around in the dot-com World Around the Corner, Stephen Schleicher, http://www.digitalwebcast.com/2001/08_aug/editorials/money.htm

Broadband Homes To Number 90Mil Globally By '04, Report, Dick Kelsey, April 30, 2001, http://www.newsbytes.com/news/01/165112.html

Why are 75% of US Households Still Offline?, Elizabeth Parks, Parks Associates, December 4, 1998, http://www.parksassociates.com/inthePress/press_releases/press_releases_surfing2.html

What Drives High-Speed Internet Services Home?, Elizabeth Parks, Parks Associates, December 4, 1998, http://www.parksassociates.com/inthePress/press_releases/press_releases_surfing1.html

Broadband Access @ Home 2001, Attaining Critical Mass in the Residential Market, Parks Associates, http://www.parksassociates.com/reports&servcies/multiclientstudies/bb@home2001/broadband2001.html

The State of Residential Broadband, Jason Marcheck, Outside Plant Magazine, March 2001, http://www.ospmag.om/past_issues/features/01_03_features/01_03_the_state.htm

Computer Ownership and Internet Access: Opportunities for Workforce Development and Job Flexibility, Employment Policy Foundation, Technology Forecast, January 11, 2001, http://www.epf.org

Will Content Drive Broadband Uptake?, Craig Liddell, July 9, 2001, http://australia.internet.com/r/article/jsp/sid/10673

Slow Economy Driving Streaming Media Onto Corporate Networks, Cahner's In-Stat, August 14, 2001, http://www.instat.com/pr/2001/mb0107sm_pr.htm

The UK Broadband Report, Dan Oliver, .net magazine, February 2002, pp. 47–52

Multimedia service quality measurement systems, http://www.genista.co.jp/solutions.htm

Perceptual Quality of Service measurement, http://www.genista.co.jp/technology. htm

Perceptual Signal Fidelity, http://www.genista.co.jp/Technology/technology-QoS.htm

Objective assessment of audio streams, http://www.genista.co.jp/Technology/audioqos.htm

Dual VQ, A Software Tool to Measure Telephone Voice Quality, White paper, Genista Corporation, Tokyo, Japan, 2001

Video QoS, A Software Tool to Measure Digital Video Quality, White paper, Genista Corporation, Tokyo, Japan, 2001

WISPA, Quality of Service Analyzer for Broadband Wireless Internet Access, White paper, Genista Corporation, Tokyo, Japan

Perceptual Video Quality and Blockiness Metrics for Multimedia Streaming Applications, S. Winkler, A. Sharma, and D. McNally, White paper, Genimedia, Lausanne, Switzerland

Perceptual QoS Management, Definition, Application and Performance, White paper, Genista Corporation, Tokyo, Japan, 2001

Industry Ramps Up Voice over IP Deployment, Ann R. Thryft, September 6, 2001, http://www.planetanalog.com/story/OEG20010906S0076

Why Will Streaming Media Make Money?

Electronic Commerce and New Ways of Working: Penetration, Practice and Future Development in the USA, Jack Nilles, JALA International Inc., Los Angeles Bonn, November 1999, http://www.ecatt.com/ecatt/country/us/inhalt.htm

TIA Responds to the FCC Section 706 Report on the Deployment of Advanced Telecommunications Capability, Ray Mileva, January 28, 1999, Telecommunications Industry Association, Arlington, VA, http://www.tiaonline.org/pubs/press_releases/1999/99-12.cfm

Education Market to Continue Lead in Networking Technology Adoption, Cahner's In-Stat Group, October 2001, e-mailed Market Alert extracted from the report "High Speed Learning: Bringing Broadband to the Education Markets," http://www.instat.com/catalog/cat-mtb.htm#mu0109sg

Microsoft's Mandate, Enterprise Case Study, Max Bloom, Streaming Media Magazine, April 2001, Vol. 1 No. 3

Upsides—Downsides

How Significant Is Streaming Media?

Streaming Media, AT&T Information Research Center Report, July 2000, http://irc.att.com

Facts and Figures About Our TV Habit, http://www.tvturnoff.org

Sample Ad Rate Guide, Ad Resource: Internet Advertising and Web Site Promotion Resources, http://adres.internet.com/adrates/article/0,1401,9251_198601,00.html

What is the Real Average CPM of Web Advertising?, David Halprin, October 18, 2000, http://www.emarketer.com/analysis/eadvertising/20001017_web_cpm.html

Number of non-US Web Surfers to Skyrocket by 2002, Nancy Weil and Kristi Essick, August 18, 1998, http://www.cnn.com/TECH/computing/9808/19/surfers2002.idg

Is the Fiber Glut for Real?, Bruce Gain and Darrell Dunn, December 10, 2001, http://www.ebnews.com/story/OEG20011210S0066

The Changing Future of Broadband, Ron Mudry, http://www.broadbandindustrynews.com/html/bl_mudry_progress.shtml

At-home High-speed Users Up 121%: Nearly 18 Million Used Broadband in July, Sid Ross, August 24, 2001, Editor and Publisher Online, http://209.11.43.220/editorandpublisher/headlines/article_display.jsp?vnu_content_id=1023176

More Analysis Results of the EURESCOM project P903 "ICT uses in everyday life", P903 Newsletter May 2001

Enhancing Security: Can the Internet Help?, Robyn Weisman, October 11, 2001, http://wireless.newsfactor.com/perl/printer/14070

Internet Holds Up Under Stress After Terrorist Attacks, Tim McDonald, September 12, 2001, http://ww.newsfactor.com/perl/printer/13463

As U.S. Was Attacked, Citizens Turned To Technology, Paul Greenberg, September 12, 2001, http://www.ecommercetimes.com/perl/printer/13465

Streaming Media Europe 2001, Jon Silk, November 4, 2001, http://europe.streamingmedia.com/r/printerfriendly.asp?id=8023

Invitation to Seminar: Content & Applications for Broadband & Digital TV: Technologies, Markets & Player Strategies, Worldwide Market Analysis & Strategic Outlook 2001-2006, http://www.arcgroup.com

Digital Media CyberTrends, John McCann, Fuqua School of Business, Duke University, http://www.duke.edu/~mccann/q-media.htm

Technology CyberTrends, John McCann, Fuqua School of Business, Duke University, http://www.duke.edu/~mccann/q-tech.htm

Digital Marketing CyberTrends, John McCann, Fuqua School of Business, Duke University, http://www.duke.edu/~mccann/q-mkting.htm

Internet in Media Time, Stacy Lawrence, May 1, 2000, The Industry Standard, http://www.thestandard.com/article/0,1902,14571,00.html

21 Million Broadband Customers in November, http://www.tomshardware.com/technews/technews-20011212.html

What Could Go Wrong?

Porno for Pockets, Jenn Shreve, Wired Magazine May 2001, p. 76

The Future Will Be Fast But Not Free: You Want Broadband. You'll Get It. You'll Pay For It. You'll Like It, Charles Platt, Wired Magazine May 2001, pp. 120-127

Telechasm: Can We Get to the Future From Here? First We Have to Get Telecom Out of the Stone Age, Frank Rose, Wired Magazine May 2001, pp. 128–135

RealNetworks Cuts 15 Percent of Workforce: Layoffs Follow Disappointing Second Quarter Results, Jose Alvear, July 26, 2001, streamingmedia.com online, http://www.streamingmedia.com/article.asp?id=7737

Broadband's Long Haul, Andrew Cohen, http://www.worldlink.co.uk/stories/storyReader$709

Petition Aims to Change Broadband Landscape, Jon Silk, November 30, 2001, streamingmedia.com online, http://europe.streamingmedia.com/r/printerfriendly.asp?id=8089

Broadband Is Dead: Sad but True, Broadband Isn't Making the Cut, but Bob Has Reason to Believe This Isn't a Bad Thing at All, Robert X. Cringely, http://www.pbs.org/cringely/pulpit/pulpit20011011.html

Report: Firms Cut Costs with Web Services, but Neglect Other Benefits, Lori Enos, August 30, 2001, http://www.ecommercetimes.com/perl/printer/13229

Battle for Broadband Britain Heats Up, Ian Lynch and William Eazel, July 10, 2001, http://www.vnunet.com/Print/1123784

UK Fails to Take Up Broadband, Computeractive staff, July 18, 2001, http://www.vnunet.com/Print/1124041

Welcome to DSL Hell: Cable Modem Provides Redemption, Charlie White, http://www.digitalwebcast.com/2001/10_oct/editorials/cw_editorial_dslhell.htm

Email A Postcard Written in Pencil, Special Report, Lawrence Rogers, Software Engineering Institute, Carnegie Mellon University, Pittsburgh, PA

Signal 2 Noise: The Broadband Fiasco, David Nagel, DTV Professional, http://www.dtvprofessional.com/2001/08_aug/editorials/signal_2_broadband.htm

Has Broadband Been Dealt a Fatal Blow?, Niki Scevak, June 8, 2001, http://australia.internet.com/r/article/jsp/sid/10492

Adventis Flashpoint: Breakthrough Improvements in Telco Provisioning, Blaik Kirby and Patrick Pugh, http://www.adventis.com

Streamlining the Internet-Fiber Connection, Sudhir Dixit and Yinghua Ye, IEEE Spectrum Online, October 12, 2001, http://www.spectrum.ieee.org/WEBONLY/publicfeature/apr01/opcom.html

This Year's Model, Alison Campbell, November 26, 2001, streamingmedia.com online, http://europe.streamingmedia.com/r/printerfriendly.asp?id=8070

Yesterday I Couldn't Spell Systems Administrator; Now I Am One!, Larry Rogers, http://www.cert.org/homeusers/ira_sysadmin.html

Nifty Gadgets at Comdex, Henry Norr, November 19, 2001, http://wireless.newsfactor.com/perl/printer/14864

Conclusion

Microsoft CEO Predicts Rosy Future for Streaming Media, Patricia Jacobus, CNET News.com, December 12, 2000, http://news.cnet.com/news/0-1005-200-4119264.html?tag=rltdnws

Executive Summary of Putting "Media" into Streaming Media, http://www.yankeegroup.com/webfolder/docmanager.nsf/00/5AE6C31C5A1E18718525

BIBLIOGRAPHY

Achbar, Mark. *Manufacturing Consent: Noam Chomsky and the Media*, Black Rose Books, 1994, ISBN 1-551640-02-3.

Allen, David. *Getting Things Done*, Viking Press, 2001, ISBN 0670899240.

Cairncross, Frances. *The Death of Distance*, Harvard Business School Press, 2001 ISBN 157851441X.

Davenport, Thomas and John Beck. *The Attention Economy: Understanding the New Currency of Business*, Harvard Business School Press 2001, ISBN 157851441-X.

de Soto, Hernando. *The Mystery of Capital: Why Capitalism Succeeds in the West and Fails Everywhere Else*, Basic Books 2000, ISBN 0-465016146.

Downes, Larry and Chunka Mui. *Unleashing the Killer App*, Harvard Business School Press 2000, ISBN 1578512611.

Dyson, Esther. *Release 2.1*, Broadway Books 1998, ISBN 0-7679-0012-X.

Evans, Philip and Thomas S. Wurster. *Blown to Bits*, Harvard Business School Press 2000, ISBN 0-87584-877-X.

Gilder, George. *Telecosm: How Infinite Bandwidth Will Revolutionize Our World*, Free Press 2000, ISBN 0-684809303.

Godin, Seth and Don Peppers. *Permission Marketing*, Simon and Schuster 1999, ISBN 0-684856360.

Herman, Edward and Noam Chomsky. *Manufacturing Consent: The Political Economy of the Mass Media*, Pantheon Books 1988, ISBN 0-679-72034-0.

Howard, Michael and David LeBlanc. *Writing Secure Code*, Microsoft Press 2002, ISBN 0-7356-1588-8.

Moore, Geoffrey A. and Regis McKenna. *Crossing the Chasm*, Harper Business Books 1999, ISBN 0-066620023.

Pank, Bob (editor). *The Digital Fact Book,* 10 ed., Quantel 2000, available from http://www.quantel.com/dfb.

Paxman, Jeremy. *Friends in High Places: Who Runs Britain?*, Penguin Books 1991, ISBN 0-14-015600-3.

Peppers, Don and Martha Rogers. *The One to One Future*, Doubleday 1997, ISBN 0-385485662.

Pilger, John. *A Secret Country*, Jonathan Cape Ltd 1989, ISBN 0-224-02600-3.

Popcorn, Faith. *The Popcorn Report*, Harper Business Books 1992, ISBN 0-887305946.

Seybold, Patricia B. *Customers.com*, Times Books, 1998, ISBN 0-8129-3037-1.

Shepard, Steven. *SONET/SDH Demystified*, McGraw-Hill 2001, ISBN 0-07-13618-6.

Symes, Peter. *Video Compression Demystified*, McGraw-Hill 2001, ISBN 0-07-136324-6.

Tapscott, Don. *Growing Up Digital*, McGraw-Hill 1998, ISBN 0-07-134798-4.

Wolf, Michael J. *The Entertainment Economy*, Times Books 1999, ISBN 0-8129-3042-8.

INDEX

Note: Boldface numbers indicate illustrations.

ABOUT THE AUTHOR

MICHAEL TOPIC is a professional innovator and product design engineer, running his own consultancy, Imaginative Engineering Ltd. Originally trained in electrical, electronic, and computer engineering, in his native Australia, he now lives in the UK, with a cat, a family, and a large collection of electric guitars. Michael has spent 18 years creating digital media tools for creative and demanding customers. During his career, Michael has lead design teams creating Internet streaming media encoders, audio and video nonlinear editing systems, synthesizers, and music sequencers.